KNOWLEDGE AND SECRECY IN AN ABORIGINAL RELIGION

OXFORD STUDIES IN SOCIAL AND CULTURAL ANTHROPOLOGY

Oxford Studies in Social and Cultural Anthropology represents the work of authors, new and established, which will set the criteria of excellence in ethnographic description and innovation in analysis. The series serves as an essential source of information about the world and the discipline.

OTHER TITLES IN THIS SERIES

Organizing Jainism in India and England
Marcus Banks

Society and Exchange in Nias
Andrew Beatty

The Culture of Coincidence: Accident and Absolute Liability in Huli
Laurence Goldman

The Female Bridegroom: A Comparative Study of Life-Crisis Rituals in South India and Sri Lanka
Anthony Good

Of Mixed Blood: Kinship and History in Peruvian Amazonia
Peter Gow

Exchange in Oceania: A Graph Theoretic Analysis
Per Hage and Frank Harary

The Interpretation of Caste
Declan Quigley

The Arabesk Debate: Music and Musicians in Modern Turkey
Martin Stokes

KNOWLEDGE AND SECRECY IN AN ABORIGINAL RELIGION

IAN KEEN

CLARENDON PRESS · OXFORD
1994

Oxford University Press, Walton Street, Oxford OX2 6DP
Oxford New York Toronto
Delhi Bombay Calcutta Madras Karachi
Kuala Lumpur Singapore Hong Kong Tokyo
Nairobi Dar es Salaam Cape Town
Melbourne Auckland Madrid
and associated companies in
Berlin Ibadan

Oxford is a trade mark of Oxford University Press

Published in the United States
by Oxford University Press Inc., New York

British Library Cataloguing in Publication Data
Data available

Library of Congress Cataloging in Publication Data
Keen, Ian.
Knowledge and secrecy in an aboriginal religion / Ian Keen.
(Oxford studies in social and cultural anthropology)
Includes bibliographical references and index.
1. Murngin (Australian people)—Religion. 2. Arnhem Land (N.T.)—Religion. I. Title. II. Series.
BL2630.M87K44 1994 299'.92—dc20 93–14380
ISBN 0–19–827900–0

1 3 5 7 9 10 8 6 4 2

Typeset by Best-set Typesetter Ltd., Hong Kong
Printed in Great Britain
on acid-free paper by
Biddles Ltd., Guildford and King's Lynn

In memory of my brother, Michael Keen

PREFACE

This book is about control of the transmission of religious knowledge on the part of Yolngu of north-east Arnhem Land, during the mid- to late 1970s, focusing primarily on men. It examines that control along roughly three dimensions: first, the ownership of forms and dissemination of information among group-centred social networks; second, religious secrecy dividing people by gender and age; and third, transformations in which more universalistic forms were preferred in order to forge relations among wider social networks of Aboriginal people and with the non-Aboriginal community, and in which religious secrecy separating men and women was, at least temporarily and in a qualified way, breached.

In describing these dimensions of the control of religious knowledge the book confronts the issue of culture and community. Yolngu seemed to participate in an unbounded though structured network of social interaction and relationships, rather than a clearly bounded group or community as the bearer of a more or less homogeneous 'culture'. If this is right, then describing the shared understandings that make discourse, even contested discourse, becomes problematic. The book begins with the assumption that agreement about meaning should not be taken for granted; meanings are contested and negotiated.

The book is about religious secrecy, but it contains few secrets. Although Yolngu men told me that I could convey secret religious matters to other anthropologists, it would be a breach of my obligations to them if I were to publish religious secrets in a work of this kind. I have omitted all but the barest indication of what goes on in the men's secret-ceremony grounds, and write of the structure rather than the content of secret religious knowledge. Because the book includes disputes and some personal details, the identity of individuals and groups has been obscured by the use of synonyms and other means, but I have not tried to disguise the identity of the main Yolngu settlements. The personal names are pseudonyms, based on the Yolngu practice of naming people after ancestral beings and their analogues; most are the names of species associated with the appropriate patrimoiety. To be consistent with the use of pseudonyms, I have omitted illustrations of group designs as well as photographs of individuals.

I.K.

Canberra, Australia

ACKNOWLEDGEMENTS

My first debt is to the Yolngu people with whom I worked for a protracted period during 1974–6, and for shorter periods subsequently, most recently in 1987. I am especially grateful to Ba:riya, the late Banhdharrawuy, the late Dja:wa, Letngawuy, the late Marangguy, the late Binyinyiwuy, Djikaṉmirriwuy, Yambal, Peter Mondjingu, the late Dha:thangu, the late Bonguwuy, Wuḻuḻu, Matjuwi, Sylvia Mulwanany, Gaḻarrwuy Yunupingu, Roy Marika, Joe Djalalingba, and many others. Many mission and other staff at Milingimbi were supportive during fieldwork periods; I thank especially Ken Nowland, Beulah Lowe, Roger Sigston, Fran and Lang Powell, Brad and Sue Harris, David Morgan, and Richard Trudgen.

Nicolas Peterson, Frances Morphy, Howard Morphy, and Nancy Williams have been close associates in Arnhem Land research over the years, and have been stimulating and supportive colleagues. Robert Bos and John Rudder have kindly allowed me to cite unpublished reports and other manuscripts. I thank Don Gardner, Leslie Devereux, David Martin, Chris Gregory, Nicholas Thomas, and Franca Tamisari, who have read and criticized parts of various drafts, Margaret Jolly for her encouragement at a critical time, and colleagues at the Australian National University anthropology seminar who have commented on versions of several chapters. Comments by anonymous reviewers have been especially helpful, and I am grateful to Tim Barton, Peter Momtchiloff, and Helen Gray of Oxford University Press for their encouragement and for guiding the book through the editing process. Libby Keen read the whole draft and made many valuable suggestions; I thank her also for constant support throughout the project. Discussions with researchers in other disciplines, especially Stephen Harris, Joe Reser, David McClay, and Michael Christie, have been very fruitful. I am grateful to Debbie Baker of the Commonwealth Archives in Darwin for her guidance, and Kate McAllan for assistance with research and proof-reading. The hospitality of the Institute of Social and Cultural Anthropology, the Pitt Rivers Museum, and Linacre College in Oxford during the final stages of writing are much appreciated. I thank Imogen Keen for the drawings and Winifred Mumford for preparing the maps and figures.

Information presented in Map 2 is drawn from Frances Morphy's 'Djapu, A Yolngu Dialect', in R. Dixon (ed.) *Handbook of Australian Languages*, and L. Hiatt's *Kinship and Conflict*.

Fieldwork and archival research has been supported by the Australian National University, the Australian Institute of Aboriginal Studies, and the University of Queensland. The Northern Land Council has provided the opportunity of continued research in north-east Arnhem Land; I thank especially Jeannie Devitt.

CONTENTS

MAPS, FIGURES, AND TABLES

MAPS

FIGURES

TABLES

ORTHOGRAPHY AND PRONUNCIATION

The reader can gain an idea of how Yolngu words are pronounced by giving the consonants the values they have in English, and the vowels Italian or Spanish values. The underlined consonants are 'retroflex', pronounced with an 'r' sound, so than *n̲* is like the 'rn' sound in 'corn' as pronounced in many parts of the United States, and so on. The letter *g* is hard; and *ng* at the beginning or in the middle of a word has the sound of 'ng' in a word like 'sing'. The letters *ny* are like the 'ni' in 'onion', not the 'ny' in 'many'. There are two 'r' sounds, the double *rr* is rolled, as in the Scottish 'r', and the single *r* is pronounced as in 'round'. An apostrophe denotes the glottal stop, as found in London dialect 'bu'er' (butter). The letters *u* and *o* are pronounced approximately as in 'hot' and 'born' respectively (not as in 'shut' and 'font'). The letter *e* denotes a long sound, approximately as in 'feet'.

The orthography employed in this book is adapted from the Gupapuyngu orthography devised by Beulah Lowe. Most of the Yolngu words and phrases transcribed in this book are from Dhuwal and Dhuwala dialects. (See F. Morphy 1983: 12–25 for further details.)

Vowels:

	short	long
high front	*i*	*e*
high back	*u*	*o*
low	*a*	*a:*

Stops and consonants:

	bilabial	dorso-velar	alveolar	post-alveolar (retroflex)	lamino-dental	lamino-alveolar
stops	*p*	*k*	*t*	*t̲*	*th*	*tj*
	b	*g*	*d*	*d̲*	*dh*	*dj*
nasals	*m*	*ng*	*n*	*n̲*	*nh*	*ny*

The cluster *n* + *g* is written *n.g* to distinguish it from the velar nasal *ng*

Liquids and semi-vowels:

	alveolar	post-alveolar (retroflex)
lateral	*l*	*l̲*
rhotics	*rr*	*r*

rr is a trill and *r* is a continuant

KIN GLOSSES

Throughout the book I employ glosses of Yolngu kin terms which require some explanation, for the specialist and general reader alike. As in other Aboriginal systems of kin classification, terms for close relatives are extended to denote genealogically more distant relatives, and terms denoting genealogically related kin are extended to refer to persons who are not related genealogically. Glosses of kin terms in the text do not distinguish these dimensions, but simply render the term as the product of more elementary terms. Where it is important to do so, I indicate the distance of the particular relationship in the text. The glosses employed are these:

M	mother, but a person's 'mothers' include mother's sisters
F	father, but a person's 'fathers' include father's brothers
wC/ZC	child, woman speaking; sister's child, man or woman speaking
mC/BC	child, man speaking; brother's child, man or woman speaking
D	daughter; hence mD: daughter, man speaking; wD: daughter, woman speaking
S	son; hence mS: son, man speaking; wS: son, woman speaking
Z	sister
B	brother
eZ	elder sister
yZ	younger sister
eB	elder brother
yB	younger brother

Relative products are indicated simply with strings of these glosses, for example, MBD = 'mother's brother's daughter'. A slash indicates alternative meanings; for example, MM/MMB shows that *ma:ri* refers to a person's mother's mother or mother's mother's brother.

In Yolngu languages the term used by a woman to denote her child (both son and daughter) is distinct from the term used by a man to denote his child (son and daughter): a woman's child is her *waku*, while a man's child is his *ga:thu*. Moreover, a man or woman applies the same term to refer to or address the children of his or her sister as she does, so that a person's sister's child is that person's *waku*. Similarly, a man or woman applies the same term to refer to or address the children of his or her brother as he does, so that a person's brother's child is his or her *ga:thu*. For these reasons I gloss the term *waku* was 'wC/ZC', i.e. 'child, woman speaking, sister's child'; and I gloss the term *ga:thu* as 'mC/BC', i.e. 'child, man speaking, brother's child'. In some contexts the gloss is more specific, e.g. *ga:thu* (mS).

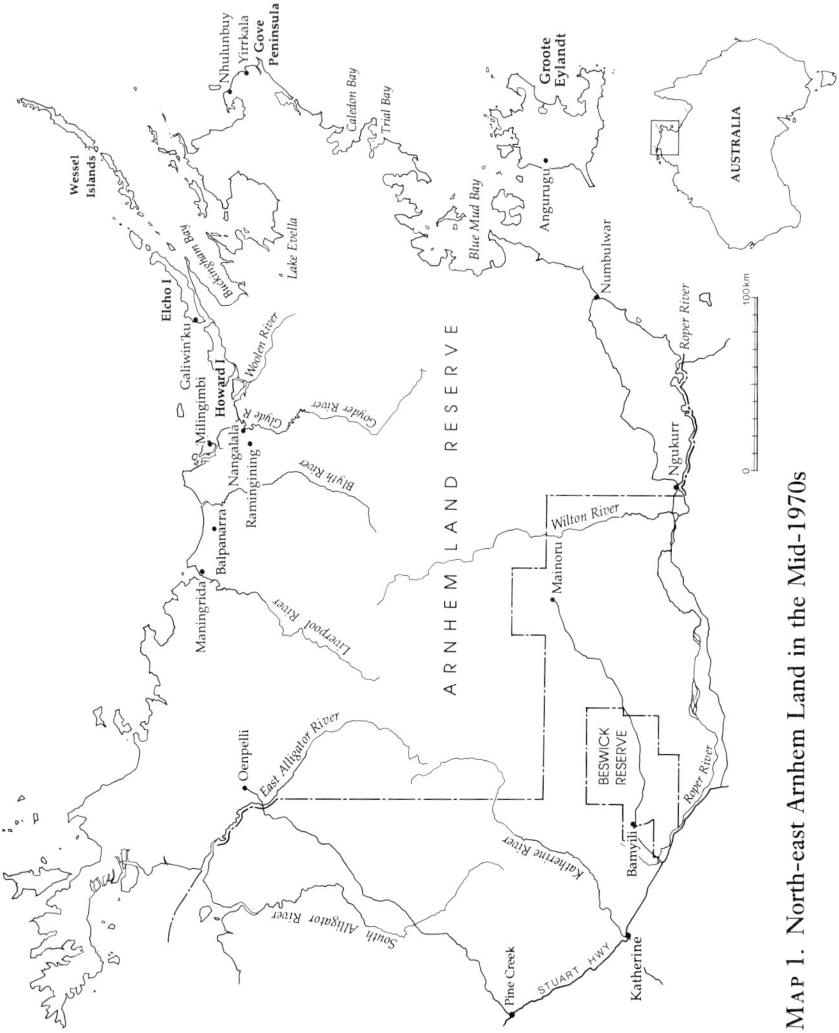

Map 1. North-east Arnhem Land in the Mid-1970s

Introduction

A quarter of a century ago Stanner (1963: 164–5) took issue with the view that Murinbata society or culture was a unified system or organic whole. In Murinbata social life 'there were only workings towards system and transient captures of unity'. Nevertheless, with some reservations, anthropologists have consistently treated the social life and institutions of people of north-east Arnhem Land in the Northern Territory of Australia (Map 1) as more or less uniform, so that it has been thought appropriate to generalize about the social organization, marriage system, art, religion, and land tenure of a region whose inhabitants have been labelled by an arbitrarily chosen proper name: Murngin, Wulamba, Miwuyt, or Yolngu.[1]

Yolngu religious beliefs and practices during the second half of the 1970s and early 1980s were not homogeneous. While co-operating with others, people of each patrifilial group had their own body of religious practices and myths and their own perspectives on the meanings of religious forms. People of each group (or more accurately, kin network centred on a group) tried to control other people's access to knowledge about their own interpretations of publicly displayed religious forms. Men and women also had somewhat distinct views of religious meaning—men excluded women from legitimate access to, and expression of, secret religious knowledge rather more than women excluded men; older men and women controlled access to esoteric religious knowledge on the part of young people. From the mid-1920s

[1] Warner (1937) writes of the 'Murngin', R. M. Berndt (1955) of 'Wulamba', Shapiro (1981) of 'Miwuyt' marriage, and H. Morphy (1977), Keen (1978), and Williams (1986) among others of 'Yolngu'. Several of these works mention some variation, and recent treatments, such as Howard Morphy's (1984) description of a mortuary ceremony, stress variability and the contextual nature of what is said and done. Speakers of languages of the Yolngu family now refer to themselves, and are referred to by others, as 'Yolngu' in contrast with 'Balanda', a light-skinned person. The racial category derives from the word *yolngu* which means 'person' or, in some contexts, 'man', and extends to other dark-skinned people.

Yolngu people were introduced to Christian beliefs and rituals, some of which were transformed and incorporated into Yolngu cosmologies and, later, ritual practice. These relative Yolngu perspectives on religious meaning, and the control of access to religious knowledge comprise the main topic of this book.

But what is meant here by 'religion' and 'knowledge'? Yolngu did not isolate a domain and label it with a term synonymous with 'religion', for religion was 'embedded' in wider practices (Bloch 1985: 74). I use the word to denote the categories, beliefs, and practices which referred to or invoked *wangarr* ancestors and related beings. Things connected with *wangarr* such as objects, designs, songs, dances, and species were all *maḏayin* (sacra, often translated as 'sacred'), and religious practice was *rom*, 'right practice' or 'the way'. Religion did indeed penetrate all aspects of Yolngu social life.

What of knowledge? Yolngu control of knowledge was structured according to the senses of their concept *marnggi*. This word can be translated 'know' or 'knowledgeable' but it had something of the sense of 'can' or 'be able' as well. Thus, as with a similar Maori concept, knowledge was not separable conceptually from power (Parkin 1985: 49). A person became knowledgeable of something by experiencing it; to demonstrate that one was knowledgeable entailed saying something or acting in some way. Knowledge, however, implied a right: a person who could not in the circumstances legitimately claim to be old enough or of the right group or gender to say or do something, such as recount a certain myth, would deny knowledge, at least in contexts where his or her actions were monitored.

The control of knowledge therefore had two aspects. First was the possibility of a person gaining experience of various kinds—seeing and hearing events and objects, hearing explanations. Second were constraints on a person's disposition to demonstrate the results of their experience and show that they were knowledgeable, by telling a story, for example, or performing a dance. Expression of such a disposition was context-dependent.

The chapters of Part I of the book (Chapters 1–4) explore the ways in which Yolngu defined groups, and how groups of the same moiety shared negotiated religious forms but were differentiated by the details of forms created in religious practices as well as their total complements of sacra. They describe how groups differed in their interpretations of ancestral events, the ancestral constitution of country, and the meaning of ceremonies. The second part of the book (Chapters 5–7) examines initiation, revelation, and religious secrecy: the ways in which Yolngu people framed and controlled the transmission of secret religious meanings, and the ways in which people became knowledgeable. We will find a constant interplay between, on the one hand, the definition of sacra (myths, songs, ceremonies, designs, and objects) connecting and differentiating groups, and, on the other, religious secrecy separating old and young, men and women.

Just as Yolngu religious practices varied across the region, so they had been in a state of continual transformation since long before European and Chinese colonization of the north during the nineteenth century. Later incursions by pastoralists, as well as the penetration of missionaries and mission organizations into Arnhem Land, had a major impact on Yolngu social life. In response to these changing conditions Yolngu later tended to favour universalistic religious forms which transcended divisions between groups and moieties, and which linked Aboriginal people (Yolngu) to light-skinned people (Balanda). In addition to this move towards inclusiveness, the power of older men which rested on the constitution and control of secret religious knowledge was eroded in some contexts, while Christianity challenged the very separation between secret and public, 'inside' and 'outside'. These and other transformations in the control of religious knowledge are considered in the third part of the book (Chapters 8 and 9).

The events and practices described in the following chapters occurred and were current in the mid- to late 1970s and early 1980s. It is for that reason that I write of them in the past tense rather than an 'ethnographic present'. The whole is also based very much on the point of view of Yolngu men with whom I have worked, and so does not pretend to give a 'complete' picture of Yolngu religion, even if such a thing were possible. Aspects of the analysis will have implications for gender relations, but any attempt to assess the balance of power or status between men and women must be regarded as tentative.

A study of the control of religious knowledge on the part of Aboriginal people gives rise to two analytical problems. The first has to do with heterogeneity of beliefs and practices, and relativity in interpretations of the social world; the second is the relationship of the control of religious knowledge to other aspects of power relations.

Heterogeneity in Culture and Practice

To speak of Yolngu practices is not to imply that there was a clearly defined population all of whom identified themselves as 'Yolngu'. Speakers of the nine related Yolngu languages (the 'dialect groups') did not think of themselves as one ethnic entity, although the languages were related.[2] Not all those called Yolngu by linguists and others identified themselves in that way: some Djinang speakers denied being Yolngu, for their word for 'person' was *yol*, not *yolngu* (Elliott 1991).

Yolngu social practices partook in a mosaic distribution of variant forms across Arnhem Land. Ethnographies that treat of cultural types show that Yolngu or 'Murngin' practices combined exogamous patrifilial groups, exogamous patrifilial moieties, circumcision, and an asymmetric mode of kin classification (Warner 1937). The Gidjingali and their neighbours also organized themselves into patrifilial groups and moieties, although they enforced exogamy less strictly, and combined these with a symmetrical Aranda-like mode of kin classification (Hiatt 1965). Further west, people were also identified by patrifilial group and used a kin terminology with some Aranda-like features. Matrilineal moieties were important in social organization rather than patrilineal moieties, although these were present, and the people did not circumcise their young men (Altman 1987; R. M. and C. H. Berndt 1970; Taylor 1987).

While this book will reveal variety within Yolngu social worlds, it may be that the similarities among the worlds constructed by speakers of Yolngu languages were greater than among the worlds of any one of these and those of their nearest non-Yolngu-speaking neighbours. Indeed, grammatically, Yolngu languages formed an enclave of suffixing languages surrounded by neighbours speaking prefixing languages. Even if the gist of this paragraph is true, however, it does not make 'the Yolngu' a bounded 'community'.

[2] Linguists refer to the suffixing languages of the region as the Yolngu (Yuulngu) or Murngic family of languages (Yallop 1982: 53).

So, this simple account can be qualified by reiterating the absence of clear cultural boundaries. First, transitional forms were to be found between the Yolngu heartland and their neighbours (see Borsboom 1978*a*; Elliott 1991). Second, people who had different practices in such things as marriage arrangements, yet who frequently interacted, had translation procedures to translate the terms of one system of (say) kin classification to those of another (Hiatt pers. com.). Third, speakers of different languages (or who were identified with given languages even if they did not habitually speak them) often intermarried and lived in the same residence groups, and many individuals were bi- or multilingual. And fourth, people who spoke neighbouring or even rather distant languages, such as Gidjingali and Rembarrnga, joined with Yolngu-speakers in the enactment of certain practices such as the *garma* genre of public ceremonies and major revelatory rites, using similar forms.

Indeed, not only did speakers of neighbouring languages possess similar domains of the social world, but some at least had vocabulary in common, and there was much lexical diffusion between some Yolngu languages and those with distinct forms of grammar (Heath 1978). Moreover, Yolngu engaged with their neighbours in common practices which entailed some commonality of language as well as multilingualism. For example, articulation in marriage arrangements may have been fostered by the common term *ma:ri* (*anmari* in Gidjingali) for a group which was a man's source of wives' mothers (Hiatt 1965: 42), and common subsection or 'skin' terms enabled Yolngu to form quasi-kin-based relations with neighbours.

Furthermore, the ethnographic record shows continual communication of resources as well as information across broad regions, and the diffusion of forms of practice such as the Karwadi-Gunapipi type of ceremony described in Chapter 8. Thus some forms of practice have been reproduced within a (more or less) localized but open population while others have been transformed, discarded, or taken up.

This story can be further modified by pointing to variation in the way in which speakers of different Yolngu languages did things, even in the case of supposedly similar ceremonies, such as Djungguwan. To pick some other obvious indicators of 'internal' variation: Yolngu folk-categories of language types oversimplified the range of variation (F. Morphy 1977); the details of kin classification varied, especially between coastal and inland people; and certain ceremonies had currency only in some regions. In so far as ceremonies could be said to have more or less invariant programmes, a type of ceremony such as Djungguwan, performed throughout the region, varied in details of style. People west of the Glyde River have lately taken on the Yabuḏurrwa ceremony introduced from the south, and some local networks embraced Christianity with greater enthusiasm than others. Finally, just as some coastal groups of the Yirritja moiety had taken up Macassan themes in their ritual (R. M. Berndt 1948*a*), so members of some groups have recently revealed a more entrepreneurial spirit than others.

A map of variation in the constitution of domains of the social world must be qualified by the recognition not only that their elements are subject to vagaries of interpretation, but that these broad domains are open in content. Some of the aspects which anthropologists have described and analysed, such as the semantic relations among kin categories, or structures apparently underlying myth, give the appearance of determinacy. But while Yolngu people would recount explicit norms about the way in which a person should behave towards another in a certain relation, it is not possible to write a rule book for the content of behaviour in relation to kin, or even for who counted as a kinsperson. This is so even though it is possible to characterize something of the quality of Yolngu social relations and interaction by contrast with people with very different ways of life.

To describe variation in forms of religious practice and belief would not be of very great interest in itself. But I suggest that relativity of perspectives was at the heart of Yolngu religious and other practices. Homogeneity implies agreement about meaning, resulting from a common history, common conditions, or coercion. Yolngu and their neighbours negotiated shared languages of forms of practice, but deliberately created differences to constitute and distinguish groups, and interpreted shared (and negotiated) religious forms differently. (Commonality of form was a matter of judgement and shifted with a person's perspective.) They constantly produced new variations of old themes and innovative interpretations of old forms in response to unique circumstances.

Furthermore, systematic ambiguity was one basis for the constitution of religious mystery and secret knowledge.

The control of religious knowledge, therefore, involved not just a flow of information and interruptions in that flow, but a certain relativity of perspectives on the world. It was recent fieldwork in Arnhem Land (in 1987) that led me to realize that people of different groups, or even of the same group, had very different perspectives on social reality of a kind that are often concealed in ethnographic representations. When mapping the significance of country to Aboriginal people, in order to document an Aboriginal land claim, for example, one frequently obtains the interpretation of a few people deemed to be the most closely connected to it, or to have the strongest claims. While some differences of opinion may arise over 'boundaries', the results appear homogeneous and systematic. In this study, however, I talked to as many people as possible in the time available who had an interest in a certain area, including those who shared songs connected with the country as well as those who were explicitly in dispute. Profound differences emerged over the identity of country, and hence the extent of any one group's territory. Members of each group tended to rank their own connections to forest country highest, although knowledgeable people did tend to agree about the location of sites with mythological significance. What this exercise showed was the relativity of perspectives on what has been called 'totemic geography' (Strehlow 1970), suggesting that there were no objective facts of rights in land. The question then arose, whose reality prevailed in practice and why?

This relativity of perspectives, together with the evident heterogeneity in Yolngu practices, gives rise to a problem of how to conceptualize what it is that is shared for communication and cooperation in practices to be possible, without built-in assumptions of homogeneity.

Assumptions of Uniform Culture

It is hardly original to stress variability; the point was made long ago by Leach (1954) and has been reiterated more recently by Pelto and Pelto (1975), Drummond (1980), Sider (1986), Keesing (1987), Strathern (1988: 9), and the Comaroffs (1991), among others. Nevertheless, whereas some recent theorizing about language and society is stimulating in relating structure to events and agency (e.g. Bourdieu 1972, 1977; Sahlins 1985; Fish 1989; Giddens 1979; Barth 1987), a limitation in these works from the point of view of this study is the pervasiveness of the idea of some kind of bounded class, group, or community as the bearer of a culture or set of dispositions, an emphasis that persists in spite of long-standing critiques of structural or cultural holism.

Among the post-structuralists, Bourdieu defines *habitus* as the structure of shared dispositions that both generates and is reproduced and transformed by 'objective structures' of the social formation. *Habitus* is the attribute of a social class in class-based societies, and, in 'undifferentiated' societies, of a group (Bourdieu 1972: 181–3). The relative homogeneity of *habitus* is the result of identical objective conditions of existence transcending individual intentions, and determined in the final analysis by the economic base of the social formation (1972: 185; 1977: 80). There is a paradox here, for *habitus* engenders an infinity of practices adapted to new situations (1972: 209), yet it is both outcome and origin of the 'objective conditions' of a group or class. But if dispositions are strategizing and innovative (1977: 73, 76), how can the *habitus* of a group or class be homogeneous, and how is it that definite classes and groups are reproduced?

The concept of culture, which remains the central anthropological theoretical object, tends to imply uniformity, and the concept tends to imply boundaries between one 'system' and another.

Among theorists of culture, and consistent with the practice theory of Bourdieu, Sahlins acknowledges that people are known to 'creatively reconsider their conceptual schemes'. Yet there is a tension between Sahlins's recognition of relative viewpoints and a conception of a cultural totality. The genesis of conceptual contradictions, he argues, can be accounted for 'precisely as partial or situational views on the global order taken from some interested standpoint', including that of the ethnographer. This process 'assumes some determinate relation to the cultural totality'. Structure 'properly refers to that totality: the system of relations between categories, without a given subject' (1985: xv–xvi). The anthropologist can place himself or herself in the position of 'transcendent subject' to 'see history working through the interested selection of social agents among the numerous logical possibilities' that are present in a cultural order (1985: xvii). It is just this conception of a cultural or structural totality that is problematic for Yolngu ethnography.

Where it occurs, commonality in cultural knowledge, Lindstrom rightly argues, 'requires a political accounting' (1990: 11). But he seems to assume a shared culture on Tanna; indeed, 'pre-modern' social orders presume the existence of shared culture, he suggests, even though they also presume conversational economies of power (1990: 14).

Writing in the field of literary and legal theory, Stanley Fish remarks that a 'code' is not a list of specific maxims or a closed set of rules that can serve as a self-executing decision-procedure, but rather a general project 'whose implementation involves the continual discovery of its own content, a discovery that is at the same time the accomplishment of its own alteration' (Fish 1989: 151). Yet the 'engine of change' is an 'interpretive community' sharing common assumptions (1989: 150). It is possible to extrapolate from Fish's position a heterogeneous and diffuse interpretive community; if the latter is an 'engine of change', then different members of the community must use codes somewhat differently, for presumably not all members discover new properties of a mode of practice or are exposed to new phenomena at the same time. (Moreover, it is clear that Fish regards American society as consisting of a multiplicity of overlapping interpretive communities.)

From a different intellectual tradition, that of transactional anthropology, Barth's recent analysis of variability and innovation in Mountain Ok belief (Barth 1987) concludes that the total field of Ok cosmological speculation confronts us not with a coherent model of reality but instead 'a field of raucous variation'. The body of shared items of cosmological knowledge which cohorts of males obtain through joint initiations is a movable, impressionable, and ever-changing inheritance, only approximately shared in the group. It does not generate a unified system of knowledge agreed by all, but a range of understandings 'sufficient so its members can be moved by the same symbols and thoughts'. Even within a local group many conceptions are only loosely connected, and carry unresolved ambiguities and enigmatic contradictions. The kind of coherence characteristic of the cosmological tradition of a locality is intuitive, agglutinative, and variable in its nature rather than logical and deductive (1987: 79). Barth sees the source of this variation as the relationship between innovative individuals and resources within the social and cultural context (1987: 77). He nevertheless writes of a 'collective cultural tradition' within which knowledge and ideas are distributed (1987: 77, 79). Is the model, then, one of a cultural whole within which variation occurs? If not, how can 'collective' be construed?[3] (See Herdt 1990: 367.)

A homogeneous *habitus* or structure of culture requires a class, group, society, or community to bear it, or a transcendental subject to constitute it as a theoretical object. In the latter case, like the *conscience collective* of Durkheim, culture is analogous to the unitary mind of a single agent. Challenges to such assumptions about shared culture have arisen out of discussions of 'post-modernism'.

Falk Moore (1987: 730) suggests that the assumption that 'a plurality of limited systems of meaning are likely to intersect' promises a better understanding of the permanently transitional condition of the 'post-modern world'. It may also help in understanding the trajectories of some social universes which have recently moved into a more direct relationship with global social processes. As Gerholm remarks when writing of rituals (1988: 194–6), plurality of perspectives, fragmented cultural universes, distinctive private experiences, and the reinvention of tradition may well be representative of times and places other than 'the post-modern world'. In Keesing's view (1987: 41), views of cultures as collective phenomena, and of symbols and meanings as public and shared, must be qualified by a view of knowledge as distributed and controlled. If cultures are texts to be 'read', then they are differently read by men and women, young and old, experts and non-experts. Moreover, cultures constitute ideologies disguising political and economic realities as cosmically ordained.

[3] Barth follows a Parsonian path in maintaining a distinction between social organization and a distribution of cultural elements which 'animates' a particular social organization and 'infuses it with its constitutive qualities' (1987: 78). This latter distinction, in conjunction with Barth's views about variation and absence of coherence among concepts and beliefs, seems to generate a contradiction; for if conceptions are only loosely connected and ambiguous, yet in some way 'constitute' social organization, how is it that Ok social organization can be so *clearly* defined in terms of local groups, gender differentiation, authoritative initiators, and oscillation between public and secret phases of ritual?

For Keesing (1987: 163, 165) the diversity of Kwaio knowledge in religious matters renders premises about culture as a system of shared meaning 'deeply problematic'. Culture as text may represent a dangerous reification in which coherence and systematic integration of systems may be partly an artefact of analysis.

Scholars, like Keesing, who have combined cultural anthropological concepts with those of a broadly Marxist framework have also questioned the conception of culture and social order as unified wholes. Among these, Sider argues that in situations of class conflict the notion of shared values provides little help, other than as a starting-point in understanding either how upper-class 'cultural hegemony' is imposed on a populace or how oppositional cultures are formed and asserted. By the same token, societies should not be assumed to have a single social structure and a single culture, for there may be, at any one time, several different structurings of the social order. What the social structure is may be open to contention and confrontation on the part of participants, not only in 'dramatic struggles' but in the everyday exchanges of social life. Sider does seem to assume the existence of shared values and beliefs among foraging peoples, however (Sider 1986: 109, 110).

In an extended discussion of conceptual schemes which include the notions of culture, hegemony, and ideology, the Comaroffs suggest, in response to post-structuralism, that indeterminacies of meaning and action, and of events and processes in history, have to be taken seriously. Culture is not an overdetermining, closed system of signs, but 'a set of polyvalent practices, texts, and images that may, at any time, be contested'. Always implicated in culture, consciousness, and representation, power is 'an often elusive and diffuse force'; 'the meaningful world always presents itself as a fluid, often contested, and only partially integrated mosaic of narratives, images, and signifying practices' (1991: 17, 27). Further, in colonial and many other contexts the semantic landscape contains a plurality of cultures or systems of symbols, values, and meanings (1991: 28).

I suggested that uniformity of culture implies agreement about, or the imposition of, meaning. But I want to take the cue from these scholars, and start not with the notion of a community or society bearing a culture, belief system, or system of knowledge, but with the attitude that no degree of agreement over meaning is to be taken for granted, and that uniformity of 'culture' across populations or continuities over time, where they occur, demand explanation no less than discontinuities and change. The problem is, how do we conceptualize what must be shared for communication and interaction to be possible, without presupposing homogeneity of culture? I shall sketch an approach, beginning with posited features of language.

Indeterminacies in Practice

What we call 'practices' and 'institutions', and describe as if they were objective entities, are constructed as objects within intersubjective discourse. The framing of many actions, entities, categories of person, states of affairs, and relations is, in part, brought into being by the constitutive use of language (Searle 1969). The enactment of practices, institutions, and relations thus entails a constant interplay of utterance and interpretation, misinterpretation as judged by a speaker (or bystander), and deceit. Forms of expression and application of expressions are constantly subject to new, inventive, ill-informed, or aberrant recombination and extension. (Just what is ill-informed or aberrant, though, is a matter of judgement.) But more fundamentally, perhaps *no* linguistic expression has an invariant interpretation; its 'fit' with the world is to a degree uncertain and contestable among users (Johnson-Laird 1983: 201; Sahlins 1985: ix; Fish 1989: 150). As Parkin (1982: xlv) expresses a similar idea, 'culture *is* people interpreting, an activity which necessarily implies at least a degree of conceptual freedom'. Or, as Keesing (1987: 162) puts it, concepts, tropes, images, myths, and rites are 'deeply ambiguous'. All this is not to say that some utterances or other communicative acts are not more ambiguous or indirect than others.

Even given a shared grammar, some of the same terms can have different meanings for different speakers. At a higher level, such as the application of explicit rules or the interpretation of entities based on metaphor, there is even more room for ambiguity and disagreement. There is also plenty of scope among speakers of the same language for incomprehension and misinterpretation, because languages are not unitary; indeed, the vagueness discussed by Johnson-Laird, the malleability suggested by Fish, and the conceptual freedom indicated by Parkin, imply non-uniformity of language meaning. Further, no one is competent in all ways of talking in 'the same' language. This is obviously true of people living in industrial societies (Giddens 1979: 72–3), but it is true also of foragers. Since many practices and institutions are linguistically constituted, brought into being out of pre-linguistic elements by the invention and application of linguistic categories, implicit definitions and explicit rules, the indeterminacy and contestability of language implies that language-dependent practices are also to a degree indeterminate, interpretable, and contestable in their nature.

While meaning is to an extent always indeterminate, people must communicate with sufficient precision for everyday practical purposes. Furthermore, the *expression* of difference presupposes shared linguistic and socio-linguistic conventions, concepts, and ontologies in terms of which it can be expressed. In order to avoid the assumption of a culture-bloc, I suggest heuristically that within the Aboriginal domain, people who were linked by the greatest frequency and density of interaction and (as aspects of the former) by relations of socialization which tended to reproduce a similar *habitus*, were in the greatest concord over the fundamentals of concepts, language use, and other modes of action—grammatical structure, vocabulary, conversational strategies, spacing behaviour, and bodily demeanour, as well as values and attitudes. Nevertheless, we find gradual shifts in the constitution of practices as we move from one language to another across north-east Arnhem Land, as we have seen; moreover, an individual could switch if necessary from one language or mode of practice to another, or translate the terms of one variant into those of a different one.

Conversely, some limited ways of talking which are confined to specific practices cross-cut language boundaries (Grace 1987). People of different regions who spoke very different languages such as Dhuwal and Rembarrnga could come together to enact ceremonies and perhaps engage in ways of talking specific to that practice. (My knowledge of Dhuwala'mirri is of a similar order: because I have researched religious matters in most detail I am able to converse quite well in Dhuwala about them, but I would not be able to talk very fluently about, for example, fishing.) Such participants from different regions could communicate to a sufficient degree for the matter in hand, and were linked by a chain of participants who were able to engage in more complete discourse.

Uncontested in the social life of north-east Arnhem Land were the outlines of broad domains in the social world. The most inclusive categories in north-east Arnhem Land included *gurrutu* (relationships, kinship), *ba:purru* (relations to ancestors, country and groups), *mala* (groups, 'clans'), *wa:nga* (country) and *madayin* (things and activities connected with the *wangarr* ancestors). These domains were not simply bundles of kinds of action and social roles, I think, but more complex semantic fields. They were not supposedly exclusive domains of activity, such as 'politics' and 'religion' in Australian society, but interwoven. Nor were they congruent with the anthropological and sociological concept of institution, which implies routinized practices carried out or recognized by the majority of members of a collectivity (Giddens 1979: 80; see also Nadel 1951; I. J. Cohen 1989: 39). Speakers of Yolngu languages held the definition of domains of the social world to a degree in common with the speakers of neighbouring languages, so that there was no closed Yolngu community. The existence of, and the general assumptions underlying, such domains were probably wholly unquestioned, although aspects of the definition and content of these broad domains was variable and highly contestable; it is these aspects that are the concern of this book.

Whatever the degree of concord, the potential for interpretive difference was built in, especially to the extent that action is language-dependent. Concord at one level, say the structure of grammar, did not entail agreement at another. Moreover, sharing of a practice of one kind, such as a particular type of ceremony, did not entail other common practices, such as mode of classifying kin. Acting out of a variety of motives and interests, people struggled over whose interpretations of terms, whose readings of rules, whose beliefs would prevail.

However, as Parkin points out, conceptual freedom is constrained, at least temporarily, by the extent to which an interpretation is acceptable to those with whom a person talks; discourse thus cannot be separated from power (1982: xlv). Fish (1989) puts it the other way round: people struggle to impose their readings on others. Similarly, if all meaning were potentially open to contest, and all power were potentially unfixed, the Comaroffs ask, how is it that history keeps generating hegemonies that seem able to impose some order and stability over long periods? How can relatively small groups gain and sustain control over large populations, drawing them into consensus with dominant values? And how is it that some meanings, practices, and concepts are meaningless, impracticable, or inconceivable (1991: 17)? Arguments and struggles to fix 'meaning', they suggest, are seldom equal; authority establishes itself in an apparently established order of things—'order' in the dual sense of 'command' and 'conditions of being' (1991: 18; see also Keesing 1987).

These questions lead me to the second problem in the analysis of the control of religious knowledge: its relation to power more generally.

Ideology and Hegemony

The Comaroffs approach the politics of meaning by attempting, like Sider, to integrate concepts of culture, ideology, and hegemony (Gramsci 1971). Culture is 'the space of signifying practice, the semantic ground on which human beings seek to construct and represent themselves to others' (J. and J. Comaroff 1991: 21). Hegemony and ideology are two modalities in a 'cultural field'—the two dominant forms in which power is entailed in culture. Humans have the capacity 'to shape the actions and perceptions of others by exercizing control over the production, circulation and consumption of signs and objects'. Ideology, which is in the agentive mode and explicit, is the world-view of a group; 'an articulated system of meanings, values and beliefs' borne in explicit manifestos and everyday practices, self-conscious texts, popular styles etc. But a kind of non-agentive hegemonic power proliferates outside the realm of institutional politics, hiding itself in the forms of everyday life; 'the silent power of the sign, the unspoken authority of habit, may be as effective as the most violent coercion in shaping, directing, even dominating social thought and action' (1991: 22, 28). The Comaroffs finally define hegemony in terms similar to Bourdieu's (1977: 169) notion of *doxa*; hegemony refers to

> that order of signs and practices, relations and distinctions, images and epistemologies—drawn from a historically situated cultural field—that

come to be taken-for-granted as the natural and received shape of the world and everything that inhabits it. (J. and J. Comaroff 1991: 23)

The more successful a dominant group is, the more their identity will disappear into the domain of the hegemonic, the tacit and uncontested (1991: 20).

Although the Comaroffs contend that inevitably some 'symbols and meanings' are not hegemonic, it seems problematic to try to sort those tacit aspects of 'culture' which are hegemonic from those that are not. Should we conceive of them as additional to a politically neutral bedrock? Moreover, the very distinction between tacit/hegemonic and explicit/ideological is questionable, for what is in some contexts implicit may in others be made explicit, as when correcting a child's use of language, countering a potentially threatening innovation, or coming up against very different practices. Moreover, to regard the tacit as hegemonic seems meaningless unless it can be shown *how* the structure of a grammar, the classification of persons, or notions of causality underpin power relations, or may emanate in some way from a dominant group or category.

Bourdieu's concept of *doxa* can be read as a reworking of hegemony in relation to what he refers to as 'ancient society' and 'undifferentiated' forms of social order. Systems of classification which reproduce, but conceal, 'objective' divisions by sex, age, or position in relations of production secure the 'misrecognition', and hence the 'recognition' of (i.e. adherence to) the arbitrariness on which they are based. In the extreme case, as in 'ancient societies', where the subjective principles of organization correspond most closely to the objective order, the natural and social world appears as self-evident—the experience of *doxa*. The instruments of knowledge of the social world are, objectively, political instruments 'which contribute to the reproduction of the social world by producing immediate adherence to the world, seen as self-evident and undisputed, of which they are the product and of which they reproduce the structures in a transformed form' (1977: 164). In relatively undifferentiated social formations in which the prevailing classificatory system encounters no rival or antagonistic principle, the political function of classifications is most likely to pass unnoticed. In contrast, orthodox and heterodox beliefs, which imply awareness and recognition of the possibility of different or antagonistic belief, emerge in the context of class struggle (1977: 169).

In Bourdieu's scheme *doxa* does not simply correspond to the tacit, as in the Comaroffs' interpretation of hegemony. The main function of modes of classification is to conceal the real 'objective' structure of social relations. 'Objective' social relations can only mean those described by the analyst in the light of social theory; I would say, rather, that evaluations of power relation or exploitation by an analyst are 'subjective', because they are necessarily made in the light of relative and contestable values embedded in that theory. This is not to say that ideological assertions or modes of belief may not be obfuscating, or that people do not attempt to conceal their actions and motives behind a veil of deceit, but that deceit does not imply the existence of a parallel 'objective' world of social relations behind the 'misrecognized' ones. It does not require the language or the values of a social theory in order to be expressed.

The social world of north-east Arnhem Land did not conform to the ideal type of less differentiated social orders proposed by Bourdieu, in which instituted discourses reduplicate the self-evidence of the world and affirm the group's adherence to that self-evidence (Bourdieu 1977: 167). Yolngu were aware of alternative beliefs within their own world, denoted by expressions such as 'a different path', and people affirmed that their own group's beliefs were, to use Bourdieu's expression, 'right opinion'.

The concept of ideology has been applied to Aboriginal religion by Bern (1979) in his general model of the 'pre-colonial Aboriginal social formation'. Religion was an ideology by means of which power relations were reproduced, and gender categories were quasi-classes. The organization of society was maintained through the control of labour, achieved through a religious ideology according to which men were in control of the conditions of reproduction and through 'totemic geography' which gave men control of land; mature men were the dominant category. Men were represented in religion as primary in reproducing the conditions of life and in the control of fertility and production; men controlled the definition of religious 'wealth' which they accumulated; they defined and controlled women's productive and reproductive power, and accumulated wives; and they controlled younger people through initiation.[4]

Bern's model has been criticized for its universality and for its unequivocally unilateral notion of power (Keen 1989*a*: 31), as well as for its distorted view of gender relations (Merlan 1988). Hamilton (1980, 1981*a*) challenged the universal applicability of Bern's model. Aboriginal women's status varied as the outcome of differences in their degree of autonomy in the religious life; 'homosocial' organization in the eastern Western Desert, together with the matrilineal organization of grindstone technology, imposed limitations on male power to a degree not evident in Arnhem Land.

More fundamentally, if we take ideology to be a set of beliefs, whether mystifying the realities of power or not (Boudon 1989: 23; Keesing 1987: 141–2), this is to reduce complex religious practice to propositional content. It is clearly appropriate to inquire how the realities of power may be disguised, legitimated, or made possible by certain beliefs, but I am more in sympathy with a view of ideology as a set of practices rather than a category of consciousness, a view which tends to dissolve 'ideology' as a separate aspect of power (Althusser and Balibar 1970; Abercrombie *et al.* 1980: 170).

The notion of an ideology of domination (and a dominant ideology) presupposes a division of a population into dominant and subordinate groups, categories, or classes, and hence a simple asymmetric notion of power. Yolngu social life did not entail a class structure, and until the arrival of missionaries, and within the Aboriginal domain thereafter, no small group sustained con-

[4] Bern goes on to argue that the dominant category was organized beyond the local group as a regional network, but men competed for control of religious property in segmentary relations. Ultimately men were able to control others through the use of force, the threat of sorcery, and through control of access to ritual. Women's ritual, on the other hand, provided neither the ideology nor the organization to challenge men's domination. Part of Bern's argument appears to owe much to Bourdieu; the taxonomies of a mythico-ritual system at once divide and unify, Bourdieu suggests, 'legitimating unity in division, that is to say, hierarchy' (1977: 165). Not only is the division of labour and power between the sexes legitimated by a mythico-ritual system dominated by male values, but the social structuring of temporality symbolically manipulates the boundaries which define age-groups and the limitations imposed at different ages. Social representations of the different ages of life, and of the properties attached to them, express and help to reproduce the power relations between age-classes in which men of mature age are dominant (1977: 165).

trol over large populations in north-east Arnhem Land. Yolngu social and political relations took the form of a complex web, in which action was organized on the basis of focused networks, gender and age categories, and patrifilial identity, rather than classes, communities, or corporate groups. We need to attend to the politics of religious knowledge along these other dimensions of power relations as well—between groups and group-centred networks, between individuals, and between Yolngu and the wider community. It follows from this view that concepts of ideology and hegemony become problematic within an Aboriginal domain, for they have no clear source. They are more salient, perhaps, when one analyses the relationship between Yolngu and the non-Aboriginal community (cf. Trigger 1992).

I have suggested (Keen 1989*a*) that the opposition of domination–subordination, even supplemented by categories such as persuasion–coercion and compliance–resistance (Guha 1983), cannot represent the subtle interplay of relatively egalitarian relationships, or indeed the political dynamics of formally asymmetrical relationships such as leader and follower. As Foucault points out, power relations 'are rooted deep in the social nexus, not reconstituted "above" society as a supplementary structure whose radical effacement one could perhaps dream of' (Foucault 1986: 222). To live in society 'is to live in such a way that action upon other actions is possible—and in fact ongoing' (1986: 221). In other words, power relations are ubiquitous; an aspect of all social interaction. There is no primary and fundamental principle of power which dominates society down to the smallest detail; rather, the 'forms and specific situations of the government of men by one another in a given society are multiple; they are superimposed, they cross, impose their own limits, sometimes cancel one another out, sometimes reinforce one another' (1986: 224). In the case of the modern state, however, power relations have come more and more under State control to be progressively 'governmentalized' in a narrow sense of the word (1986: 224), and in any social order a prevailing direction of power relations can be discerned.

Rather than trying to tease out the ideological and the hegemonic as subclasses of culture from the totality of Yolngu religious practice and the control of religious knowledge, I propose to consider the latter as what Foucault (1980: 194) refers to as *dispositif* ('apparatus'), a heterogeneous mode of discursive and non-discursive practice which incorporates various technologies or techniques of power, and which is a conglomerate of practices with a variety of motivations, origins, and purposes. Nevertheless, ideology and hegemony remain useful characterizations of aspects of practices. Some of these 'techniques' were related to pervasive qualities of Yolngu social interaction concerning the property of language which Steiner (1975) calls 'alternity'.

Religious Secrecy

Discussion of the control of religious knowledge in the anthropological literature, manifested most recently in Herdt's (1990) contribution to the field, tends to discuss 'secrecy' and 'secret organization' as a special mode of organization with its own unique properties. Thus secrecy imparts value, promotes in-group solidarity, gives rise to an alternative reality and ontology, and so forth (see Chapter 7). In north-east Arnhem Land there was no separate domain or institution of 'secret organization'; rather the control of religious knowledge was an elaboration of a pervasive obliqueness in social interaction.

What Steiner calls 'alternity' (1975: 222) is as much an inherent aspect of language as indeterminacy of meaning. The capacity of language to put forward propositions about the future is part of a more general capacity to frame non-factuality and counter factuality, or the 'fictional', and a capacity for concealment (Steiner 1975: 160, 231).[5] Bourdieu's concept of 'misrecognition' (*méconnaissance*) can be seen as an aspect of this more general condition. Misrecognition is not so much intentional deceit as a pervasive failure to make motives and conditions explicit: such as self-interest in Kabyle exchange and the arbitrary nature of an ideology justifying domination (Bourdieu 1977: 171).

Modalities of obliqueness and deceit, what might be called 'opacity' in communication, were pervasive features of Yolngu social interaction and characteristic of some other Australian Aboriginal peoples (Kendon 1990: 452–4). As with the Aboriginal

[5] I suggest that alternity is the product of the property of human language which Hockett calls 'displacement', which makes it possible for us to refer to objects and events that are remote in time and place from the act of utterance itself (Hockett 1958: 579; Lyons 1977: 80).

English of south-east Queensland people (Eades 1982), the language of reasons was non-specific, consisting more in the use of conjunctions than in that of causal terms; people were not under a social obligation to answer questions but might remain silent without rudeness; and they elicited information not predominantly by the use of direct what/why/where/who questions but by offering information to be corrected or corroborated. Eades relates this obliquity to the public nature of social life and the consequent maintenance of private social space or, we could say more generally, autonomy. By the same token, to shape and guard the communication of information at the level of group interaction is not confined to formal religious matters, but is the stuff of everyday interaction (Sansom 1980).

The constitution of guarded religious knowledge was founded in ambiguity. I posit a general structure in the control of religious knowledge both between groups and age/gender categories, resting in the ambiguity of the meaning of enacted forms. Yolngu religious practices involved participation in shared though negotiated forms (of song, dance, design) by people of different groups and regions with rather distinct beliefs and even modes of social organization, who both contributed segments specific to their own identities, and interpreted these ambiguous forms in their own ways (cf. Leach 1954: 4). A similar structure was the basis of the control of the transmission of religious knowledge between old and young, men and women; people enacted many forms in public, but some restricted access by others to the many-layered interpretations of those forms. As Sutton (1987) has recently expressed a similar structure in his discussion of Aboriginal song, 'mystery' was in the hands of the few, who deliberately used devices of obscurity and ambiguity in the control of religious ideology.

The foregoing discussion implies both a distinct domain of Aboriginal practices and continuity of forms of practice through time. However, the same theoretical stance from the point of view of which practices are ambiguous and contested, implies also that they have an inherent potential to change. People 'creatively reconsider their conceptual schemes' (Sahlins 1985), an interpretive community is a 'moving project' in which the work of transforming the landscape into material for that project transforms the project itself (Fish 1989: 150), and it is more appropriate to regard a 'code' as an ongoing process rather than as a set of stable elements of culture which endure through time, and 'whose implementation involves the continual discovery of its own content, a discovery that is at the same time the accomplishment of its own alteration' (1989: 151).[6] Nevertheless, by no means all transformations in Yolngu religious beliefs and practices have been the result of endogenous processes (indeed, the very idea of 'endogenous' is suspect in the light of the foregoing discussion). Many have arisen as the direct result of the increasing incorporation of Yolngu social life into that of the wider community. Trajectories of change are traced in several chapters; the remainder of this chapter sets out to establish the wider context of these transformations, especially the imposition of mission and government control.

The Social and Historical Context

The 3,000 or so Yolngu who lived on the wedge-shaped peninsula of north-east Arnhem Land during the mid- to late 1970s formed a linguistic enclave, albeit modified by linguistic diffusion. They spoke suffixing languages (of the Pama-Nyungan language family) while their neighbours to the south and west spoke prefixing languages (of the non-Pama-Nyungan language family). Yolngu languages were more closely related to those of Cape York Peninsula in northern Queensland across the Gulf of Carpentaria and to those of the people of the semi-desert than to those of their neighbours. Speakers of Yolngu languages lived in the three major coastal settlements of Milingimbi, Galiwin'ku (formerly Elcho Island Mission), and Yirrkala, the two smaller townships of Ramingining and Gapuwiyak (formerly Lake Evella mission), and at many 'outstations' or homeland centres established especially since the 1970s by people returning to their traditional countries (Map 1). Many more outstations have been established during the 1980s. Yolngu moved from a life supported wholly by hunting,

[6] The scope of a code, Fish argues, may be extended to render intelligible an occurrence it could not have predicted, so that the code, which is inseparable from the practices it enables, has been augmented and modified and so changed (Fish 1989: 151). Sahlins (1985: ix) refers to the process in which concepts acquire new connotations as 'the functional reevaluation of the categories'.

gathering, and fishing, to engagement in a kind of farming village, then to a partial dependence on transfer payments.

Compared with the Aboriginal people of many parts of Australia, Yolngu were protected from gross destruction of their population and way of life following British colonization of Australia by the unsuitability of the region for the cattle industry, by an enlightened mission-policy, and, with the exception of the Gove bauxite mine, by the fact that the strongest push in mining exploration came after the conversion of the reserve to Aboriginal Land under the Aboriginal Land Rights Act (1976). This gave Aboriginal people the power to deny permission to mining companies to explore and mine on Aboriginal land.

The establishment of missions in north-east Arnhem Land was not the first Yolngu experience of relations with non-Aborigines. Men came annually in prahus from Macassar (in what is now Sulawesi) on the north-west monsoon to live on the beaches and to gather and process trepang (or sea cucumber), returning on the south-east winds at the end of the rainy season. These annual visits began perhaps in the sixteenth century and continued until curtailed by Australian authorities at the beginning of this century (MacKnight 1976). Many Yolngu religious traditions reflected their relations with the Macassans. Songs told of the Macassans and their life in the Celebes, as well as peoples of 'Badu', the Torres Strait Islands (R. M. Berndt 1948*a*). Ceremonies represented the practices of Macassans, including their rituals. The subjects of these songs and ceremonies were not merely historical or typical figures but *wangarr* ancestors of the human Macassans. People of the groups who sang about the Macassans regarded Macassar as a land of the dead to which spirits may return; and groups with the Badu traditions believed the islands to be the home of Yolngu spirits of the dead. These 'Macassan' ceremonies were still current in north-east Arnhem Land in the mid- to late 1970s, but some people reinterpreted them as representations of missionaries and their activities. Through the exchange of personal names Macassan names entered the Yolngu lexicon, along with other words. Some Macassan place names, such as that of Kopanga, at the mouth of the Blyth River, and that of Gambung, on Langarra (Howard Island), remained in use, for the Macassans applied their own names to the landscape of north-east Arnhem Land and declared certain rocks sacred, as sites for offerings to sea spirits (R. M. and C. H. Berndt 1954: 45).

Yolngu opposition was one of several factors which prevented the successful establishment of the pastoral industry in north-east Arnhem Land. The earliest British attempts to settle the north took place in 1824 and 1827, but while much of the Northern Territory was 'taken up' by land speculators in the decades following the establishment in 1863 of the administration of the Northern Territory by the South Australian Government, by no means all the land was occupied by whites. It was the construction of the overland telegraph between 1870 and 1872 that is said to have stimulated the gold-mining boom, pastoral activities, and attempts at agriculture (Bauer 1964; Powell 1982). At least two attempts were made to establish cattle stations in north-east Arnhem Land from 1885, but the projects failed because of the distance from the main market, difficulties mustering in that terrain, unsuitable grasses, and the opposition of the Yolngu manifested mainly in the form of cattle killing (R. M. Berndt 1954: 98–101; Cole 1979: 80).

The South Australian Government administered the Northern Territory until 1910. When the Commonwealth took over in 1911, the Northern Territory administration was headed by the Administrator, who was directly responsible to the Commonwealth Minister for the Interior, and later to the Minister for Territories. By that date there were some 3,000 non-Aborigines living in the north, although only a few explorers and would-be settlers had made forays into north-east Arnhem Land (R. M. and C. H. Berndt 1954: 76, 92–4, 98). Legislation between 1910 and 1920 was 'protective' and restrictive in effect (Donovan 1984: 19, 21–3). The Northern Territories Aboriginals Act of 1910 specified conditions of employment of Aborigines, made the Chief Protector, who headed an Aboriginals Department, the guardian of every Aboriginal person, and provided for the declaration of reserves and for Crown land to be leased to missions. Marriages between Aborigines and non-Aborigines were controlled, and the Act also provided for relief and care of the old and infirm, the sick and the indigent. A 1911 Ordinance gave the Protector the power (which he could delegate) to take any 'Aboriginal' or 'half caste' into custody. Later legislation provided for the payment of Aboriginal labour. In spite of these provisions, effective control rarely extended beyond Darwin, and indeed the office of Chief Protector was vacant between 1914 and 1927. Economies established during and after the War meant that the police were in practice responsible for the 'protection' of Aborigines, but in areas remote from non-Aboriginal settlement, such as north-east Arnhem Land, missionaries became agents of the State.

In response to conflicts on the Roper River and in Arnhem Land between Aborigines and would-be settlers, and following discussions with government, the Methodist Overseas Mission (later incorporated into the Uniting Church) established three mission stations across north-east Arnhem Land. After a false start at Galiwin'ku, the first was established at Milingimbi in 1923 (M. McKenzie 1976: 7; Cole 1979: 114).[7] In contrast with the Macassans, the missionaries had the aim of transforming the Yolngu way of life.

Yolngu did not simply tolerate the presence of the missionaries, say for economic reasons; rather the coercive power of the State was necessary for the establishment of the mission settlements. The missionaries who first settled at Milingimbi island were armed, and in the early decades missionaries communicated with Darwin Police by radio and telegram over reported killings and spearing of stock. However, it seems clear that T. T. Webb, who was Superintendent from 1926, preferred to minimize outside interference and to deal with violence in his own way. In his annual reports Webb unfailingly recounted a 'peaceful' year. In doing so he was hiding the violent conflict between migrants from the east and the more local Burarra people.[8]

Patrol officers moved through Arnhem Land from time to time during the two decades following the Second World War, sometimes in the company of police, in order to investigate such matters as violence, the abduction of women by Aboriginal men, and alleged murders.[9] The police eventually established a more immediate presence at Maningrida government settlement to the west of the Yolngu region, begun in 1957, and at the mining town of Nhulunbuy.

[7] See also Review Report Milingimbi Institution North Australia District, *Methodist Overseas Missions from Establishment to 31st December, 1949*, F1 1953/266, AAD.

[8] See Milingimbi Station Reports, F1 1953/266, AAD.

[9] See Memo G. Holden to District Superintendent, 20 Sept. 1953, F1 54/81, AAD; F. H. Moy to Administrator, 26 Sept. 1950, F315 49/393AIII, AAD; S. H. Kyle-Little to Director of Native Affairs 31 Jan. 1948, F315 49/393AI, AAD.

The imposition of State control profoundly transformed Yolngu social life in at least one respect. Direct intervention by missionaries, police, and patrol officers, aided by some Yolngu men and women and backed up by the promulgation of a Christian ethic of forgiveness, eradicated overt blood-feuds. Missionaries and police prevented or took steps to punish not only killings of non-Aborigines by Aborigines, but also Aboriginal killings of other Aboriginal people. And missionaries and police intervened not only in homicide but in spearing which did not lead to death, and in disputes that might lead to violence and death.

The presence of the mission as a resource at Milingimbi led to the more permanent presence in the Glyde River region of Yolngu whose countries were in the Galiwin'ku and Buckingham Bay regions to the east. This movement led to conflict, especially with the Burarra-speaking people of Milingimbi and Cape Stewart. Yan-nhangu speakers of Milingimbi and the other islands of the Crocodile group were married and allied through ceremony both to other Yolngu and to the Burarra. In the early years of the mission, fights between Yolngu and Burarra men were 'frequent and fierce, resulting in numerous casualties, many of which required lengthy treatments'.[10] Webb apparently encouraged the settlement of conflicts by means of the Makarraṯa conflict-resolution ceremony.[11] Children of people who had migrated to the mission, and whose own countries were far away, learned about the country in the vicinity through foraging trips, and visits to relatives. Many gained vicarious experience of their own lands and waters through stories and ceremonies, but were attached to mission country by the dogma of spiritual conception and by life experience.

In spite of its reliance upon the coercive power of the State, the Methodist Mission was relatively benign, largely as a result of the enlightened views of John Burton, the new General Secretary of the Methodist Overseas Mission, and T. T. Webb (M. McKenzie 1976: 40). It elected to gain its ends through persuasion rather than force and without radical cultural surgery such as a dormitory

[10] See Review Report Milingimbi, 1949, F1 1953/266, AAD. What is called the Gupapuyngu ('Gobaboingo') faction in the Report were predominantly speakers of Dhuwal, Dhuwala, and Djangu, belonging to groups such as the Daygurrgurr Gupapuyngu, Djambarrpuyngu, and Wan.gurri.

[11] See Review Report Milingimbi, 1949, F1 1953/266, AAD, p. 2.

system, banning of ceremonies, prohibition of arranged and polygynous marriages, and enforcement of the use of English. However, some individual superintendents and ministers made sporadic attempts to ban ceremonies and encourage monogamy, and some mission workers regarded Yolngu religion as Satanic.[12] Early relations with the missions gave the appearance of being largely voluntary, although the provision of tobacco was an indirect mode of coercion because of its addictive character. From their point of view Yolngu exchanged periodic labour for resources of tobacco and food, and perhaps the fear of incursions and reprisals attracted people to the security of mission settlements. More women than men lived on the missions before the Second World War, perhaps in order to escape oppression by men. The number of more or less permanent Yolngu residents grew to about 200 in 1949, while some 250 visited from time to time.[13]

The exchange relations between Yolngu and missionaries became transformed as succeeding generations born at the mission acquired a *habitus* somewhat different from that of their parents (as many as four generations, traced through women), conditioned in part by the routines of a farming community, dominated by clock and calendar, by Christian discourses, and by schooling established by the missionaries. The Methodist doctrine of non-interference in Yolngu social structure and ceremony ensured that intervention in socialization was not as radical in Arnhem Land as in other parts of Australia. Nevertheless, a substantial proportion of children received schooling from the 1950s, learning new skills, a new language, different concepts, and a distinct view of the world. This is not to say that Yolngu children internalized the complete structure of concepts and values of Australian culture, for they interpreted the information and behaviour to which they were exposed through a grid of fundamental Yolngu concepts and attitudes, as recent research in education has shown (Stephen Harris 1984; Christie 1984).

What children missed by daily attendance at school was perhaps as important as what was gained. Removed from the daily business

[12] One white person at Galiwin'ku regarded Yolngu culture and especially religion as evil. Another thought that bark paintings were vehicles for demons and burned them (Bos 1988*a*: 125; R. M. and C. H. Berndt 1980: 4).

[13] See Milingimbi Mission Station Report for year ending 31 Dec. 1931, F1 1953/266, AAD.

of camp life, children regularly spent the days with much larger numbers of peers than formerly, including those of the opposite sex. People learned and adopted new ends and new routines, especially those related to settlement services, maintenance, and administration, interpreted and modified in the light of Yolngu concepts. Money was woven into the gift economy, and similarly transformed.

The establishment of mission-settlement communities produced a new mode of variation in culture. One axis of this variation was the distinction young people drew between town and 'bush', regarding 'bush people' (*d̲iltjipuy yolngu*) as rather inferior. Another was the difference engendered by a sort of founder effect and the influence of key individuals. In this way Elcho Island mission settlement, among whose first residents were Yolngu who identified strongly as Christians, became something of a centre of Yolngu Christianity and innovation.

The missionaries had established what was in effect a farming village, but it was by no means self-sufficient. The economy was subsidized by a small government grant, contributions to the Methodist Overseas Mission, the payment of extremely low stipends to the missionaries and lay staff, and payment in kind to Aboriginal workers. With the aid of community workers and advisers, aspects of this settlement organization survived the withdrawal of mission control from 1974, although the farming economy did not.

Arnhem Land people largely lost control of the movement of outsiders within Arnhem Land until the passing of the Aboriginal Land Rights (Northern Territory) Act 1976. Moreover, during the first half of the century the Northern Territory Administration attempted to prevent too many Arnhem Land Aboriginal people from moving out of Arnhem Land to settle in Darwin. Individuals, however, could move in and out of the mission regime more or less at will. Loss of the power of movement was greatest during the Second World War, when non-Aborigines were evacuated from the north and the army strictly controlled the movement of Aboriginal people.

Perceptions of the relative power of Aboriginal, white, and Japanese people must have been greatly affected by experiences during the war. With the Japanese bombing of Milingimbi, Yolngu living there had some direct experience of the military power of an industrial society. More positively, some Yolngu entered into new kinds of relations with whites, as employees in army camps and participants in a reconnaissance unit under the anthropologist Donald Thomson.[14]

A major shift in government policy was framed before the Second World War and implemented in the 1950s. After the war Aboriginal affairs were still under the effective control of the Commonwealth, but the scale of government administration increased by an order of magnitude. The staff of the Northern Territory Welfare Branch, which replaced the Department of Native Affairs, increased to nearly 1,000 by 1971, compared with a Native Affairs staff of 64 in 1953.[15] Consistent with 'assimilationist' policy, according to which Aboriginal people would be accorded the same rights and privileges as other citizens, and were expected eventually to live like non-Aboriginal Australians, the rights of other citizens were gradually extended to Aboriginal people.[16] Aborigines received social security benefits in 1958, the right to vote in 1962, and the right to drink alcohol outside the reserve in 1964. For 'tribal' people of full Aboriginal descent the process of assimilation was to be gradual, and entailed a continuing curtailment of legal adulthood: the Northern Territory Welfare Ordinance (1953) authorized the government to place Aboriginal people under its care as 'wards'. Reserves were to be temporary places of preparation for full participation in Australian life (Cole 1979: 90). The new policy led to radical changes in the relationship between government and missions, which were to receive more funds from government and have a more clearly defined role, acting as a buffer between 'the tribal state and detribalization'.[17]

Through the provision of transfer payments, a 'training allow-

[14] See Milingimbi Station Reports, F1 1953/266, AAD.

[15] The Department of Native Affairs remained under the Administrator and Government Secretary. Provision was made in 1947 for a Legislative Council for the Northern Territory, but this body had powers only to frame ordinances (Donovan 1984: 166). When the Labor Party was defeated in the 1951 federal election the administration of the Northern Territory became one of the functions of a new Department of Territories, and remained so until 1968, when it again came under the Department of the Interior. The Northern Territory Welfare Ordinance was passed by the Northern Territory Legislative Council in 1953 and proclaimed in 1957.

[16] See Memorandum Hasluck to Fadden, 24 July 1952, F1 49/21, AAD.

[17] See *Northern Standard*, 16 Dec. 1938, p. 4c.

ance', and award wages, Yolngu became more involved in the money economy. At the same time, the monitoring and control of movement, births, health, and mortality increased (for example through the compilation of the register of wards), as did the presence on the mission settlements of a variety of government agencies. Among the more profound agents of change was systematic subsidized schooling, established in 1952, and taken over by the government Department of Education after 1966.[18] Through residential colleges, and especially through education received in southern cities, a few individuals became extremely proficient in white practices, especially as administrators and ministers of religion.[19] Medical authorities also intervened in Yolngu lives in a radical way. Yolngu valued and sought out medical care which was more effective than indigenous techniques, but Australian agencies exerted control of health through coercion as well as persuasion, for example forcing the Yolngu to adapt ways of disposing of the body after death.

Evidence of the relative powerlessness of Yolngu in the control of the overall conditions (as opposed to the quotidian details) of their lives was the excision in 1963 of part of the Arnhem Land reserve (proclaimed in 1931) for bauxite mining, and the building of the mining town of Nhulunbuy, which reached a population of about 5,000. Sympathy for Yolngu of the Gove area was an important aspect of the public sentiment which led to the defeat of the conservative Coalition Government, and the election of the Whitlam Labor Government, with its policy of 'self-determination' for Aboriginal people (Stewart Harris 1972).

The policy changes initiated by Whitlam and continued in many respects by the Coalition Government under Malcolm Fraser, had radical effects on Yolngu lives. In this era Yolngu were to feel under increasing pressure from outside agencies upon which they were dependent, but without the former support of the missions. A separate federal Department of Aboriginal Affairs was created in December 1972, community welfare funding increased, and the Federal Government established the Woodward Commission

[18] Review Report Milingimbi Mission, 1952, F1 1953/266, AAD; Letter Moy to Chairman of MOM, 18 Mar. 1953, F1 54/81, AAD; Review Report Milingimbi, 1949, F1 1953/266, AAD.

[19] See Milingimbi Council minutes 23 Nov. 1974, NTRS 40 Box A7398, Drawer 8.5, NTAS.

to report on possible land-rights legislation. In November 1974 a fully elected Legislative Assembly of the Northern Territory sat for the first time, although the Territory was not accorded statehood. The monolithic Welfare Branch was dismantled at the beginning of 1973 and its functions distributed among the Departments of Health, Education, Labour, etc.

With the passing of the Aboriginal Land Rights (Northern Territory) Act of 1976, Arnhem Land Aboriginal reserve became Aboriginal land under inalienable freehold title held by an Aboriginal Land Trust on behalf of those with a traditional right to use and occupy the land. This form of tenure did not include the coastal waters.[20] Traditional Aboriginal owners as defined in the Act had the power to veto proposed mining exploration and extraction, a power which so far has been exercised throughout north-east Arnhem Land. One effect of the Act, together with the withdrawal of the mission, has been to give local land-holders more authority in the administration of communities.[21]

Whitlam Government policies encouraged Aboriginal communities to administer their own affairs. The missionaries had incorporated Aboriginal leaders (mostly men) into the administration of the stations from the 1950s, first informally and later as 'village councils'. During the 1960s elected community councils were formed, and the Federal Government made provisions for incorporated Community Councils to receive and administer grants (M. McKenzie 1976: 171; Northern Territory Report 1957–8, 1964–5).[22] Consistent with these changes in government policy, the Uniting Church shaped its own policy of with-

[20] Complementary legislation enacted by the Legislative Assembly governs access to Aboriginal land, and provides for Aboriginal communities to apply for seas to be closed, although closure affects only those without a current fishing-licence.

[21] We can see a trend during the 1980s for the power of the larger groups whose forebears migrated to the mission stations to be eroded, so that Council leaders now tend to be people of local groups. Another effect of the Land Rights Act has been to enable the Northern Land Council, which has the duty to look after the interests of traditional Aboriginal owners, to intervene in local politics. The all-Aboriginal Council includes members elected by Yolngu, and the bureaucracy consists for the most part of non-Aboriginal people, and Aboriginal people of mixed descent. The articulation of Land Council with local politics is exemplified in the action in the early 1980s by the then Chairman, who intervened to determine traditional Aboriginal ownership of the Koongarra area in the Alligator Rivers region, which was disputed, in favour of a pro-mining faction.

[22] See also Letter Milliken to Superintendent of Milingimbi 27 Dec. 1962 (NTRS 40 Box A7400 NTAS).

drawing from the direct control of Aboriginal settlements, and of acting solely in an advisory and pastoral capacity where requested (United Church 1974). Thus by the 1970s Yolngu were increasingly involved in the administration of settlement affairs, while still relying on the advice and participation of whites as well as their own skills in literacy and numeracy.

Settlement administration did not give power over the affairs of the population of a more widely defined space, however, except indirectly through such means as the provision of funds for outstation resources, or an outlet for craft production. Councils were responsible for such matters as housing, settlement maintenance and public health, schooling, public order, and the admittance of visitors. These new powers gave rise to uncertainties and to conflicts within communities, especially between those whose authority rested on the control of religious practice and those with new skills.[23]

Yolngu also took over the local church. The missionaries seem not to have forced Christianity on to Yolngu, but to have inculcated Christian beliefs through persuasion and schooling. As important, perhaps, was the adoption (and transformation) of Christian concepts and practices by certain individuals as a corollary of close and interdependent social relations with missionaries and lay workers. As well, certain modes of Christian discourse went hand in hand with involvement in institutions such as village councils.[24] The last white minister left the community at the end of 1978, and a Yolngu man was inducted as pastor in the same year. This man began training for the ministry in 1980. A Christian Fellowship with Yolngu leaders met daily during 1980.

The Milingimbi community adviser expressed hopes in 1973 that the Milingimbi economy would be based on fishing, craft, sewing, and a canteen, and that companies would be formed, but these hopes proved to be unfounded.[25] Some Yolngu became motivated to start up their own capitalist enterprises by the early

[23] See Milingimbi Village Council Minutes 1970, 1971, 1972, 1973, NTRS 40 Box A.7398, Drawer 8.2.1, NTAS.

[24] In 1977 Milingimbi Church membership was 69, including 10 white people. See Milingimbi Parish Council minutes, NTRS 40 Box A.7398, Drawer, 8.5, NTAS.

[25] See Report on Milingimbi funding, 29 Oct. 1973, by John Van der Wall. NTRS 40 Box 7397, Drawer 8.1.1, NTAS.

1970s if not earlier.[26] One of the few enterprises that succeeded for a while was the hiring and showing of 16 mm. films, for which families charged admission. During this period, ostensibly 'Aboriginal' organizations, with a high level of white involvement, proliferated. These included an 'advancement' association, housing association, and a pastoral company.[27] With the withdrawal of the missionaries from a supervisory role the organization of settlement work all but collapsed by the late 1970s, although many people were still officially employed and drew wages.

In the middle of the decade the all-pervasive day-to-day activity in which most adults were engaged throughout the settlements was card-playing, one of the means by which money was distributed round the population. At Milingimbi the daily routine of gambling was interrupted from time to time by alcoholic binges which erupted in quarrels and fighting. Alcohol was not available in the reserve but the increase in cash income with the training allowance, and especially the introduction of award wages for Council workers, together with the increased frequency of air services to and from Darwin and Gove, led to more alcohol being brought into the community. During the brawls women, few of whom where drinkers, would take refuge in the houses of school teachers, missionaries, and health workers to escape the violence of drunken men.[28] Petrol sniffing by some children was also

[26] One man applied in 1972 for a lease to carry out timber milling, gardening, and fishing on the Woolen River, but this came to nought. See Letter K. Nowland to B. Clarke 3 Nov. 1972, Letter K. Nowland to Chairman of Lands Board 3 Nov. 1972, NTRS 40 Box A.7397, Drawer 8.1.1, NTAS. People living on the Howard Island Outstation in the mid-1970s were also anxious to begin a cypress-pine logging industry in 1975, although their hopes did not come to fruition.

[27] Milingimbi Progress Association received a grant from the Australia Council to assist artists and craftsmen at Milingimbi and the mainland communities. Milingimbi Community Advancement Association introduced the sale of refreshments at the film show, and attempted to start a weekly social evening for young people. The Ramingining Housing Association, with both Europeans and Yolngu on the executive, had a programme of building 10 houses in 1973–4. A grant of $30,000 made in 1973 assisted the establishment of the Murwangi Pastoral Company, managed by a white adviser and employing Yolngu stockmen. The Company was never viable, and at best provided some meat, rather irregularly, for the local community. Its Yolngu workers enthusiastically adopted the style of dress of Aboriginal stockmen—cowboy hats, checked shirts, jeans, and leather riding boots.

[28] Milingimbi people tended to be divided into pro- and anti-alcohol factions which cross-cut patrifilial group membership. The anti-alcohol faction dominated the Community Council, which attempted to control the import of liquor with the aid of the Darwin Police, whom they asked to search luggage at Darwin Airport. Council

regarded as problematic in this period. More recently the enactment of the Northern Territory Liquor Act assisted community councils in banning the import of alcohol.[29] As well, Yolngu have adopted the drinking of kava on a large scale, which has replaced alcohol to some extent, but after the period with which this book is concerned. Yolngu men have found that they can combine kava drinking with ritual, for unlike alcohol it does not lead to aggressive behaviour.

An increase in the frequency of air transport (about three flights a week in 1976 through Milingimbi), together with the higher disposable income, enabled Yolngu to travel more widely and more often. It also brought an increase in the number of visitors to Arnhem Land communities, including government officials, doctors and dentists, police, academics, social workers, and educationists (M. McKenzie 1976: 167).

About 770 Yolngu were living at Milingimbi in the mid-1970s, with another 300 or so at the mission outpost of Nangalala and the new township of Ramingining on the mainland nearby. Escape from the disruptions of settlement life was one motive for the move out of settlements back to traditional lands; political rivalry was another (see Gerritsen 1982). The policies of the Whitlam Government fostered a proliferation of new 'outstations' or 'homeland centres': small communities of people living on land to which they have close traditional attachment.[30]

members even took it on themselves to search the bags and cases of people arriving at Milingimbi airstrip. See Milingimbi Town Council minutes 31 May 1974, NTRS 40 Box A.7398, Drawer 8.5 NTAS; Milingimbi Town Council minutes 16 Nov. 1974, NTRS 40 Box A.7398, Drawer 8.5 NTAS.

[29] In June 1979 the Liquor Commission advised the Milingimbi Council that it had agreed to declare the area a restricted area under the Liquor Act (Milingimbi Council Minutes 29 June 1979, NTRS 40 Box A.7398, Drawer 8.2.1, NTAS).

[30] 'Outstation' is a term from the pastoral industry, and refers to an outpost of the cattle or sheep station in the country remote from the main station buildings. Settlement on the missions was probably never absolute, e.g. Harold Shepherdson, the Superintendent of Elcho Island Mission, serviced at least four bush communities by light aircraft from the mid-1950s (M. McKenzie 1976: 175). By the mid-1970s several outstations had been established out of Milingimbi on the mainland, but numbers burgeoned through the 1980s. People maintained links with major settlements by four-wheel-drive vehicle, light aircraft, and boat, which enabled them to combine foraging with links to the money economy. Income was derived from transfer payments as well as from the sale of art and craft, and could be spent at the main centres on food and consumer goods. Grants to help with their establishment were available from government from 1970. See Correspondence, NTRS 40 Box A.7399 NTAS.

Throughout this period religion remained a dominant force in Yolngu social life. The following description of Yolngu religious beliefs and practices of the 1970s and early 1980s begins in the same way that Yolngu introduce newcomers to their religious 'law' (*rom*)—with stories about ancestors.

PART I

The Same but Different

Trees and other plants were a rich source of metaphor in Yolngu thought. In the songs of Cloudy Water group the roots of Poison Pea point in the direction of other Dhuwa-moiety groups. These are connected through Mistletoe Tree, which the Djang'kawu sisters planted at each Dhuwa-moiety group's country.

A distinction between outward forms of practice displayed by the group-centred network to other groups, and content supposedly transmitted only with the permission of group members, approximates the structure of Yolngu religious communication. Two or more groups possessed a form in common such as the programme of a complex ceremony, albeit differentiated in detail, but each could include specific elements of its own, and each attached different significance to its sequences of dance and song. They could thus co-operate in a performance while each retained its particular identity, ancestry, and connections with country. The contrast between shared form and differentiated content has to be modified to the extent that people disagreed about the proper 'forms' of religious practice, invented new ones, and incorporated (and transformed) new forms into old forms. Gerholm (1988) refers to such shared forms as the 'hard surface' of ritual or, following Lewis (1980), the 'ruling' which is there for all to observe, while each participant walks away from the ritual having experienced something different. In north-east Arnhem Land, however, it was not simply a matter of a contrast between public form and private experience, but of shared forms current among one public, and distinctive interpretations among a variety of less inclusive publics and individuals.

A similar ambiguity and relativism prevailed in the organization of groups which held country and religious property. There was no uniform group-constitution, for members often disagreed about the relationship between subgroups if the group was divided in that way, and individuals gave different and incompatible accounts of the genealogical structure of a group. Ambiguities pervaded the construction of ancestral worlds, and each group had its own perspective on ancestral events within shared (if ambiguous) categories and assumptions about ancestral powers. People of distinct groups and kin networks contested the definition of country and rights in it, expressing difference in terms of a more or less shared language of ancestral connection and ceremony. Groups of the same moiety formed multiple, cross-cutting connections by sharing similar (but contested) ritual forms, while maintaining their differences.

These matters are the concern of this chapter and the three that follow. We begin with stories about *wangarr* ancestors.

1

The Same Story

Objectified Stories

The identity and actions of spirit ancestors (*wangarr*) were recounted in innumerable myths. On the face of it Yolngu myths, unlike accounts of everyday events, were established forms, more or less reproduced from generation to generation, but while their content was obviously less transient than stories of everyday happenings, this stability was only relative. In reproducing myths, people transformed them, the transformations expressing particular relations, interests, and concerns.

Two young Cloudy Water women dance in a sand-sculpture as the Djang'kawu sisters

Perhaps we should regard Yolngu stories about ancestral beings not as text-like, but as representations of events. My point is that if one thinks of an entity, a myth, and of this entity as like a text, then the problem arises as to what counts as a variation of 'the same' myth rather than different myths? If we think in terms of recounting ancestral *events* rather than a text, there can be innumerable speculations about ancestral events involving ostensibly 'the same' ancestors. Still, Yolngu did objectify stories (*dhawu*), categorizing a tale as 'such and such a group's story'. Depending on their status, tellers were supposed to ensure that they were monitored by others with rights in a story, so that narratives about *wangarr* ancestors were, up to a point, group products. But groups, especially the larger ones, were not united in following invariant orthodoxies. Even greater diversity arose between stories about 'the same' ancestors told by members of different groups. Accounts of ancestral events were relative to the perspective of the teller, which was to good measure a group perspective, and to the context of the telling.

Terms for the Construction of Ancestral Realities

Stanner reconstructed a typical conversation with an Aboriginal man explaining the significance of a totemic place:

> Conversations follow rather like this: 'There is my Dreaming [place]. My father showed me this place when I was a little boy. His father showed him.' . . . What had the father said? 'He said: "Your Dreaming is there; you want to look after this place; you don't want to let it go [forget, be careless about it]; it is from the first [totemist] man."' . . . What did the father do there? 'He used to come here every year with the old men, the wise men; they used to do something here [hit, rub, break off pieces, brush with green leaves, sing]; that way they made the [totem] come on, come back, jump up, spread out.' *How* did that happen? *What* is in the place? 'We do not know. *Something* is there. Like my spirit [soul, shadow, invisible counterpart]; like my brother [father, father's father, mate, friend, helper]; like my Dreaming [naming the totem entity].' Will he think more? What else did his father say? That there was something in the Dreaming-place? The dark eyes turn and look intent, puzzled, searching. 'My father did not say. He said this: "My boy, look! Your Dreaming is there; it is a big thing; you never let it go [pass it by]; all Dreamings [totem entities] come from there; your spirit is there; we do not know what; *something*."' There is a struggle to find words, and perhaps a lapse

into English. 'Like engine, like power, plenty of power; it does hard-work; it *pushes*.' (Perhaps now the anthropologist begins to understand; to fit his own abstract, discursive symbol-structures to that understanding.) (Stanner 1979: 135; the brackets and emphases are Stanner's, I have omitted some of his commentary.)

A theist might argue that there is a spiritual reality towards which Aboriginal belief, as expressed in this passage, is groping. Stanner himself seems to have been of this mind, although the sense of his final comment in the above passage is that Aboriginal people have beliefs that are determinate but not easily *understood* by someone who thinks using 'abstract, discursive symbol-structures'. My view is rather different: it is that these beliefs were *constituted* in the use of imprecise terms. Individuals were free to fill in some of the more concrete details in imagination if they wished, but much in Yolngu mythology seems to have been deliberately ambiguous. Religious mystery was created through ambiguity.

Comparisons with what other ethnographers have recorded show that, while they were inherently imprecise and constantly under revision, elements of Yolngu languages about ancestral worlds transcended time and space. Yolngu constructed ancestral worlds from a shared lexicon and stock of metaphors, a limited set of expressions and enactments the efficacy of which lay, in part, in their indeterminacy. In explaining ancestral events to me men would set the events in a time-frame (I can think of no better expression) of *baman'*, 'long ago', emphasized with a particle—*baman' birr!*, 'very long ago'—or explained with a spatial metaphor—*barrku birr!*, 'far off'. The temporal category *baman'* contrasts with the present (*dhiyangu-bala*, 'now') and the past of living memory, or of remembered human ancestors. To further emphasize the distance, a man would invoke genealogical distance: 'My father did not know the *wangarr*, his father did not know, *his* father did not know; long ago!' Men who instructed Warner during the 1920s evidently used similar expressions, for he took the word *baman'* ('bamun') as the proper name of an ancestral period in rather the same way that more recent commentators have construed *wangarr* as a time-category (Warner 1937: 566).

Warner defined *wangarr* ('wongar') as 'a general name applied to the totemic spirits', while 'in the time of Wongar' referred to the mythological period 'Bamun' (1937: 568). According to Howard Morphy (1984: 17) and Williams (1986: 28), *wangarr* refers to a distant time in the past and to the spirit forms that existed; *wangarr* (time) exists in the present, too, in the signs and spirit-traces that endure. This is misleading in that Yolngu did not have the equivalent of an abstract category 'time', and so did not play this language-game of parallel domains of time. Yolngu languages add the instrumental or ergative suffix *-yu/-thu* to an entity or person to denote the time-frame associated with it, as in *d̠ambuy-wal̠uy* ('head' + INST-'sun' + INST), 'mid-day', so that *wangarryu* can be translated '*wangarr*-times' (Rudder 1991). *Wangarr* ancestors existed and were active long ago; their traces and powers remained, and people explained some of the *wangarr* were still alive and active beneath the waters and earth. This does not mean that Yolngu believed that past time continued in parallel with the present.

Stories about *wangarr* ancestors, together with their coeval human 'forerunners' (*ngurrunanggal*), did accord the beings extraordinary properties and powers, such as huge size and the ability to bring about a tropical thunderstorm, and they contained fabulous events, such as a man being burnt in a fire and becoming the moon. But more subtle constructions lay alongside these simple inventions. In one type of construction the names of non-human entities displaced human nouns and names: hollow tree walks from the hills to the creek, dingo converses with the Macassans about the merits of houses and cities. In another, the description of a normal activity of a non-human species can be read as an analogue of a human action, thus 'honeybee stung Nowa the Macassan after he refused to go back to his own country' can be interpreted as 'Honeybee (singular or plural) speared Nowa the Macassan after he refused to go back to his own country'. While people sometimes explained that the *wangarr* were people, mythic discourse did not require a simultaneous translation in thought from actions and properties of non-human entities to actions and properties of human beings.

A further subtlety consists in what was left unsaid, and what was left enigmatic. We shall encounter an example later in which the identity and gender of a figure was left unclear by a switch in the proper name of the protagonist of the story from 'Djang'kawu' to 'Wal̠awal̠a'. In this example, the story of the Djang'kawu, does the second name refer to another *wangarr*? If so, is the *wangarr* female like the first, or male? Or is the name an alternative proper name for the same two sisters? Gender is also ambiguous in the narrative when the protagonist is referred to simply by the second-person pronoun *ngayi*, 'he/she', which leaves gender unmarked. Ambiguities of a similar order arise from the fact that Yolngu nouns are not marked for the plural.

Uncertainty as to identity and the idea of metamorphosis seem to be inextricably linked in many examples of Yolngu myths and interpretations of song and ritual. I recall reading a transcript of a story at the Milingimbi literacy-centre in which a father takes his son hunting. In the climax of this simple story the boy mistakes a termite nest for a wallaby or kangaroo. Transmutation occurs in many stories about *wangarr*; after a disaster of some kind a being transmutes and 'becomes *wangarr*', or some object used by the being is transformed into a feature of the country. Ambiguity or polysemy and transformation of identity are conflated in the concept of 'turning' (Jorgenson 1990). Yolngu men used the expression 'turn' to denote the moment of transformation, or simply a pause:

> Djuranydjura the Dingo travelled by canoe from Honeybee mainland country to Langarra (Howard Island). With his wife and children he tried to set off for Galiwin'ku (Elcho Island) in the bark canoe. He pushed and pulled the canoe to get if off the bank, pushed and pulled, but it sank. Djuranydjura went up and down grieving for his wife and children. . . . Rocks stand there now.

In the following passage, translated from a mixture of Dhuwala tongue and English, the Honeybee-group leader told me about the relations between the *wangarr* ancestors, the *maḏayin*, and living people:

> Here am I at the beginning, a man, a man; and at the beginning I am an animal. I turned from being Honeybee: here am I a man—that is my *maḏayin* now. I am a Possum too, in the beginning a long time ago, a Possum. Later I turned, and I am a man. Here I sit, a man, like this. I turned each time: Honeybee, Possum, Catfish. I turned and I am a man who speaks the Dhuwala tongue. That is the story: here I am a man, just the same. Here I sit, a man. That is all. From this animal, that animal, that animal, that fish, that animal, whatever animal—I am a man. I have Emu, perhaps I have Catfish, Possum, Darter: there is the design, and here is the waterhole. From there—from the Catfish, Possum, Honeybee—I have the *maḏayin* [sacra] now, everything; I am a man. I

have a song there, I have a song there, a song there and a song there because we *come* from there. I am from there. I am a man. That is all.

The leader here identified with the *wangarr* ancestors of his group, which were both persons and animals. (One of these, Honeybee, was the ancestor proper who created the group, but all were *wangarr*.) Each 'turned', and there were both men and *ma̲dayin*. It follows that he possessed the *ma̲dayin*, including the songs related to the ancestors. Other people said more simply that the *wangarr* were people and turned into animals, or that they were people and the animals came after them. 'Turn' could also denote the relationship between exoteric and esoteric meaning—between 'form' and 'content'. Here a semiotic relationship also implied a substantive transmutation.

The above are some of the conceptual, linguistic, and rhetorical elements which went towards the framing of ancestral worlds. But other semiotics were just as important in giving ancestral realities substance. People pointed to evidence of ancestral actions in features of the country, expressed fears of ancestral forces immanent in certain places, and followed procedures to tap these ancestral powers in the enactment of ceremonies. In rituals, visible and tangible form were given to *wangarr* and other beings.

Categories of Spirit-Being

Elsewhere I have tried to put together typologies of spirit-beings in Yolngu cosmology (Keen 1978). Such an attempt tends to force overlapping and somewhat indeterminate categories into types governed by unequivocal criteria. A typology might look like this:

- Yolngu conceived of *wangarr* ancestors as beings with human form, but having some of the properties of the beings or entities whose names they took, such as Rock or Honeybee, and as having extraordinary powers. They were active long, long ago in 'far off' times. They camped, foraged, made love, quarrelled and fought, and bore children, somewhat like humans, but they were involved in extraordinary events and were transformed by them, perhaps into species such as jabiru or entities such as the moon. Some *wangarr* engendered the ancestors of human groups. These were a group's *gulu'kulungu* ancestors, but there was little uniformity in beliefs about the nature of ancestral procreation: some people affirmed that it was parthenogenetic, others that it was heterosexual.

- *Motj* were also *wangarr*, but the term denoted particularly the dangerous serpents which were believed to lie in deep waters and to cause tropical thunderstorms.
- *Ngurrunanggal* were human beings who were the coevals of *wangarr* and saw or interacted with them.
- Speakers of Dhuwal and Dhuwala dialects conceived of *mokuy*, literally 'ghosts of the dead', as libidinous and cannibalistic 'trickster' spirits who inhabited the forests and were the subject of the *garma* genre of songs and of erotic stories. Warner (1937: 66) was told that these beings were *wakinngu*, without kin, and that they broke all laws. People might hear them playing the drone-pipe and laughing in the forest. Yolngu sometimes referred to them as 'the devil' or 'devil-devil' in English. Some myths relate how the *mokuy* acted in concert with the *wangarr* in distant times. People sang of the *mokuy* during mortuary ceremonies, identifying them with the deceased. *Mokuy* also denoted 'corpse' and the malign spirits of the recently dead who inhabited trees and especially thickets of vine jungle. *Birrimbirr*, on the other hand, were that aspect of the soul that returned to the waters of the group of the deceased.

This kind of typology falls into difficulties, for the classes of *wangarr* and *mokuy* are not clearly discrete. For example, I have heard the Djang'kawu sisters referred to both as *wangarr* and as *mokuy*. People of the Yirrkala region did not apparently refer to the subjects of public songs as *mokuy*, but used this term to denote the 'followers' of beings with proper names (Williams 1986: 33). Similarly, concepts of *mokuy* and *birrimbirr*, which have occasioned lengthy discussion on the part of anthropologists about differences in their meaning (Warner 1937: 443–7; Williams 1986: 32–3; H. Morphy 1984: 40–1, 1991: 106) were not clearly discrete. *Mokuy* had the particular sense of 'corpse' as well as 'spirit of the dead', but the terms could be substituted one for the other, or combined as *mokuy-birrimbirr*.

Through these ambiguities the discourse and ritual action of distinct individuals and groups could mesh, but interpretations did not have to converge. There was a more or less common language of ancestors and other spirit beings, but it was neither uniform nor immutable. Yolngu had constructed *wangarr* modelled after people of the Torres Strait (R. M. Berndt 1948*b*) or the annual Macassan visitors to the coast, whose visits probably began in the sixteenth century, and they speculated about the relationship between Christian spiritual beings and the *wangarr*. For example Dji<u>n</u>iyi<u>n</u>i Gondarra wrote in 1980:

> Church leader have been encouraging me to write an Aboriginal theology of the land. When we become Christians, we see more clearly our relationship with the land and with God. It was God who entrusted the land to our ancestors. We were living in a land of plenty, like first creation people. We had our own technology, out own social laws, our own pattern to follow. Life was so beautiful before.
>
> I'm sorry to say it, but I picture oppressors, both yolngu and balanda, coming into our garden of Eden like a snake. Satan used the snake as his instrument to tempt God's people ant to try to destroy God's plan for his people. The bad influence came in breaking our relationship with God, with man, and with land. (Gondarra 1980: 8–9)

It is indicative of the tenacity of indigenous beliefs that people speculated about the interrelations between Christian deities and the *wangarr*, rather than dismissing one or the other *tout court*. Many Yolngu have seen the imposed European law as coming from God, and encoded in the Bible. Even those who placed the Christian God at the top of a hierarchy of powers held on to certain of the causal and logical relations among *wangarr*, places, persons, and sacra, even if they saw *wangarr* powers rather differently.

The beginnings of moral dualism may have appeared in the late 1970s. Yolngu beliefs were not strongly dualist: there was no great divide between, on the one hand, beliefs and practices concerned with the *wangarr* ancestors that I have glossed 'religion', and, on the other, 'sorcery' beliefs and practices. The body of the dead and the ancestral bodies in the form of traces in country, designs, and objects, were both believed to be sources of power. People thought that they could draw on ancestral powers to benefit themselves and harm others, and threats of sorcery defended the integrity of secret religious knowledge. A move towards moral dualism may have occurred at a meeting of Christians at Galiwin'ku. Participants rejected the word *ma<u>l</u>agatj* as the name of God on the grounds that it implies agressiveness, and emphasized the peaceable nature of *wangarr* (see Chapter 8). The degree to which beliefs about the *wangarr* ancestors were deemed relevant had also changed by the mid-1970s, especially amongst some younger people educated in schools and colleges outside the region. Indeed, older people rightly saw these institutions as a threat to social reproduction.

Yolngu patrifilial groups (*mala*, *ba:purru*, often translated as 'clans') owned stories about ancestors and *ma<u>d</u>ayin*, and they possessed areas of country and elements of ceremony. The named, exogamous groups were divided between the exogamous patrilineal moieties, Dhuwa and Yirritja. We shall see in the following chapters that just as stories about ancestors, the definition of place, and the outlines of religious forms were to varying degrees ambiguous and contested, so was the constitution of groups. The following stories about the Djang'kawu belonged to Dhuwa-moiety groups.

The Relativity of Truth: The Djang'kawu Sisters

A Cloudy Water Story

In stories told by people of the Cloudy Water group, as I shall call it, the name 'Djang'kawu' referred to two sisters, a mature woman who had borne children (*gungman*), and a nubile girl (*wirrku<u>l</u>*). Someone explained that the name Djang'kawu was the name of their *ba:purru*, 'group', as distinct from their two personal names. People with related traditions generally agreed that long ago *wangarr* called Djang'kawu travelled west along the coast of north-east Arnhem Land from an island in the sea off Gove Peninsula in the east, described as a land of the dead. (A young man who had seen aerial photographs of the Gulf of Carpentaria expressed scepticism to me about the existence of Burralku.) They went on a bark canoe or a raft made of a roll of paperbark (*djutu*) through the sea, and overland on foot, following the journey of the sun, and indeed they were said to be like the sun. Each woman carried two digging-sticks or walking-sticks called Ga<u>n</u>inyi<u>d</u>i, one in each hand, with which they gathered food, speared the ground to make freshwater springs, and paddled the canoe. Each also carried a woven dilly-bag (*bathi*, a small, egg-shaped basket, coiled from pandanus palm leaves) frilled with *wa<u>n</u>a* ('arm') feathered cords,

which were the public equivalent of men's secret *rangga* sacred objects. As they travelled they saw and named places, plants, animals, birds, fish, and other creatures. They bore the first Dhuwa-moiety people, named them, and gave them languages. At each Dhuwa group's country they created freshwater springs in the salt-flats and mangroves and on the beaches along the coast by piercing the ground with their sticks. The theme of fresh water mixing with salt water recurs throughout songs of the Djang'kawu, for the springs bubble up into the salt water at high tides. The sisters met or were followed by other *wangarr* on the way, and they also met men who deprived them of their dilly-bag and sacred objects, forever gaining for men the right to control the religious law, according to Yolngu tradition, a tradition that some have questioned. Some of their actions resulted in, or centred on, permanent topographical features, especially rocks and trees regarded as permanent, or perpetually replaced. Many of these features were equivalent to *rangga* sacred objects, implying that the *rangga* were transformations of features of the country, and implying also that representations of objects, places, and ancestors in painted group-designs were permanently connected to the land.

The whole category of people comprising all the groups taken to be descended from the children of the Djang'kawu constituted one *ba:purru* (translated 'group' in some contexts), named Ma:tjarra or Bungunbungun (cf. R. M. Berndt 1952: 86; Dunlop 1984). The status of the sisters as *wangarr* ancestors of an indefinite number of groups of the Dhuwa moiety through their journey from place to place was similar to, if not modelled on, that of a serially polygynous woman who, as one husband died, married a man of a different group. The difference is that in the story the Djang'kawu gave birth to children of their own moiety and procreated through parthenogenesis.[1]

What follows summarizes a public exposition of the myth, as told to me by the Cloudy Water leader, and one of his brother's sons.

[1] That the ancestors came from so far away reflects the geographical separation which increases with each ascending generation of matrilineal ancestry. People mapped uterine-kin connections on to the patrilineal and exogamous groups and countries as wholes. A person's mother's mother's mother's (*waku*) country was quite likely to be far from their own country, although the marriages of some people of groups in the Buckingham Bay region linked them in cycles of affinity, which could bring a person's own country close to that of their MMM's group (see Shapiro 1981: 133).

The Djang'kawu travelled from Burralku, an island in the sea towards the sunrise, to Sunrise country. There they created the Sunrise group. They carried *ma<u>d</u>ayin* (sacred objects) to give to all the Dhuwa groups.

They created Driftwood, Blow-fly, Dilly-bag, and Digging-stick groups, who have Red Kangaroo *ma<u>d</u>ayin*.

They created a Black-capped Tern group at its country.[2]

Then they went to the place where Shark travelled on the other side of the country, also Black-capped Tern group.

The two sisters travelled inland to make the forest, carrying the *ma<u>d</u>ayin*, giving languages, creating groups, and leaving songs.

They put the *ma<u>d</u>ayin* objects into the ground at Thunder group's place, and made spring waters.

Then they went to a second Thunder-group country, made places and people and put in the *ma<u>d</u>ayin*—but they carried *ma<u>d</u>ayin* still.

The two sisters created groups at a third Thunder-group country and other places.

Then on to Wa:<u>l</u>angur and 'the place of lightning' at Cloudy Water country. At Wa:<u>l</u>angur the Djang'kawu hung up the dilly-bag, and went to gather crabs and mangrove-worms. When they returned and looked for the dilly-bag they could not find it; men had stolen it and taken it back to the 'inside' ceremony ground where they were dancing, so they could not get it back. They remained at Wa:<u>l</u>angur where they created springs, created the Cloudy Water people and collected food.

The Djang'kawu—really someone called Wa<u>l</u>awa<u>l</u>a—heard his or her *ma:ri* (mothers' mothers and/or mothers' mothers' brothers) singing at Galiwin'ku. They crossed to the country at Galiwin'ku and said 'sing like this *ma:ri*, slowly' (in *bi<u>l</u>ma*-song style). Then they asked their *ma:ri* if they had water; and made springs with Ganinyi<u>d</u>i digging-sticks. The Galiwin'ku country and Wa:<u>l</u>angur are *ma:ri'manydji* (in a relation of MM/MMB–wDC), and Cloudy Water group has the Ma<u>d</u>ayin ceremony.

The Djang'kawu left all the *ma<u>d</u>ayin* at Wa:<u>l</u>angur and went towards Milingimbi just carrying digging-sticks.

They created water-holes and left *ma<u>d</u>ayin* at Stringybark country at Langarra.

And they went on to Mangrove country and Lorikeet country. At Lorikeet country the Djang'kawu gave the Nga<u>n</u>marra Mat and Red-collared Lorikeet to that group, and went on travelling west to Oenpelli and beyond.

In more private settings Cloudy Water men described the creation of sacred objects in more detail, or outlined just how the

[2] Black-capped tern is not a species but a generic term for a number of species referred to by Yolngu as *djarrak* (Rudder 1977).

sisters reproduced. Thus, for example, the *djutu* paperbark raft of the Sisters 'turned' not only into a rock, but also into a sacred object which was the subject of a painted design. This story established a fusion of identity between a sacred object, an everyday object used by the *wangarr* ancestors of which the sacred object was a transformation, and a feature of the landscape. The rock was a transformation both of the canoe and the sacred object, which 'followed' this transformation. In this way a type of sacred object had a causal connection in Yolngu belief with the *wangarr* ancestors and a particular place. These connections, signified by the concept of *l̲ikan* ('elbow', 'joint') explain why holding a sacred object and related designs were grounds for a claim to possess land: they showed the connection between a group's ancestor/s and the country (see Chapter 3).

In imparting information in the form of stories, an individual was accountable to others. Very often a man would ensure that someone else with rights in a story was present as 'witness' (expressed thus in English; or simply to 'see', *nga:ma*, and 'hear', *nha:ma*). This gave some protection against being accused of having said the wrong thing, or having gone beyond one's prerogatives in giving information. The exceptions to this rule were some older men, who were less accountable, although even they tended to relate tales belonging to their own and closely related groups. Men also stressed the truth of what they told me, saying that they knew it was true because their father or other senior male relative told it to them, and because they listened to their fathers. Others, they said, were not knowledgeable, or got the stories a bit wrong (*djarrpi*, 'crooked') because their fathers did not teach them.

Notice how group-centric the above telling of ancestral events was. The men who told it to me placed the central episode in which men stole the dilly-bag at their own country, and denied that the sisters carried *mad̲ayin* objects to groups with areas to the west of their own. Moreover, certain episodes are contradictory: the tellers insisted that the sisters carried no more *mad̲ayin* objects after losing them at the Cloudy Water country Wa:l̲angur (a pseudonym), yet asserted that the sisters gave *mad̲ayin* to each group. Some people resolved this contradiction by saying that the sisters carried only their digging-sticks and that these were transformed into *mad̲ayin* objects. Groups had their own perspectives on ancestral events.

People of many of the other groups linked by this story located that central episode on their own countries. The versions of the ancestral events which I translate below are those of three Dhuwa-moiety groups whose countries ranged from Buckingham Bay in the east to the Glyde River estuary in the west. The countries were not contiguous, separated as they were by other Dhuwa- and Yirritja-moiety land or waters, but some narratives about the journey of the Djang'kawu refer to them in succession. I allude briefly also to the Sunrise story. This group's country was far to the west of Cloudy Water country, on the Gove Peninsula. A later chapter will show how these different views of similar events are reflected in co-operation in the performance of ritual. Common to all depictions of the journey is that the two sisters travelled from east to west following the sun to a succession of group countries. They arrived at a named place, named the group, put the sacred objects in the ground, and created the group.

The central episode in this Cloudy Water story describes events at Wa:l̲angur in more detail. In this episode men who were forerunners (*ngurrunanggal*) of the group stole the dilly-bag full of sacred objects from the sisters. The result, according to both men and women, was that men gained control of the Mad̲ayin ceremony (named after *mad̲ayin* 'sacra') and other rituals and women became the workers, gathering food for the men. By implication the men gained control of women's reproductive powers:

The Djang'kawu hung up the dilly-bag in the tree and went to get nerite and telescope shellfish and crabs. The men came [from a dance ground in the forest, where they were performing a Mad̲ayin ceremony, across the lagoon and swamp,] and stole the dilly-bag full of *rangga* sacred objects, taking it back to the ceremony ground.

'Hey! Where is the dilly-bag?' said the sisters. Then they heard '*binbir*', the men dancing at the Mad̲ayin ground. 'Maybe our elder brothers have it.' They put their fingers in their ears [so as not to hear the secret calls in the men's dances] and went to see.

'They stole it, and now they are dancing,' they said. 'We will go back, work, and get food—the men can do the dancing.' (If men had not stolen the dilly-bag the women would still dance, and we men would have to cook!)

The men said, 'We'd better get it!'

Sitting at the men's ground during a Mad̲ayin ceremony, Wandhuma, an aspiring Cloudy Water leader, remarked to me, 'If men had not stolen the dilly-bag the women would be dancing here, and we men would be sitting at the hearth!'

Cloudy Water people whom I knew placed this central episode at their country and they denied that the sisters carried sacred objects to groups with places to the west of their own. However, people of many of the other groups on the same mythological track located a similar central episode on *their* countries.

Variation in Cloudy Water Stories

A brother's son of Wandhuma, also of the Cloudy Water group, told a version of the Djang'kawu story to Robert Bos which differed in some particulars from the one recorded above (Bos 1979*b*). According to Wulakitj, the man who stole the dilly-bag from the sisters was the male group-ancestor Wa_lawa_la, who saw the dilly-bag hanging on the tree, took it, called out in a loud voice and began to dance. The sisters heard the singing and called out, 'Look, he has our ceremony dilly-bag.' The Djang'kawu saw a sorcerer who was crouching down watching.

Wulakitj was something of a Christian evangelist who later became a pastor. In his view the Djang'kawu were peace-makers. It was the Macassans who came later and taught the Yolngu to be 'sword fighters', and taught them about alcohol. Wulakitj also brought the Djang'kawu and Wa_lawa_la very close genealogically to the living. Wa_lawa_la was the father or father's father of the two named human ancestors who were the *ma:ri'mu* (FF) of the leader and the *ga:thu* (FFF) of Wulakitj and his siblings. The Djang'kawu 'put Wa_lawa_la in charge'. Thus this man thought of the Djang'kawu as having lived six or seven generations before his own.

Differences among people of the same group sometimes led to delicate negotiations. A person was likely to be affiliated to, and have responsibilities in, groups of the same moiety with incompatible interpretations of ancestral reality: the person's own and mother's mother's groups, for example. People of the same group with different kin-networks might have to negotiate the appropriate events to relate to, say, an anthropologist.

Wandhuma frequently interposed the episode in which someone called Wa_lawa_la crossed to Galiwin'ku. The leader of another lineage of the group, who lived more or less permanently at Galiwin'ku, asserted more recently that it was the Djang'kawu who changed their name to Wal̲awal̲a when they travelled across to the island (John Rudder pers. com.). Either way, the episode was quite irreconcilable with other stories current among groups connected with Galiwin'ku and the islands to the north-east (see R. M. Berndt 1976: 152). According to the people of these places the Djang'kawu travelled south-west along Galiwin'ku Island and then on to the mainland. The Cloudy Water story seems to have been intended to ratify a claim on the part of Wandhuma and his lineage to that part of Galiwin'ku. The last members of the Two Women group of Galiwin'ku died within living memory and, according to Wandhuma and to some of the sons and daughters of women of the deceased group (the *waku*—wC/ZC), they asked the Cloudy Water people to look after the *mad̲ayin* (sacra) and the place, which was in a *ma:ri* (MM/MMB) relation to the Cloudy Water group, the proper relation for a claim to succeed to a vacant area (see Chapter 3). Many Cloudy Water people referred to Two Women country on the island as their own country, and many of the group's members lived there. By recounting this episode the Cloudy Water men in effect denied the independent status of Two Women country, and incorporated it as their own. The name Wal̲awal̲a was one of the *l̲ikan* names of the Cloudy Water group associated with their mainland country Wa:l̲angur, and with the group. (*L̲ikan* names or *bundurr* were names of a *wangarr* ancestor related to a particular place and group, and called out in ceremonies.) Someone said that the name was the language of the group. Another mature man of the same lineage as the leader commented to Robert Bos, 'We take over everything now, land, *mad̲ayin*, everything, because [Wal̲awal̲a] went there to help' (Bos 1979*a*: 5).[3]

The difference between Wandhuma's delineation of the Djang' kawu myth and accounts current in other groups led to a delicate situation when two men of the Cloudy Water group, one of whom had a Galiwin'ku group as mother's mother, jointly told me part of the story. When telling me the Djang'kawu story, sitting one afternoon above Milingimbi beach, Djikad̲a, one of Wandhuma's brother's sons, checked with the older man Ngat̲ili about whether the sisters themselves crossed over to Galiwin'ku, or whether someone called Wal̲awal̲a went over:

[3] A leader of a Wessel Islands group did record a version of the journey of a Cloudy Water man to Galiwin'ku similar to Wandhuma's story (Yule 1980: 30).

Djikada: Right, those two started to walk from Wa:langur—or did they come down to the water that way?

Ngatili: Yeah.

Djikada: Yes, Wa:langur first and then they went to Galiwin'ku?

Ngatili: Yes.

Djikada: To Galiwin'ku, and they heard . . .

Ngatili: They stayed, stayed there.

I.K.: At Galiwin'ku?

Djikada: At Wa:langur.

Ngatili: They sat down at Wa:langur.

Djikada: And those two went from there . . .

Ngatili: They didn't go.

Djikada: They did go, in the daytime.

Ngatili: Wait. Those two stayed there. That Walawala was there at Wa:langur, Walawala sat there, and yes he went, all right, and called out from there that way—this is your story. That, whatsit, that little island . . .

Djikada was perfectly familiar with Wandhuma's constantly reiterated story about how Walawala heard the songs from Galiwin'ku and travelled across to give Madayin songs to the islanders, and create the springs. He checked with Ngatili partly because of this man's seniority by age, but partly I think because of Ngatili's links with adherents to a different view of the Djang'kawu journey. But, as I said, people of other groups had even more divergent ideas about ancestral events.

A Thunder Leader's Story

The younger of the two brothers who were acknowledged leaders of the Thunder group told me his story on a visit to Cloudy Water country with myself and Wandhuma, the Cloudy Water leader. After recounting various events Buwata, a man in his fifties, came to the central episode in which the dilly-bag was lost at his country. But he narrated this part of the story in a rather allusive way, perhaps out of deference to the Cloudy Water leader, for we were sitting at Wa:langur, Cloudy Water country, where, from Wandhuma's point of view, the dilly-bag was lost.

We join Buwata's tale at M____, Thunder country, where the sisters encountered men, warriors:

> 'What shall we do?' they said. And they . . . made a group, the first people, like that. 'Ah, good.'

'Off you go, work, collect cycad-palm nuts,' [they said to the men.] 'Here you are, here's your dilly-bag. And you, your dilly-bag.' And the people got food.

The Djang'kawu sat at their ceremony ground while the others worked—here a dilly-bag, there a dilly-bag.

Then all those first people were thinking, 'Hey! Why are those two women sitting there while we're working away with the stones pounding cycad-palm nuts? What are they sitting about for? They're just hanging around.' They talked in this way among themselves.

The two sisters sat in the shade, and the mob of men worked and gave them food. 'You, go and get more food over there, we'll give you the dilly-bags, you carry the dilly-bag to get this, and that, and that; the sun's going down.' And they just sat there.

One of the sisters said, there in the cycads, 'Sister, let's go and get some shellfish.' The men saw them. 'Putting off work!' they said.

The two sisters hung up the dilly-bag and off they went. They got nerite shellfish, angular ear, telescope shells, and crabs.

Those two carried two things at Marapay, the dilly-bag and firewood. There was the dilly-bag and there were the warriors. The sisters got some firewood and they heard:

'Shhhhhh——Gunbur Mongungala waaaaa, yu!, waaaaaaay.' [The men called on the ceremony ground.] 'We'll pick up more shellfish' they said, and went to another place.

'This way, shellfish'—'Right.' They got more, went on, still not satisfied, got other kinds, collected more—'How many?'

They went around, climbed a tree, and looked this way and that.

'Enough now?' (And that same tree still stands—we'll go and see it later.) They went on until: 'We're getting near that place.'

'Aaaeeeee' [they heard.] Walking and sticking in their walking-sticks they neared the camp. They came up to an ant-hill, put in their hands, then,

'Sister, leave it, we're getting rather near the men.' So they left that place and said, 'We'll go!' So off they went to [places in the forest] and that place with a dead man's name. There they came out [of the forest].

They walked about fretting and weeping. Bustards were standing around, and they dug up *gomili* [*Clitoria australis*] food.

Buwaṯa's story then took the sisters on to Wa:ḻangur, Cloudy Water country. Wandhuma, the Cloudy Water leader who was listening laughed and said to me, 'It's the same story!' referring to his own group's Djang'kawu myth. It is remarkable how allusive some of the references are. Buwaṯa made no direct mention of the theft, saying merely, 'There was the dilly-bag and there were the warriors'. He merely reported the sister's remark, 'Sister, leave it, we're getting rather near the men', and the fact of their 'fretting and weeping'.[4]

A Sunrise Story

In the Sunrise-group version of the events, as documented by R. M. Berndt (1952), the name 'Djang'kawu' referred to two sisters and a brother who procreated though hetero-sexual intercourse. When Buwaṯa finished his story, Wandhuma commented on Sunrise belief:

> This [Thunder and Cloudy Water group] story is from the old people, Day—— and Dji—— [the apical male ancestors of the Cloudy Water group]. That story told by [Sunrise men] is on a different path. They [the Djang'kawu] were women, those two who came from the east. The truth is from those two old people, in the beginning, a long time ago.
>
> We know, we younger people, and our fathers knew. Those two are just women, no man! It is a bad thing [*ya:tj*] that they tell that story, and a lie. In the beginning, *women*.

Buwaṯa added, 'Yes! Their "penis" [man] went back, the uncircumcised one. Yes, there was no man.' They went on to explain how the women procreated alone. Buwaṯa here accepted the Sunrise version as valid taken in relation to their own country, and reconciled it with his own. Wandhuma seems to have merged Cloudy Water and Thunder-group stories into one in contrast with that of the Sunrise group when he spoke of the story coming from Cloudy Water ancestors.

How people reacted to differences of myths depended, I suspect,

[4] Buwaṯa recorded a more complete version for a document on land rights (Yule 1980): 'Yolngu men didn't have any authority before this time, before the day Djang'kawu came with the sacred Mindjalpi [dilly-bag]. But on that day they brought it, they cut some baltarr (forked sticks) and hung the dilly bag on them. Then they went off to look for some shellfish and crabs. The men were watching all this, and as soon as the two sisters had left, they decided to take over the dilly bag with its power. The original dilly bag didn't have a lid; it was completely closed in. So the men cut the top off and began to sing. As soon as the Djan'kawu [*sic*] sisters heard the marrwaṯ (special songs), the younger sister said: "They took out bag from the forked stick." "Never mind," said the older sister, "Let them have the power, because we call them wa:wa (brother)". So that's how the men got the authority over the land. And from that time the men hunted the women away from the sacred dilly bag. So today, the dilly bag which belongs to the men in each mala (clan) is the proof that that mala has the authority over that land—because the dilly bag came from Djang'kawu. This dilly bag here is proof of ownership of—my homeland centre at Buckingham Bay.'

on the distance of their relationship, and whether a representative was present or not. Wandhuma was scathing about Sunrise beliefs but tolerant of Buwaṯa's, at least in his presence, for these two groups co-operated closely in ritual. Men would press me to come to them for their group (or mother's group) stories, warning me that other people tended to mix them up.

A Mangrove Story

The man who recorded a Djang'kawu story from the point of view of Mangrove group was a leader of the lineage of the Honeybee group of the Yirritja moiety living on the mainland. Mangrove was his mother's group. Ḏan.gi told this story to a researcher and in the presence of a young married man of the Thunder group, so having a witness with direct rights in the tale.[5]

In Ḏan.gi's telling the Djang'kawu started their journey by paperbark raft (*djutu*) at Burralku, the island of the dead, and came through a series of group countries along the coast and coastal islands, including Sunrise, Thunder, and Cloudy Water countries. Ḏan.gi then brought the story to a Mangrove country on the east bank of the Glyde River, then to the country on the west bank where the men stole the dilly-bag:

> They sat at my 'mother' place. They came out from there at daybreak—no, it wasn't daybreak. They walked, covered up, hitting the clapsticks and singing. They were cold.
>
> 'Oh, We've no husbands! Just the two of us.'
>
> 'You sleep there and I'll sleep there.'
>
> 'No, come on, we'll sleep in one place.'
>
> 'No.'
>
> 'We've no husband.'
>
> 'Never mind, you sleep there, and I'll sleep here.'
>
> 'Oh very well.'
>
> Dawn broke. They got firewood and water. They made the sacred water. 'This is a sacred place, Dha:ngura,' still speaking Djinang language. 'We must dry out the things at this shade, here at [these places].' They hung up the things, dried them, and went to get shellfish, cooked them, got more Nonda shellfish, came out from the mangroves, and put them in the dilly-bag.
>
> They dug up the shellfish but they mixed up the firewood, Dhuwa and

[5] I am grateful to Dr J. Reser for making a recording of this narration available to me.

Yirritja—those mangroves that stand there, Gukawu and Lanyupi. They should have used Yirritja wood [to cook Dhuwa shellfish]. They hit the wood and broke it up.

The older sister took Nyu<u>l</u>ungu<u>l</u> shellfish [a bivalve] and sucked it out of the shell, like this.

'Hey! sister, what are you sucking? It's just mud! That's because you got the wrong firewood—mixed it up!' [the younger sister said.]

They went on, and caught some fish. After that, in the afternoon, they speared Witchetty Grubs, the black ones—all this in one day! Then they reached G——, and Little Kingfisher [man] had got there already, the day before, at that sacred place. He was puffed out. And there was a discussion, and after that the two thieves left.

From there where they saw the fish, the Black Butcher Bird went silently—this is all significant, all Dhuwa, all meaningful. Those two—you see they followed the way the women went and got to the sandy place—they were men. They [the men] got closer, and they got muddy because they had to bend down like we did [when we visited the place] on to the ground. He bent down, the man did.

The sisters went to get fish, and they [the men] came behind and took the dilly-bag and then sang, following the Ma<u>d</u>ayin ceremony. They were good at performing their ceremony. Both of them sang it, and then they cleaned themselves up. They really knew the song, and followed it at the secret ground, loving it! They went 'arrrr, arr, arr, waaaaa!' [the first part of the Gapila fish dance] at the ceremony ground.

They [the sisters] finished with that place, and went on. It was late afternoon. 'Where shall we go first?'

'First to Dh——, it's a rich place.' So they went there.

'Sister, let's get some vegetable food.'

'What food?'

'*Dharruma* [*Portulaca oleracea*], *milwini*, it's Yirritja [moiety].' They got it, put it in the basket and carried it.

'Wood for a pipe, sister!' So they cut some wood. 'It stands in the stringybark forest, the bush.'

They crossed the creek at . . . the point where there's a round, deep place with turtles, and where the leeches bite. They crossed over, and the leeches did bite the younger sister.

<u>D</u>an.gi then took the story to the west where they went around a place with a dangerous snake, on to an Emu place, then a place where they encountered Little Kingfisher. They crossed to Milingimbi island, where they ate Sting Ray and saw Woolly-butt in flower (red in colour) and Red-collared Lorikeet, then on to places along the coast to the west, where the leeches bit and where they spoke Burarra language.

Comparisons and Contrasts

There are some telling comparisons to be made between these narratives. They have a common framework of the ancestral journey and the same elements appear, but some are given special significance. In the Thunder-group story the contrast between those who 'work' and those who have control of ceremonies is emphasized in that, before they lose the dilly-bag, the Djang'kawu sit around and make the men work. They make thunder and rain by spearing termite queens—'Thunder' (*wul̲ma*) was one name for the group.

In the Mangrove-group version the episode in which one of the sisters complains about the lack of a husband is reminiscent of a section of the Gunapipi ceremony, of which the Mangrove people were owners, and seems, if my interpretation is right, to hint at homosexual intercourse and perhaps parthenogenetic procreation. The episode about mangrove firewood is a marker of the group's identity. The group's name refers to that episode, which can be taken as a play on sexual morality; 'cooking' shellfish with the same moiety firewood produces only mud, a metaphor for the results of incest. Indeed, the relations of sexuality to moiety are dominant themes in the Gunapipi ceremony (Chapter 8).

In this narrative the men who stole from the Djang'kawu were two Black Butcher Birds, whereas in Cloudy Water group tradition this Bird called from Mistletoe Tree (*Exocarpos latifolia*) as the tide came in. D̲an.gi alluded to the similar episode in Thunder and Cloudy Water stories, for he mentioned that the Mad̲ayin ceremony was going on at both places. The reference to the Black Butcher Bird men getting muddy may well have been an interpretation of the body-paint worn for the related dance in the Mad̲ayin ceremony, an interpretation specific to this group's version. Cloudy Water men interpreted the same dance as Little Kingfisher.

Taken together the stories show how common elements of ceremony are subtly transformed from group to group: the sisters' digging-sticks were of different timber in the countries of different groups; the name of the paperbark raft changed; the sisters spoke different languages from place to place; they performed specific actions at each group place which differentiated the group from others and reflected the other content of each group's religious traditions. From east to west the male sexual role in creation is increasingly weakened in the myths, from sibling incest (R. M. Berndt 1952: 18) through unsuccessful attempts on the part of men to have intercourse (Warner 1937: 338), to parthenogenesis on the part of women who carried phalluses with them in the form of digging-/walking-sticks, to homosexual intercourse perhaps. Further west the myths imply that leeches were the impregnating agent.

Members of each group recognized that other groups lay on a 'different path', but attitudes to these differences depended on the closeness of the relationship. Wandhuma was scathing about the Sunrise myth, but joked about the similarity between the Thunder version of events to his own. Peterson (pers. com.) recalls how on one occasion people of the Thunder group living at Mirrngadja in the mid-1960s returned from Nangalala, outraged at what they had been told about the stealing of the dilly-bag. Two young women and their mother, who were ritual experts, poured scorn on what was to them an incorrect story.

In the above examples people of different groups disagreed about whether there were two sisters or two sisters and a brother, though agreeing on the general structure of the story and an overlapping vocabulary of symbols. The central and apparently unique event occurred at the country of the person relating the story, but people were of course aware that details told by others were incompatible. The details of the ancestral events differed depending on the group relating the story, and among different tellers of the same group. The male ideology was that a man followed his father. These differences among the myths of distinct groups were recorded in group names and alluded to in other groups' stories. People appeared tolerant of the differences among close consociates, but less tolerant of more distantly related people. More contentious, perhaps, were the differences in ancestral journeys and ancestral identity apparently propagated to legitimate claims in another group's land. But Yolngu did not appear to seek a unitary vision of the world.

2

Group Identity, Kin, and Leaders

In the last chapter I showed how people of different groups gave rather different accounts of events in which 'the same' ancestors were involved. The chapter took the constitution of group identity for granted, and assumed that people of a group had more or less the same beliefs. The reality was more complex, however, for those who were privy to the details of a group's religious traditions included descendants of women of a group. Moreover, people of the same group sometimes disagreed about ancestral events or related matters. This chapter discusses the constitution of groups, as well as the roles of leaders, and those aspects of the classification of kin and the organization of marriage which will help in understanding succeeding chapters.

Cooking turtle on the beach

Yolngu Patrifilial Groups

In the mid- to late 1970s patterns of social interaction in conjunction with the definition of groups and social relations in Arnhem Land formed a structure of overlapping, interlocking, and open social networks rather than a segmentary structure of clearly defined groups. These indigenous forms were extended and modified by recently imposed settlement organization. In spite of the continuity of networks across the region connecting speakers of Yolngu languages with others, there were apparently some lesions where the density of interaction and communication, indicated by intermarriage, was thin (White 1976; White *et al.* 1990).

Anthropologists have applied the terms 'clan' or 'sib' to social entities which the Yolngu constituted in discourse and action, and referred to as *ba:purru* and *mala*. They have represented these as exogamous patrilineal descent-groups, sometimes divided into subgroups, and clustering into sets of a higher order of inclusiveness, indicated by such terms as 'linguistic group', 'phratry', 'clan-set', 'song-set', and 'totemic union', as well as 'exogamous patrilineal moieties' and 'dialect groups'.[1]

Shapiro has argued strongly that sibs did not partake in a segmentary organization based on descent and that they were only weakly corporate (Shapiro 1981: 22). I shall go further and suggest that the identity and boundaries of groups were often ambiguous, that people disagreed about their 'internal' structure, including who was the leader, and that groups did not sort into a taxonomic hierarchy of different types of group at different levels of inclusiveness implied by concepts such as 'clan' and 'phratry'. Rather than being constituted by enclosure within boundaries, Yolngu groups and group relations, like places, extended outward from foci. Connections among groups were not those of enclosing 'sets' but extendable strings of connectedness (Keen, forthcoming). Yolngu groups were not like the corporations in Roman or English law or corporate groups of anthropological theory, but 'kinds' of people with ancestry and attributes that both linked

[1] On totemic unions see Shapiro (1981: 23). People at Yirrkala distinguished between *manikay mala*, 'song group', and *ma<u>d</u>ayin mala*, 'sacra group'. The former referred to groups with the same myth. Williams (1986) believes this kind of 'set' to have a greater scope than the latter. Keen (1978) uses the expression 'clan aggregate' and H. Morphy (1977) writes of 'clan sets'.

them to, and differentiated them from, others. For these reasons I shall simply use the term 'group' (or sometimes 'patrifilial group'), applying it in the same way as Yolngu use the word *mala*, to less inclusive and more inclusive groups alike.

Yolngu constituted *mala* and *ba:purru* through beliefs about creation and reproduction, images of ancestral journeys and traces of ancestral activity, attachment to places, and webs of connections. But let us begin with *mala*, 'group'.

Groups

A Yolngu concept of 'group' was signified by the term *mala*. This had the sense of aggregation, perhaps perceived as the spatial proximity of a plurality of elements separated to some degree from others.[2] This notion of aggregation was apparently carried over metaphorically to the sense of a 'group' of persons with similar attributes or who were connected in some way, whether grouped together in space or not. The word *mala* went together with a proper name (*Gupapuyngu mala*, *Ma:tjarra mala*) to denote a collectivity of people sharing a named *ba:purru* identity. In this usage *mala* was at least partially synonymous with *ba:purru*. A person had the same *ba:purru* identity as their 'father' (*ba:pa'mirringu*).

The word *ba:purru*, one of the words commonly translated as 'clan', may have derived etymologically from *ba:pa*, father/ father's brother, and the suffix *-wurru*, 'through' (Schebeck 1968: 42). However, people living at Milingimbi in the 1920s were 'unconscious' of a link (Warner 1937: 30), and more recently some people of Yirrkala denied this connection (Williams 1986: 70). While *mala* signified a collectivity of individuals with common attributes such as *ba:purru* identity, or a clustering of individuals (people, geese, rocks) in space, *ba:purru* denoted a relationship between *wangarr* ancestors, spirits of the dead, the yet-to-be-born, places, and elements of ceremonies associated with the ancestors and the group. It implied a shared identity through those connections, which came primarily through one's father. As Daymbalipu explained to Williams: '*Ba:purru* is something important in our lives; it is like soul, you can't see it. It means something

[2] The Yolngu-matha Dictionary defines *mala* as 'group (of people), tribe, clan, crowd, mob, set, bunch' (Zorc 1986: 167).

we share before we are born and keep on sharing during our lifetime and after we die' (Williams 1986: 66; cf. H. Morphy 1990: 316, 321).

The semantics of *mala* and *ba:purru* did not fit the semantics of taxonomic hierarchical classification and the related class semantics, for according to the linguistic usage of many Yolngu people, just as a named group such as 'Cloudy Water people' was a named *mala* or *ba:purru*, so the string of groups including Cloudy Water, Thunder, Mangrove, and others with the Djang'kawu sisters as ancestors, comprised the Ma:tjarra *ba:purru* or *mala*. The relationship here is 'fractal' rather than hierarchical. (However, some Djinang-speaking people recently averred that *ba:purru* does not apply to a plurality of groups but only to the less inclusive group (Elliott 1991).)

If we are looking for some kind of anchorage or starting-point for explaining the structure of groups, there are two possibilities: country and group names. The persistence of *ba:purru* or *mala* identity attached to persons, places, and sacra is revealed by their recurrence in anthropological and administrative documents, but groups had many alternative names, used in particular contexts and referring to particular attributes (Schebeck 1968; Williams 1986: 61 ff.). Many *ba:purru* names referred to the group's *wangarr* ancestors or to some elements associated with an ancestor. (It is necessary to use real names here to make the point.) Daygurrgurr, for example, was a name of the Eel-tailed Catfish relating to the Hollow-log mortuary-ceremony of that group; the *ba:purru* name Ba<u>l</u>amumu referred to the waters connecting Djapu to other groups (Williams 1986: 68). Other names denoted a group's tongue, while names such as Gurrka-wurtha, 'penis of fire', connoted the warlike attributes of a group and the ferocious aspect of *wangarr* (Williams 1986: 57 ff.). People rendered the names of other groups (as well as their own) in their own dialect, so that 'Ba:lmbi' in the Djinang tongue became 'Ba<u>l</u>mawuy' in Dhuwal and Dhuwala.

It is possible to distinguish a minimally inclusive usage of some group names from an extended use in which the identity can be divided into several identities. But it is misleading to refer to any such usage as 'clan', for the groups did not have a common structure. The following discussion refers to these least-inclusive group-identities.

As we shall see in detail in the next chapter, Yolngu identified each loosely bounded area of country denoted by a 'big name' in terms of both moiety and group names. Thus the area which I call in this book Wa:langur (see Chapter 3) comprised two main 'big-name' areas and a host of small-name places. People identified it as Dhuwa-moiety country, and as Cloudy Water country after the group of that name. Those with that identity 'held' the country and its associated sacra. Most such groups possessed several 'big-name' countries, some contiguous, others separated by as much as 60 kilometres, as well as several small areas embedded in the countries of other groups. The number of people with this least-inclusive, common *ba:purru* identity varied from one to over 200 in the mid-1970s, with an average of about fifty-five. About half of the groups were classified as Yirritja and the rest as Dhuwa.

There is a difference between these proper names of groups and names which meant 'pertaining to place *x*' and which often applied to a less inclusive aggregate of people than the group. For example, among the people who identified as Liyagalawumirr group, *Mirarrminawuy mala*, 'group pertaining to the country Mirarrmina', comprised one cluster of patrilineages while another lineage was the *Guruwananawuy mala*. Not all groups were divided clearly into constituent groups, as we shall see.

A person took his or her father's group identity (or rather identities), but because of women marrying again when widowed, or because of a person being engendered by a man other than the mother's husband, this rule was not as simple as it sounds. Where a woman remarried when her child was an infant, whether because of the death of her husband or a separation, and where her new husband's group was different from that of the child's father, the latter could claim a secondary identity with this man's group. People recorded (in oral tradition) human ancestors to about the *ga:thu* (FFF) of living adults, but groups were not defined with reference to a common ancestor. Moreover, distinct groups seem never to have been linked by common agnatic human ancestors, although they were linked through common *wangarr* ancestors and by relations through women.

Mala or *ba:purru* were moiety-specific. Dhuwa- and Yirritja-moiety categories extended to the people of a group, its countries, *wangarr* ancestors, all the associated sacra (*madayin*, including songs, designs, sacred objects, and dances), and all the species and entities after which the sacra were named. The Sun and Shark, for example, were Dhuwa moiety, while the Moon and Kangaroo were Yirritja. A person was of their father's moiety or, by default, the opposite moiety from the mother's, a rule that came into play in the rare event of children born of intra-moiety unions (Shapiro 1967; 1979: 36).

Moieties and hence groups were exogamous: a person had to marry someone of another group and of the opposite moiety. Thus a Yirritja woman married a Dhuwa man, her child was Dhuwa and was of a different group from her own; a Dhuwa woman married a Yirritja man and her child was Yirritja and also of a different group from her own. Consequently, any community of people living together and related as husband to wife, mother to child, father to child, brother-in-law to sister-in-law, and so on, would include people of different groups and of both moieties. Moiety identity was strongly exclusive; Yolngu assiduously guarded moiety exogamy and did not tolerate intra-moiety unions. In this sense the moieties were 'closed', although open in the sense that they were not territorially bounded. Speakers of the prefixing Gidjingali language, and whose country lay to the west of Cape Stewart, seem to have been more tolerant of intra-moiety unions (Hiatt 1965).

A group was not united politically in the sense that all activities were centrally co-ordinated, and people with a common *ba:purru* identity were usually dispersed, although full siblings tended to live in the same community, often near their fathers if they were still alive. Except for the smallest groups, all people with a common group-identity probably never came together in one place at the same time. Certain people who were related through matrifiliation to people of a group had strong interests in that group's affairs, although the relative powers of members and affiliated non-members were open to dispute. Furthermore, larger groups were divided by factions embedded in distinct cognatic-kin networks which linked people of distinct group-identity. Conversely, people sometimes extended *ba:purru* identity contextually to uterine kin and affines when referring to an aggregate such as those performing in a ceremony or residing on an outstation (cf. Shapiro 1981: 25). It is perhaps for such reasons that Aboriginal people use the English 'mob' so readily to denote an aggregate of persons.

Indeterminacies in Group Identity and Relations

The account so far may give the impression of patrifilial groups each owning several tracts of country, and of some being clearly divided into land-holding groups. This impression needs to be corrected, for people disagreed about the genealogical structure of a group and, where their existence was acknowledged, about whether constituent groups were uniquely associated with particular countries. Moreover, there are no clear criteria for designating some differences of identity as 'subclan' relations as opposed to 'clan' relations, for even where people agreed about the differences, the relation of constituent 'subgroup' to group identity was a relative matter.

Genealogies showed that some groups were composed of a single lineage from one man or two brothers; others included up to four lineages, each up to five generations in depth. In men's accounts a lineage sprang from one man or two close brothers (with at least one parent in common). In Yolngu terms these were groups of people with the same *ba:pa* (F), *ma:ri'mu* (FF), and *ga:thu* (FFF).

Some groups identified lineages and sublineages by names inherited in the male line marking relations to a common ancestor. In this way Honeybee men explained to me that their *ga:thu* (FFF) had two names, Wa____ and Ng____. His two sons, the founders of the two dominant lineages, and their successors took one of these names each, as well as their personal names, so distinguishing the lineages. Cloudy Water people simply identified lineages by the personal names of the ancestors of each.

Cloudy Water people often gave me very different accounts of genealogical relations, especially in the way they linked lineages to male ancestors. The older Honeybee people agreed on the general structure of their genealogical relationships but several Cloudy Water men had divergent views. In the mid-1970s the group consisted of four distinct lineages of four generations in depth (Fig. 1). Men generally agreed about the constitution of the lineages (although they differed about details) and grouped these as sublineages of lineages founded by two ancestors, but they did so in different ways (Fig. 2). There was no orthodox genealogy of this group. Indeed, there were no specialist genealogists, nor means for one person to impose their version on others. Indivi-

D Lineage

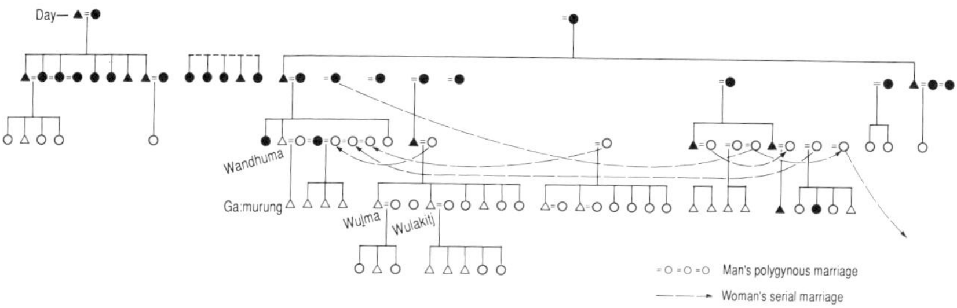

L Lineage

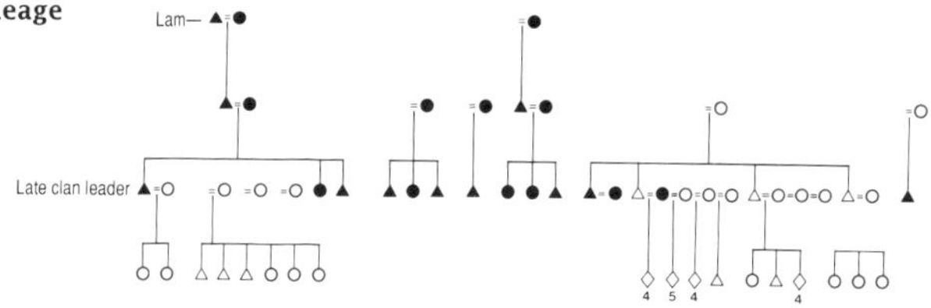

B Lineage

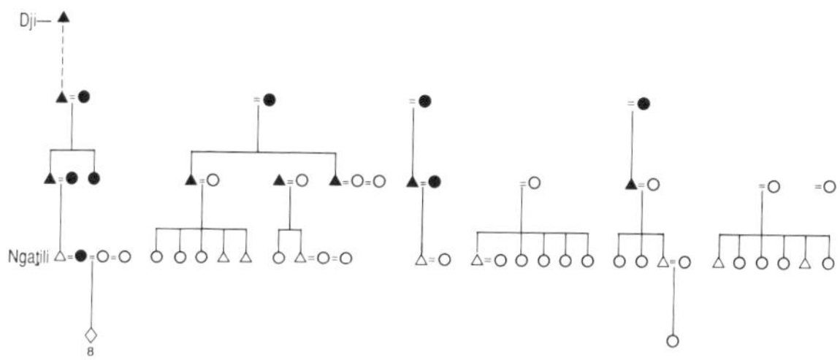

G Lineage

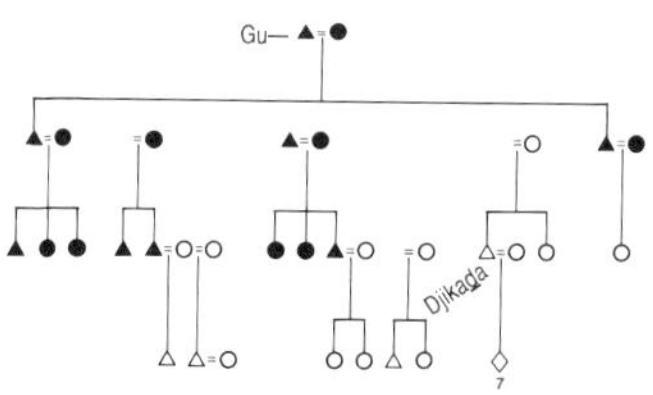

FIGURE 1. Cloudy Water Genealogy

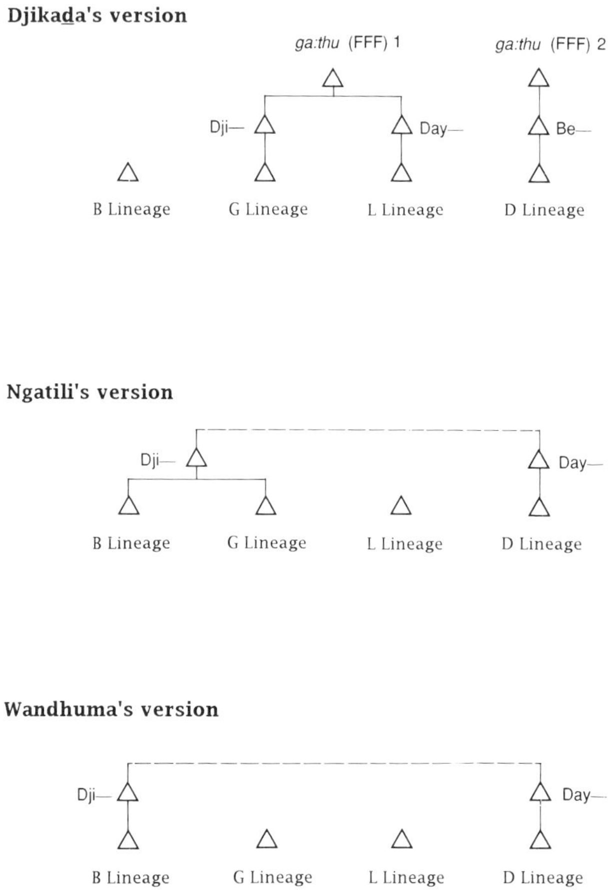

FIGURE 2. Versions of Cloudy Water Genealogy

dual versions reflected individual affiliations and interests and political relations at the time (Keen 1978: 37).

People of some larger groups agreed that their people were divided into smaller groups, each especially associated with one contiguous country. Other groups were not divided in this way. In yet other cases people with a common group-identity disagreed about the division of responsibilities and affiliations. Stringybark people with whom I discussed the matter agreed that the group had three main countries, that one lineage was associated with one country, three lineages belonged to a second, and that older people of the second country looked after the third, for its people had died out. They named each constituent group after the country as the '*x-wuy*' group—the group 'pertaining to' *x* country.

Crocodile people disagreed about the way in which its three groups, each named after its *likan* name, related to their several countries. I recorded a variety of opinions: that each of three Crocodile groups held a distinct country, as well as having rights in the others; that they comprised one group with common rights in the various countries, although each group had somewhat distinct sacra; and that people of one constituent group could sing about the country of another but not live on the land. (This was during 1987.)

'Elbow' (*likan*) names were a class of polysemous names, called out at key points in ceremonies, and which signified at once a *wangarr* ancestor, a place, and a group (see Chapter 3). The way in which men related these to countries reflected differences in whether and how they recognized group divisions.

The Stringybark leader explained to me that four of the *likan* names which he called out at mortuary ceremonies applied to one island-country, and four names to the mainland country. The first of each four was the 'leader' or 'boss' of each country. One of these 'boss' names signified the Olive Python *wangarr*, as well as the *ba:purru* group. Men used to call out the name during fights, for the ancestor 'bred' the group, the man explained.

A senior man of the Honeybee group, in contrast, explained to me that each group had but one *likan* name. His own bore the same name at all three of its places—'One group, one *bundurr*', he commented. Despite this assertion, when the same man told me the *bundurr* of other groups, he attached a different name to each country. His first claim perhaps reflected the fact that his own group was not divided into distinct land-holding subgroups.

Groups with the same *ba:purru* identity but holding distinct countries were sometimes distinguished as 'top' or up-country people (*dha:gupapuy*, mouth-nape-pertaining to) and 'bottom' or down-river people (*dha:dhudipuy*, mouth-buttocks-pertaining to).[3]

[3] According to Williams (1986: 71) some names refer to a 'clan' subgroup at a particular place and season. Her example is Marradhanggimi, 'fish-poison people', a subgroup of 'bottom' Djapu.

Such usages could refer to place of residence rather than country owned; I have heard Honeybee people who lived at Nangalala outstation referred to as 'top' or 'forest' people as opposed to the 'bottom' or 'sea-side' people of the group who lived at Milingimbi.

What distinguished a group with a proper name, such as Stringy-bark people, from its constituent groups named after country? Morphy writes that 'clan' members had common rights in songs, designs, and ceremonies, and used a common group-pattern; however, people associated with different countries often produced somewhat different sets of designs when making bark-paintings for sale, or when painting on the body in rituals (H. Morphy 1991: 49). Can we say that ideally group members had an unqualified right to the use of any land of the same group identity? In practice, although they might assert that they had the right, people would sometimes be hesitant about setting up camp in the country of a distinct group with the same *ba:purru* identity unless invited. Conversely, people would at least claim the right to use the country of a group with a different *ba:purru* identity but with the same *wangarr* ancestor, and in some circumstances enact it.

Yolngu 'groups' were named 'kinds' of people, as I suggested earlier. These kinds of people were not clearly defined political/jural entities; rather, assertions of similarity and difference were aspects of tacit agreement to co-operate, or attempts to establish relative freedom from constraints in ritual practice. There was no absolute distinction between autonomy and dependence, however. For example, men predominantly of one lineage of Cloudy Water group were able to decide to organize the performance of the Ma<u>d</u>ayin ceremony described in Chapter 6, yet did not assert a separate identity from other Cloudy Water lineages. They were obliged to invite the leader of Thunder group to participate, yet these groups had separate identities as well as linkages. No doubt genealogical relations constrained patterns of division; it seems unlikely that a group of adult siblings with a common father's father would split into separately identifying groups except with reference to place of abode.

A contrast between 'group' and 'subgroup' (or perhaps, 'constituent' group), then, can be misleading. The distinction between 'clan' and, say, 'land-holding group' or 'clan subgroup' is intended to capture the relationship between distinct aggregates of people with the same *ba:purru* identity but who identify with distinct countries, sacra (*ma<u>d</u>ayin*), and *<u>l</u>ikan* names. One of the reasons for avoiding terms such as 'clan', however, is that there was no clear relationship between such differentiation and organizational features. For example, Whale people of M____ country seem to have asserted some independence from Whale people of <u>D</u>____ country, leading some analysts to describe these as separate clans or sibs. Stringybark people associated with different countries were more unified in their activities.

But relationships between groups could change over a period. During the mid-1970s Honeybee and Lightning Snake people stressed their unity as Up-country people (a proper name), for the Lightning Snake men were supposed to have lost their songs, the Honeybee men performing on their behalf. By the end of the 1980s, however, following the death of their influential leader, Honeybee men were far less dominant in settlement affairs, and Lightning Snake people were singing their own songs (Franca Tamisari pers. com.).

Strings of Groups

Groups which were differentiated by attributes such as name, country, and elements of ceremonies, joined with others to form more extensive *ba:purru* or *mala* (groups) through shared ceremony-elements and/or common 'tongue'.

People whose countries were connected by ancestral travels formed 'one group' (*ba:purru wanggany*, *mala wanggany*) with 'one sacred object' (*ma<u>d</u>ayin wanggany*) and 'one ceremony' (*bunggul wanggany*). People would say that the songs, designs, dances, or sacred objects of two groups were 'the same' (*balanya bili*), although they might qualify this unity by saying that they were a little bit 'separate' (*ga:na*). A shared name indicated the existence of the relationship. In just such a way, Cloudy Water people, Mangrove, Sunrise, Thunder, and other groups, were 'one group' called Ma:tjarra or Bungunbungun, with Djang'kawu as their common *wangarr* ancestor. However, each country and group was connected to several such strings, which cut across each other. One of a group's alternative names would connect it to one string of groups, another name would join it to a different, though possibly overlapping, string. Moreover, the same name might be used on one occasion to denote a tongue (*matha*) and on another, a group (*mala*).

Another relation between groups was to be 'conjoint' (*dha: manapanmirri*, 'joined together'). This relationship seemed not to entail shared songs and ceremonies but to reflect the close kin-relationship between members who could combine to perform a ceremony. They 'washed' together or, like the Cloudy Water and the Black-capped Tern groups, they performed the circumcision ceremony together (see Chapter 5).

A common practice was to extend the name or names of one's own group to identify others of the same extended group. For example, Stringybark people applied their name in certain contexts to Sun, Mangrove, and Dilly-bag groups, who were 'kinds of Stringybark'. The four groups had the Wagilak *wangarr* ancestors and associated rituals in common, and the name referred to a common *maḏayin* element. This practice led to people of different groups disagreeing about what name should be used for the string of groups.

These extended groups were not 'sets' of less inclusive groups, I suggest, for they were open in definition, and all could be extended further through the exchange of ceremony elements and names, or the discovery of previously unrecognized similarities. Furthermore, which groups might be said to be joined by a common identity was relative to a perspective. A Milingimbi person would provide a different list of Sandfly groups from that given by a person from further east, although these lists overlapped. The former might list groups whose countries lay on the coast to the west and east of Milingimbi, whereas people from Yirrkala, a settlement some 160 kilometres east of Milingimbi, listed groups whose countries lay to the west and south of their community. Peoples with places between the two settlements featured in both lists. In Milingimbi discourse the name 'Up-country people', on the other hand, was generally confined to two groups, Honeybee and Lightning Snake, although the Fish people were said to speak the same tongue.

In the case of groups such as Stringybark, Cloudy Water, and Honeybee, it is possible to generalize that the more extensive groups or 'strings' in which they partook held only certain ritual elements in common. (The string metaphor is mine, but is consistent with Yolngu symbolism of group connections.) Each 'minimal' group shared only certain elements of another group's sacra. However, it was always possible for people of a formerly undivided group to claim close association with a particular country, and assert separate identity, marked, say, by a *likan* name or design belonging to that country, and implying a unique complex of connections with other countries and groups through ceremony.

The social implications of groups being said to form 'one group' varied greatly. In some cases the common identity was little more than a reflection of the fact that they shared a common ritual element. In others it implied very close ritual, social, and political ties. (See Chapter 5 for an example.) It is possible that identity in terms of more extended groups became more salient in some cases, as a result of co-residence on missions. At Milingimbi in the mid-1970s Honeybee and Lightning Snake people were more likely to identify as Up-country people, and the very inclusive Black-capped Tern group had more salience than the name of smaller groups related by this identity.

Hence Yolngu did not spend time arguing about whether a particular named entity was one kind of organization or another, translatable as a clan subgroup, a 'clan', or a 'phratry'. They did discuss and argue about whether people were 'one' (*wanggany*) and 'the same' (*balanya bili*) or 'separate' (*ga:na*) and 'different' (*wiripu*); or both the same *and* different. Judgements about unity and difference depended on context (who was present, what the agenda was) and on what attributes were at issue (resources related to country, song and ceremony, alliances with persons). Unity and difference were expressed in the use of proper names and of ceremony elements shared in complex, overlapping strings.

This pattern of ambiguity of group identity, together with the loose, cross-cutting web of affiliations, was reflected in the relation of groups to language.

Group and Tongue

Yolngu drew connections and differences between groups by juxtaposing two names denoting contrasting attributes in this way:

x *matha*, y *mala*	(x 'tongue', y group)
x *mala*, y *ba:purru*	(x group, y *ba:purru*)

Many of these names were shared by more than one group, reflecting the sharing and exchange of ritual elements. While people no doubt qualified one name with another in order to draw distinctions (and connections) between groups, what names they chose depended again on the context and the individual; the pairings did not imply the existence of cross-cutting named sets of different types, as in R. M. Berndt's theory of Yolngu groups.[4]

The Yolngu generic term for a way of speaking was *matha* (or a synonym for this word such as *ya:n*), literally 'tongue'. The degree and kind of differences between named ways of speaking that people took to be socially significant was very varied, from no discernible differences, through differences in a few lexical items or a simple phonological rule, to major syntactical and lexical differences.

Yolngu classified ways of speaking into nine categories, called 'dialect groups' by Schebeck (1968), according to the characteristic form for the demonstrative pronoun 'this/here', such as *dhuwal*, *djangu*, and *dhangu* (Map 2). A Dhuwal-speaking person or group was referred to as *dhuwal'mirr* ('*dhuwal*-having'), and so on. Since speakers of a common 'dialect group' did not constitute a sociological group, I shall refer to these ways of speaking as 'this-categories'. Dhuwal, Dhuwala, and Djangu were moiety-specific tongues (Dhuwa, Yirritja, and Yirritja respectively), whereas all the other this-categories included patrifilial groups of both moieties (Schebeck 1968: 10–11). The countries of up to ten groups associated with a particular this-category were contiguous. The exceptions were Dhuwal and Dhuwala groups whose countries were interspersed, and the Djangu-speaking groups (Yirritja moiety), whose countries were interspersed with those of Dhuwa-moiety groups, mainly Dhangu-speaking.

Many anthropologists and linguists have reported the Yolngu dogma that each group had its own particular way of speaking, often referred to by a *ba:purru* name, such as *L̲iyagawumirr matha*, *Manggalili matha*. But this ideology was not universal; people of

[4] This pairing of 'tongue' and 'group' names, as well as the pattern of sharing of names, led R. M. Berndt to posit two kinds of overlapping groups or sets, *mala* and *matha*, such that one *matha* could consist of several *mala*, or one *mala* could consist of several *matha*. A particular group could be specified as the conjunction of two sets, that is, as a *mala-matha* pair (R. M. Berndt 1955). See Schebeck (1968) for an extended discussion of this issue, and Keen (forthcoming).

Map 2. Yolngu Languages and their Neighbours

some groups did not claim to possess a distinct tongue but said that they spoke a tongue in common with one or more other groups. Some labelled their way of speaking with their group name, and extended this name to the tongues of other groups in the same this-category. For example, people sometimes said that Cloudy Water people spoke Black-capped Tern tongue after the name of a group whose tongue was of the Dhuwal this-category. On occasion, however, a Cloudy Water person labelled his or her way of speaking as 'Cloudy Water tongue' (or by an alternative name of the group). Such applications were thus to some degree relative to each group's perspective. The constituent groups of some strings of groups (or extended groups) spoke tongues of the same this-category, while others, such as Sand-fly people, did not have a common tongue. Gupapuyngu groups, for example, spoke Dhuwala (often referred to as Gupapuyngu language), whereas Liyagalawumirr or Galbanuk groups spoke Nhangu, Dhuwal, and Djinba this-categories. Although the language names were mapped on to land and groups, this was not a predominant mode of social identity, except for Djinang speakers, who often used the name as an identity marker.

Named tongues of the same this-category often differed in some respects, and, as Frances Morphy points out (1983: 6), Yolngu made a conscious effort to maintain even minor differences between such tongues. But differences within two related this-categories (Dhuwal and Dhuwala) at either end of their geographical range were greater than differences between them as spoken in the same community (F. Morphy 1977). The ways of speaking associated with groups and strings of groups did not conform, therefore, to linguistic criteria demarcating 'language' and 'dialect'.[5]

As Merlan (1981: 144) found in relation to the Mangarrayi people, there were differences between linguistic and territorial affiliation and language actually spoken. Due to many years living in a settlement in which Dhuwal- and Dhuwala-speakers were numerically dominant, people who identified as Yan-nhangu had come to speak mainly Dhuwal.

We find in the foregoing the same relativity of perspective and fuzzy boundaries in Yolngu language-use and categories of language as in group properties and structure. People named ways of speaking relative to their particular perspective. While a more or less definite aggregate of people 'agreed' on a grammar, such as that of the Dhuwal this-category, the speech of people using 'different' this-categories was actually closer than that of some speakers of 'the same' this-category. As with ceremony, those who were most closely related and who interacted with the greatest

[5] Frances Morphy (1983) notes the difficulty in applying the usual linguistic criterion of mutual intelligibility in a multilingual community.

frequency had the most in common linguistically, although maintaining differences salient to their identity.[6]

Yolngu *ba:purru* groups or *mala* were not necessarily groups of close siblings, fathers and fathers' sisters and men's children. In the case of large groups, a person might find quite distant mothers' mothers' brothers (*ma:ri*), women's daughters' children (*gutharra*), and other non-agnatic kin with the same group-identity, as well as close agnatic kin.

The Classification of Kin

Ba:purru, *mala*, and *matha* were aspects of a person's identity which he or she shared with others. Some personal names identified an individual with a patrifilial lineage, while others were more specific to an individual, though often shared with at least one other living individual. A person had several names, one of which might be suppressed if someone with the same or a similar name died. These names derived from places, *wangarr* ancestors or other beings, or *ma<u>d</u>ayin* attributes of ancestors, and were ideally given by the person's *ma:ri* (mother's mother, mother's mother's brother) or *mukul ba:pa* (father's sister) (Warner 1937: 90, 513–15). People identified others primarily as kin and using kin (or subsection) terms rather than personal names. In what follows I shall outline only those aspects of Yolngu kinship particularly relevant to the topic of this book.

A case could be argued for avoiding the use of 'kin' and 'kinship' for the Yolngu category *gurru<u>t</u>u*—contrasted with the word *mulkuru* ('stranger'), a person with whom one has no relationship—to denote the existence of a social relationship. It did not contrast with other kinds of social relationship, such as 'friends' or 'colleagues'. However, *gurru<u>t</u>u* relations were based on (although not restricted to) genealogical relationships grounded in beliefs about procreative processes; for that reason I shall retain the word 'kin' as well as 'relation' and 'relationship'.[7] For the non-specialist

[6] Yolngu did not simply map 'language' or 'dialect' on to territory or social units as in the model of the tribe associated particularly with Tindale and Birdsell (see Sutton 1978; Merlan 1981). As Sutton (1978) found among Wik-speakers, linguistic variation was isomorphic neither with ritual groupings nor marriage clusters.

[7] On the genealogical basis of kinship, see Scheffler (1973); Keen (1985). The term *gurru<u>t</u>u* means 'kin relation', *gurru<u>t</u>umirri* (kin-relation-having) means 'related by kinship', and one's *gurru<u>t</u>umirri mala* ('kin-relation-having group') is one's group of kin, or kindred. (Cf. Lowe 1976; Zorc 1986: 149; Shapiro 1981: 41; Heath 1982: 50.)

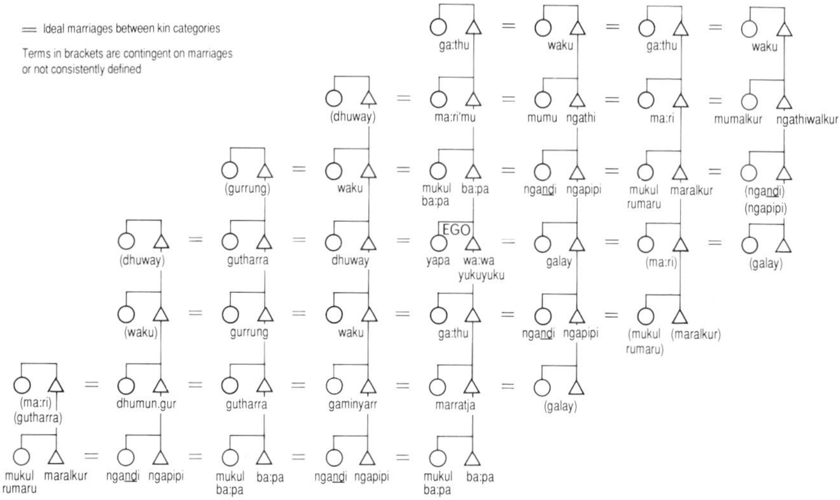

FIGURE 3. Yolngu Kin-Classification

reader I should explain some key features of kin classification and associated practices which contrast with those in cultures of European origin.

In the pre-colonial past, the Yolngu social universe was almost wholly composed of kin. Not all relations were equal, however. Yolngu distinguished between 'full' and 'partial' kin, and assessed the quality of relatedness along a dimension of distance. When referring to close kin, speakers added the suffix *-mirringu* to a term, as in *ngarraku ba:pa'mirringu*, 'my close father'. Figure 3 shows the basic set of kin terms used for reference, some of which were also used for address. Yolngu employed another set of terms for indirect reference (e.g. *nganakal*, 'your mother's brother'). Still others referred to 'avoidance' relatives (e.g. *manawiny* = woman's male *gurrung*), that is to say actual or potential affines with whom behaviour had to be reserved. There were also many synonyms for the basic terms.

If people of English-speaking societies bothered to trace kin relations beyond first or second cousins, all but lineal kin would be 'cousins' of varying degrees and remove. Yolngu, however, applied terms for close kin such as 'mother' and 'sister' to more distant kin as well. The extension of these and other kin-terms came about in a number of ways. First, through 'sibling merging', a person's mother's sister was also their 'mother' (*nga:ndi*), and their father's brother was their 'father' (*ba:pa*). This mode of extension applied to more distant generations as well—one's father's father's brother was one's 'father's father' (*ma:ri'mu*), and mother's mother's sister was 'mother's mother' (*ma:ri*).[8] It was, of course, possible to distinguish linguistically between the mother who bore you, and her sisters. The children of a person's mother's sister (their 'parallel cousins') were, it follows, the person's siblings, and the children of mother's mother's sister were 'mother' and 'mother's brother' (*ngapipi* or *gawal*). Other modes of extending close-kin terms had to do with the extension of the kin universe more generally.

One way in which Yolngu computed existence of a kin relationship where for English speakers there is none was through 'inherited' kinship. A person who is, say, your mother's 'brother', although no one knows the precise link (in terms of parent–child relations), is your 'maternal uncle' (*ngapipi*). This is so because *any* 'brother' of someone who is your 'mother' is your *ngapipi*. Consistent with this way of finding a relationship, if two people who did not know their relationship met, they could compute one by thinking of a person to whom both knew his or her relationship. Mary and Joe did not know their relationship, but both knew Arthur. It turned out that Arthur was Mary's 'brother' and Joe's 'father', so Mary and Joe deduced that Joe was Mary's 'brother's son', *ga:thu*, and she was his 'father's sister', *mukul ba:pa*.

Yolngu generally counted inherited kin and kin found through common distant connections among the 'distant' (*barrku*) or 'partial' (*ma:rr-gangga*) kin relations as opposed to 'close' (*galki*) and 'full' (*dhangang*) kin. But just who counted as 'close' or 'full' kin depended on context, although in general they were those related by direct genealogical ties.[9] Like the Macassan traders, more

[8] This form of equivalence, which is extended to other categories of relative also, is technically same-sex sibling merging (Scheffler 1978: 115).

[9] Shapiro (1981: 38–9) defines 'full' kin in terms of the relationship of groups ('sibs') to Ego's group. I found, however, that the contrast of 'full/partial' depended partly on context, for there were no strict criteria. In general the contrast marked genealogical proximity—full kin were close kin. It is certain that not all members of the group of a 'close' kinsperson were necessarily classified as close. One man at

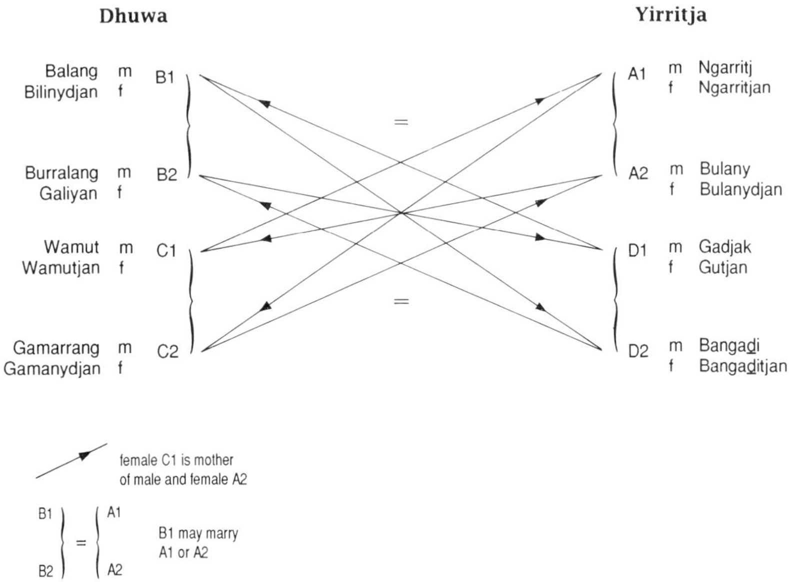

FIGURE 4. Yolngu Subsections

recent non-Aboriginal outsiders who developed a close relationship with Yolngu have often been incorporated as quasi-kin. In north-east Arnhem Land during the mid-1970s the default relationship-term for white males was *wa:wa*, older brother, or *yapa*, sister.

A third way of extending kin relations was by means of subsection names. Subsections are a variety of what used to be called 'marriage classes' by anthropologists, but which are better thought of as 'naming systems', for they were not as a rule used to regulate marriages. What the Yolngu called *ma:lk* or 'skins' comprise a set of sixteen paired names, eight for males, and eight for females (Fig. 4). In this system the names relate to one another in ideal kin-relations—*bilinydjan* is the 'mother' of *bangadi* and *bangaditjan*, and *bangaditjan* is the 'mother' of *burralang* and *galiyan*, and so on,

Milingimbi, for example, classified people of two lineages of his sister's child's group, who had very indirect genealogical links to him, as 'partial' kin. Genealogical and related spatial distance among close relatives may be assessed also. In a discussion recorded by Heath a Yolngu man referred to his *waku* (wC/ZC) as 'close' and his *gaminyarr* (mDC/BDC) as 'distant' (Heath 1980: 67).

forming two distinct cycles of names. The names are also paired in ideal marriage-relations; ideally a woman called *bilinydjan* should be the wife of a man called *ngarritj*, and by second preference, *bulany*, and so on. These ideal marriages, in conjunction with the matrifilial relations between subsection names, result in ideal patrifilial relations among subsections, again in two cycles if only first-preference spouses are taken into account. An individual was assigned a subsection name on the basis of their mother's subsection, by 'indirect descent', following the ideal relationships between the names. Thus the children of someone called *bilinydjan* were *bangadi* and *bangaditjan*, and so on.

Probably all Yolngu in the mid- to late 1970s answered to a subsection name, although they were not universal in the upper Arafura swamp-area during the late 1950s (Nicolas Peterson pers. com.), having been relatively recently introduced (McConvell 1985). Yolngu used this form of name as a mode of address in preference to personal names, and in preference to some kin terms in the case of avoidance or sensitive relationships (such as a man's *galay* who is not his wife). A person could compute a kin relationship with someone on the basis of the ideal relationship between their subsection names, or Yolngu versions of subsection names from other regions. Thus two people who were *wamutjan* and *gamarrang* could say that they were related as 'mother's mother' (*ma:ri*) to 'woman's daughter's child' (*gutharra*). Yolngu incorporated an outsider by giving him or her a subsection name, after which anyone could work out a specific relationship to that person. Because these subsection names (or equivalents) had a currency far wider than north-east Arnhem Land, two people of quite different regions could work out a relationship on the basis of their respective subsections.[10]

The above shows how the kin universe was extended to a wide network of people. One other feature of kin classification should be pointed out, and that is the way in which children and siblings'

[10] Subsections are formally incompatible with the structure of Yolngu kin-categories, and did not map on to lineages or groups in a predictable way, except in so far as they were divided between the moieties. It was in order to maximize the fit between the two systems of classification that Yolngu provided a choice of ideal marriage between each subsection and two others. Furthermore C. H. Berndt found that people at Yirrkala would sometimes change subsection to fit a marriage which, though based on correct kin-relations, was between people of inappropriate subsection (C. H. Berndt 1970*a*: 38).

children were classified. A person referred to the children of a sibling of the same sex by the same terms as his or her own children. Thus a man's son or daughter was his *ga:thu*, and a person's brother's son or daughter was also his or her *ga:thu*. A woman's child was her *waku*, and a man or woman's sister's child was his or her *waku*. In contrast, a person applied the same term to the children of a sibling of the opposite sex as that sibling did, so that a woman called her brother's children *ga:thu*, just as he did, and he called her children *waku*, just as she did.[11] I shall gloss the term *ga:thu* with the formula 'mC/BC, (man's child, brother's child) and *waku* with the formula 'wC/ZC, (woman's child, sister's child).

This characteristic of kin classification, shared with many other Aboriginal systems, meant that full siblings of both sexes had the same perspective on the kin network. With the possible exception of spouse's relatives, they could apply the same terms to the same people and so speak of 'our' *waku* (wC/ZC) or 'our' *ga:thu* (mC/BC) and so on. This characteristic is consistent also with the way in which people talked of kin relations as existing not only between persons but between groups as wholes, and among countries, *ma̱dayin*, and *wangarr* ancestors (see Chapter 3).

I am not going to describe the structure of kin classification in any detail here (but see Keen 1982). Suffice it to say that it was markedly asymmetric, with cross-cousins on the mother's side being distinguished from those on the father's side, and with first and certain second cross-cousins on the mother's side being assimilated to the same category. The only other general point about Yolngu kin-relations that I want to make here is that certain kinds of relationships entailed rules about appropriate behaviour. Brothers and sisters had to avoid all contact with each other after childhood, as did people in the relation of 'wife's mother' (*mukul rumaru*) to 'daughter's husband' (*gurrung*). And certain relationships entailed the potential to make specific claims—mothers and sisters helped a woman with child care; a man's sister's husband was supposed to support him with gifts and help in a dispute; and certain relations were involved in marriage claims.

Two aspects of kin relations are important in a discussion of the

[11] This principle of equivalence is technically parallel-cross neutralization (Scheffler 1978: 116).

control of religious knowledge. These are age-related polygyny and the mapping of kin relations on to groups, country, and *ma<u>d</u>ayin*.

Age-related Polygyny and its Conditions

The Yolngu had one of the highest rates of polygyny recorded for Australian Aborigines; the Djapu leader Wonggu had as many as twenty-five wives in the 1930s, and even in the mid-1970s two men had married ten and eleven wives respectively and the average for older men was high (Table 1). Surprisingly, marriage patterns revealed in genealogies compiled in the mid-1970s exhibited no marked change through time. However, some women were beginning to refuse to agree to leviratic marriage on the grounds that they would lose their widows' pensions. More recent genealogical data suggest that some young women have been refusing to consent to arranged marriages with much older men, and several young men have recently begun to evade the constraints of the marriage rules by seeking wives in other, sometimes very distant, Aboriginal communities. If these trends continue and grow, one would expect the conditions for widespread polygyny to break down, which would have widely ramifying effects on the dynamics of group structure, land-holding, households (or hearth-holds), and the size and scope of economic networks.

The control of religious knowledge by older men was one of the conditions under which they could be the beneficiaries of age-related polygynous marriage. I have argued (Keen 1978, 1982), drawing on proposals by Rose (1968) and Maddock (1972), that age-related polygynous marriage required that most men married later than women, and had a shorter married life. (The average age-difference in the genealogies was about 13 years, and the most extreme case was a man in his mid-sixties married to a girl in her mid-teens.) Women married serially, sometimes to younger men as a woman got older (Keen 1982: 634). Older men who benefited most from age-related polygyny could sustain the arrangements only by means other than their own use of physical force. They did so through authority derived from control of access to valued religious knowledge and practice, and from apparent access to supernatural power. This authority enabled them to channel the young men's aggressive power towards their own ends

TABLE 1. Incidence of Polygynous Marriage

Age of husband	Number of wives											Totals			Average wives per husband
	0	1	2	3	4	5	6	7	8	9	10	men	wives	husb.	
61–70	1		2	1			1		1	1		7	30	6	5.00
50–59	2	4	7	1	1	2	1				1	19	51	17	3.00
40–49	1	13	7	2	3	1						27	50	26	1.92
30–39	7	15	6	2	3							33	45	26	1.73
20–29	38	9		1								48	12	10	1.20
TOTALS	49	41	22	7	7	3	2		1	1	1	134	188	85	2.21

rather than against them, and was more a matter of young men deferring to older men's interests than the ability to issue binding commands. I would add here that the authority-sustaining effects of initiation rites were not simply brought about through an economy of religious knowledge; such rites had more direct effects on the dispositions of initiands (see Chapter 5). Furthermore, the use of physical force and threats on the part of men was an important aspect of marriage politics; in the past, men sometimes killed a husband and took his wives by force.[12]

That age-related polygyny was not in the interests of women (*contra* Rose 1968) is suggested by expressions of preference for monogamy among Yirrkala women recorded by Reid (1983). However, women were not mere pawns in a male marriage-game, for as women got older they were more able to assert control of their own lives. Some recent changes in the conditions of Yolngu lives have worked in the interests of women, especially the availability of alternatives and the imposition of external control of the use of force, which have led to the erosion of the authority of older men and consequent changes in marriage practices (see Chapter 9).

There is no need to elaborate on Yolngu marriage-practices here, for many others have described them in a degree of detail not relevant to the themes of this book (see for example R. M. Berndt 1955, 1971; Keen 1982; H. Morphy 1984; Shapiro 1981; Warner 1937). Put simply, Yolngu men and women spoke of a woman being 'given' to a man rather than the other way round (cf. Merlan 1988). Many marriages were arranged when the pros-

[12] I have suggested other conditions and consequences of Yolngu polygyny elsewhere (Keen 1978, 1982). The demographic properties of Yolngu kinship led to the delay of men's marriages, since women in whom a man had primary claims as wives were on average some 13 years younger. Older brothers had prior claims over younger brothers, leading to a few men acquiring many wives. The consequence of these marriages was that some groups grew quickly at the expense of others, with consequent competition for country and the sacra attached to them, although Chapter 3 suggests that this competition was not played out among solidary groups. There may have been a degree of positive feedback, for men of larger groups were better able to press marriage claims and succeed in elopements than men of small groups. These patterns of marriage linked a group to a few others through many marriages across several generations, linking pairs of large groups in political alliances, cross-cut by kin-based factions. The smallest named patrifilial groups of Gidjingali people to the west were much smaller than Yolngu groups, and a group was linked to many others by a few marriage ties to each group. Western Yolngu groups came somewhere in between these extremes.

pective husband was young, and even before the birth of the girl who would be the wife. A proper marriage was that between a man and woman who were in the relation of *dhuway* (the man) to *galay* (the woman), the relation of 'FZS' to 'MBD', and the preference was for marriage between genealogically close cross-cousins who were, by implication, of groups whose countries were geographically close.

Other relationships were important in the arrangement of marriages. Looking at it from a man's point of view, a man's kin on his mother's side were potential wife-bestowers: his mother's brothers (*ngapipi*) of varying degrees of closeness were potential wives' fathers; daughters of his mother's mother's brother (*ma:ri*) were potential wives' mothers (his *mukul rumaru*); and their matrikin in turn might be involved in bestowal and spouse exchange (the man's *mumalkur* and *ngathiwalkur*, MMMBD and MMMBS). The classification of kin was such that similar marriage-claims could be repeated two generations apart. A man could try to arrange to marry his MBSSD as well as his MBD (and indeed his MBSD); moreover, similar marriages connected lineages through the generations in continuing alliances. The resulting network of marriages between lineages and groups had a markedly asymmetrical character at the level of sublineages, with successive links through women doubling back and cycling among some groups, and connecting others in long strings.

The relations involved in marriage bestowal, as well as the resulting marriage-networks, were expressed as kin relations between groups, their countries, their sacra, and their *wangarr* ancestors (see Chapter 3). The kin relations with which this book will be especially concerned, and which defined rights in, and relations to, country and ceremony, included relations through both the father and the mother (Fig. 5). A person traced connections within his or her group back to father and father's sister (*ba:pa*, *mukul ba:pa*), father's father and his sister (*ma:ri'mu*), and father's father's father (*ga:thu*). Relations were traced forward to *ga:thu* (man's child/brother's child), *marratja* (man's son's child), *mukul ba:pa* (mSSD) and *ba:pa* (mSSS). Matrilineal kin-categories were important in defining rights in the country and ceremonies of other groups, as well as men's marriage-claims: *ngan̲di* (mother), *ma:ri* (mother's mother), and *waku* (mother's mother's mother). Descendants in this line were *waku* (woman's child/sister's

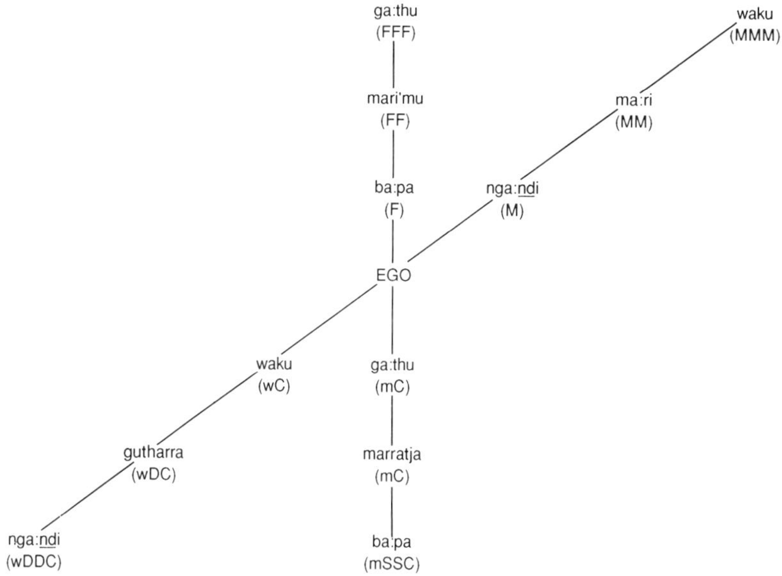

FIGURE 5. Kin Categories Applied to Groups, Country, and *madayin*

child), *gutharra* (woman's daughter's child/sister's daughter's child), and *nga:ndi* (wDDD). We shall see that the same categories expressed relations between groups, countries, and *madayin*.

Indeterminacies and Variation

Let me conclude this brief outline of *gurrutu* by stressing variation and indeterminacies; although some widespread common axioms could be isolated, there was no uniform Yolngu 'marriage system'. First, the classification of kin was less than determinate in a number of respects. One of these was no doubt the 'performative' aspect of kin relatedness (Sansom 1988): a person would emphasize or deny the existence of a real or close relationship on the grounds of a person having nurtured them, or as the result of a grievance. Then, people would manipulate categories and 'rules' in their own interests; say by justifying a sexual and domestic relationship that was wrong in kinship terms on the grounds that it was 'straight' through subsections. Similarly, when I asked one man why some of his wives were of a group other than his mother's and why he lived on their country, he said that the group and country were of his 'mother' because his mother's country lay on the same river as his wives' country.

Individuals varied in their knowledge of definitions at the boundaries of the set of axioms, for example over what the spouse of *dhumun.gur* should be called; and considerable flexibility came into play when it came to applying terms to people. Choice in reckoning a kin relation by different routes was not an outstanding feature of Yolngu kinship, since most marriages were between people in a correct relationship. Where anomalies did occur, people corrected them by applying the rule of 'throwing away the father', and computed relationships through the mother only (see H. Morphy 1991: 82 for an example). The subsection system (*ma:lk*) did provide an alternative basis for reckoning kinship which by no means always coincided with relations calculated through kin links (cf. Rumsey 1981).[13]

Speakers of Yolngu languages agreed over core axioms but constituted more distant kin differently. Residents of Yirrkala, on the extreme north-east tip of Arnhem Land, did not consistently distinguish between *ma:ri'mu* (FF/FFZ) and *ma:ri* (MM/MMB), but applied the latter term to both kinds of grandkin, and it seems that they never used the terms *mumalkur* (MMMBD) and *ngathiwalkur* (MMMBS) (Howard Morphy pers. com.). Djinba- and Ritharrngu-speakers, whose country lay at the southern end of the Arafura swamp and in the ranges, did not distinguish MMMBS and MMMBD from MF/FMB and FM/MFZ (both *natcha* in Djinba and *karmala* in Ritharrngu), and perhaps did not formalize the exchange of sisters' daughters' daughters as did Yolngu of the coasts (Shapiro 1968; Peterson 1971: 170 ff.).

A second area of variation was marriage practices. The marriages of Djinang people, living on the western extremity of the

[13] e.g. I asked two other Honeybee men how they were related to one another. One of them told me that he thought they were *maralkur–gurrung* (MMBS–FZDS), but an older man present said that the two men were brothers for they were all of the same subsection. Here, shared subsection may have been taken to be an indicator that the men were connected by an identical number of generation links to a common uterine ancestor, implying a sibling relationship as well as common subsection.

Yolngu region, appear to have been intermediate between the reciprocal marriage-exchange typical of those living further to the west who spoke prefixing languages, and the asymmetrical exchange of Yolngu people to the east. Men expressed the latter as a preference, but sometimes engaged in reciprocal marriage-exchange with neighbouring groups, including the Gunadba to their west (Borsboom 1978*a*: 24–5).

Some of the differences in the way in which Yolngu people have perceived the overall 'system' have given rise to differences among anthropological accounts of Yolngu kinship, and consequent arguments. Other ethnographers have taken Shapiro to task for attributing to Yolngu a semi-moiety system of four implicit categories among which marriages circulated (Shapiro 1969, 1981: 107 ff.; R. M. Berndt 1976; H. Morphy 1977, 1978; Keen 1978, 1986*a*). Recent research at Galiwin'ku, however, where Shapiro carried out the bulk of his fieldwork, shows that some men of that community do indeed represent marriage relations in just that way (Rudder pers. com.). Local meta-theories of marriage practice reflected, and perhaps in part guided, local variants.

If people had different and conflicting ideas about kin relations, whose accounts would prevail and why? Where marriage claims conflicted, success at making them stick or in getting away with violating the rights of others depended on the support a disputant could muster from patrifilial group and other kin.

During the mid-1970s an attempt by a young man of a small inland group to live with a woman promised to a man of a large group, and who was his *waku* (ZC), did not succeed against the wide network of people who rallied to support the offended man. On the other hand, when another man of this same large group eloped with a woman promised to a man of another small group there was little he could do but grumble. There were no overarching judicial bodies to arbitrate and enforce law. People tried to recruit support for their interpretations of *rom*, 'law' or 'right practice'.

Leaders: 'First-born'

Yolngu expressed group leadership in terms of a primary kin-construct, that of birth order. Leaders of a group or set of siblings were 'first-born' (*ngurrudawalangu*) as well as 'boss' (*bunggawa*), reflecting a pervasive ranking on the basis of age as well as

TABLE 2. Yolngu Age-Gender Categories

Non-gender specific	Male	Female
norrutj, *murrimurri* very old person (implies blind or infirm)		
worrungu, *wul̲man* old person	*yindi yolngu* 'big man', man with grey hair	
d̲ilak mature, knowledgeable person (used also to denote relative age)	*wurrwiliny* man in thirties and forties, 'mature man'	*gungman* woman with three or four children
		yothumirri woman who has borne a child
marrathulma young men and women (16–25)	*gurrmul* (sing.), *yawirriny'* (pl.) bachelors, circumcized males	*wirrkul̲* young woman, usually unmarried
	gad̲aku uncircumcized male	
yothu (sing.), *djamarrkuli* (pl.) child, children		

gender.[14] The classification of siblings marked rank in terms of birth-order and gender. While both males and females distinguished older brothers from younger brothers, males did not distinguish older sisters from younger sisters:

wa:wa	older brother, male or female speaking
yapa	sister, male speaking older sister, female speaking
yukuyuku	younger brother, male speaking younger sibling, female speaking

Male and female age-categories were distinct except for those relating to the very young and the old (Table 2). Categories of females tended to relate to the reproductive cycle, while those of males related to initiatory status and age.

Consistent with the iterative character of kin relations, that is to say the fact that one had very many 'mothers', 'fathers', etc.,

[14] See definitions of *ngurrud̲awalangu* in Lowe 1976 and Zorc (1986).

Yolngu of my acquaintance did not recognize a single office or 'position' of leadership in each group; indeed, given ambiguities in the definition of groups this is hardly surprising (cf. Myers 1986: 260). Where a woman was the first-born of a group, men sometimes acknowledged her as the 'true' leader who 'held' the sacra, but they usually referred to the oldest (or second oldest) man as the leader or 'boss' of a group, and the acknowledgement did not give the woman the right to perform a man's role in ceremonies, or to participate in men's secret aspects.

During the mid-1970s people sometimes spoke of two brothers as leaders of the Thunder group in the same breath, implying that they were the two 'first-born' and not that they shared a single 'position'.[15] Moreover, people of some groups disagreed about who actually was the leader of the group. For example, the Crocodile group was divided into three lineages or 'families' associated loosely with distinct countries. Warragalili claimed to be the leader of the group, and was acknowledged as such by some members. However, the brother of Warragalili's deceased father, who was older than Warragalili, claimed to be leader of the group as a whole. The most respected singer and religious leader for the group was of a different lineage from Warragalili and his father's brother, but was not acknowledged as leader by other people of the first lineage.

A person who had more say in a group's day-to-day affairs than an older sibling, due in part to personal qualities of energy and ambition, would be acknowledged as leader even if he was not strictly 'first-born'. Thus Yolngu at Milingimbi distinguished between the first-born male of a group, lineage, or group of close siblings, and the man with real power if this was a different person (see also Warner 1937: 63). The effective leader was *bunggawa* ('boss') or, as one young man expressed it to me, a person with a 'strong sternum' (*gumurr d̠a:l*). These remarks apply both to leaders of a group as a whole and to subgroups where these were clearly differentiated.

Age was relevant to a leader's powers, though not all were *yindi yolngu*, literally 'big men', men with grey hair. Where the leader was a young man, then the 'big man' of another group of the same

[15] Warner describes how the next oldest brother may be said to 'help' (*gungga'yun*) his older brother (1937: 63).

moiety was more likely to control the group's religious affairs. Nevertheless, some group leaders who were *wurrwiliny* ('mature men'), with the requisite degree of competence, did effectively lead the group's religious life.

Thus knowledge was important in leadership; the leader of one group might co-ordinate the religious affairs of another group where the men of that group, including the acknowledged leader or leaders, were deemed to lack knowledge about their songs and ceremonies.[16] According to Williams (1987: 44) leaders were expected to travel widely, meet with the heads of other groups to extend knowledge, and cement or extend old ties. While I do not know if this was a general expectation, it is true that prominent leaders in the Milingimbi region in the mid-1970s, especially the Honeybee leader, did travel widely to attend ceremonies (see Chapter 7 for an example).

The most important formal role that a male leader performed in ceremonies, that of 'caller of the invocation' (which included *l̠ikan* names) or 'power-man' (*djirrikaymirr*, Dhuwa; *d̠alkarramirri*, Yirritja; 'power-name-having'), was not an office but a role which any competent man belonging to a group with rights to perform the invocation could take on, when permitted by older men. (These terms seem to have been moiety-specific among Milingimbi residents but not, apparently, in eastern Arnhem Land.) In larger groups several men took on the role of power-man in ceremonies, but the leader could always perform the task unless he was infirm (in which case he became leader in name only). Yolngu men saw the acquisition of the esoteric aspects of a group's religious traditions (*rom*) necessary to be power-man as a long and arduous business. One Cloudy Water 'mature man' explained to me:

> A leader must know all the *l̠ikan* names . . . I know how to prepare the *rangga* [sacred objects], but *djirrikaymirr* must know the dances and *l̠ikan* for all the Dhuwa groups. I am still learning, and Wandhuma is a good man to teach us . . . A *djirrikaymirr* is someone who knows all the *l̠ikan* names and dances as well as preparation of the *rangga*. Men might come

[16] Williams (1987: 44) also generalizes that Yolngu people expected heads of groups and subgroups to be physically vigorous, have forceful personalities, and be capable of expressing this force in acts of aggressiveness—they should be 'dangerous' (*mad̠akarritj*).

[to the ceremony] and say, 'You know the *madayin* but do you know all the dances?' I would say, 'No.'

Women were excluded from legitimately expressing that knowledge, although they took part in decision-making, both secular and religious, and we shall see that women had more similar roles to men in Christian movements.

The leader/s of a group claimed, or were acknowledged, to 'hold' (*ngayathama*) and 'look after' (*dja:ga*) the members, the places, and the *rangga* sacred objects. Some men at least believed or theorized that the powers of such men derived in a substantial sense from the *rangga* sacred objects and hence the *wangarr* ancestors. Indeed the Honeybee leader equated the concept *dalkarra*, synonymous perhaps with *likan*, 'elbow', with power itself (*ma:rr*) possessed by a leader. Influential and respected leaders exercised rhetorical skills and used a markedly wider vocabulary than younger men and women, deploying these skills in their role as mediator in disputes and in decision-making (Williams 1987: 49). However, they did not make laws binding on the community and they directed the day-to-day activities of others only in limited contexts, such as sons helping prepare bark for painting or commanding a grand-daughter to fetch food. But any father or grandfather with children and grandchildren of the appropriate ages might act in this way.

In the sphere of religious performance, the authority of a male leader lay not so much in being able to command others to perform actions as in his agreement being necessary for something (especially a ceremony) to be carried out. One man explained to me that a leader was a person who could say, 'Let us begin the ceremony', and was the boss of the *madayin* (sacred objects and sacra). His presence was necessary for an activity to take place, juniors attributed to him greater knowledge than they possessed, and he had some control of their access to religious knowledge, including invitation to, or exclusion from, secret aspects of ceremony. In a modality typical of the relationship characterized by Aboriginal people as 'boss', leaders were expected to share resources under their control (such as ceremonies, religious knowledge, or land-based resources) in return for gifts and 'work' on the part of dependants (Williams 1987: 44).

Much religious knowledge was not formal but practical, such as the ability to sing long song-series and perform complex dance-sequences, so that if older men did not agree to perform ceremonies, then younger people did not have the opportunity to learn them. This is particularly relevant where the older men of one group controlled the religious life of another group of the same moiety. According to Williams, groups at Yirrkala without a senior ritual-leader were at a disadvantage; others might claim that the deceased leader did not teach members all he knew before his death, and make claims against the group's rights in ritual and land (Williams 1987: 44–6). However, other older men of a group, or older *waku* (wC/ZC) and *gutharra* (wDC/ZDC) of the group could also pass on knowledge (see Chapter 5).

Ambitious and able leaders of the pre-mission past, now referred to as 'kings' in English, extended their spheres of influence widely (Williams 1987: 44). Some leaders (among other men) of the pre-mission era gained a reputation as warriors and abductors of women, and indeed success in such ventures contributed to their leading large and powerful groups (see Thomson 1948: 409–10). In the mid-1970s Yolngu leaders were still among the most-married men: two Honeybee leaders living at Milingimbi and Nangalala had children by nine and eleven wives respectively. The ability of leaders to marry more wives than other men derived in part from the prior claims of the oldest of a group of brothers with rights in the same women as potential wives (Keen 1982). Such multiple marriages extended and intensified a man's social network and potential sphere of influence (see Chapter 9).[17]

Religious Control of the Economy

In spite of the separation of male and female spheres of production and, up to a point, distribution (C. H. Berndt 1970*a*), there seems to be little doubt that men, especially older men, had a greater degree of control of economic practices, and that this

[17] Among 19 group leaders the average number of wives per group leader in my data is 3.37, with a range of 0–10. This compares with an average for all men over 50 of 3.48, however 4 of the group leaders were under 45 years. There was during 1974–6 a strong correlation between group size and polygyny among leaders. Some of the most renowned leaders of the early part of this century, at least in the east of the region, were more highly polygynous: Wonggu (who was about 50 in the early 1930s) married 24 wives during his mature years, and Mun̲duku̲l married 18 (R. M. Berndt 1965: 88).

power derived from their control of religious knowledge and practice. In the foraging economy, production relations were articulated with religious roles: access to productive roles on the part of young men was co-ordinated with access to ceremonies, while older men left routine economic activities in order to specialize to a greater degree in religious activities. Women's productivity, on the other hand, tended to increase with age (Peterson 1972: 22, 1973: 184).

Ownership of country and attempts to control access to its resources were expressed in the idiom of ancestral connection, *rangga* sacred objects, and *likanbuy* designs (see Chapter 3). Consequently male leaders had the greatest ability to define, control, and compete over country, for it was they who deployed *rangga* sacred objects and the designs that demonstrated a group's connection to country. Residents of an outstation camp exercised some control over the use of the country on which the camp was situated; the oldest, and hence generally most authoritative categories of residents of bush camps in the 1950s and 1960s tended to be land-holders or mother-holders, or the spouses of these (Shapiro 1973; Peterson 1972, 1978).[18]

Access to some items of technology such as fish-traps was limited through their definition as sacred objects, restricting their use to particular group-centred networks led by older men (Thomson 1938; 1949: 62). These and other sacred objects, as well as rights in elements of ceremonies such as songs and designs, were currency in the exchange economy. Because of their value, people exchanged religious elements in the settlement of disputes, during marriage negotiations and after marriage, for other goods or services, or as an aspect of kin relations. The ability of older men to control the production, use, and exchange of sacred objects, designs, and songs gave them a strong advantage in this exchange economy.

Women during their reproductive lives, and adult men before

[18] It is possible that residence was more strongly patrilocal before people moved in large numbers to the missions and the countryside emptied with the exception of a few small communities (Thomson 1949). While people hoped to control access to their own or their mother's group's country, studies suggest that less than half the residents of bush camps and outstations in the 1960s and 1970s were members of the group on whose country a wet-season base-camp or outstation was situated (Shapiro 1973; Peterson 1972, 1978). Mother-holders formed an even smaller proportion.

marriage and in the early years of marriage, bore the brunt of customary restrictions on consumption. Older men directly or indirectly decided who had access to foods through ritual restrictions which they could impose or waive, declaring them *dhuyu* ('sacred', 'taboo'). Older men could also impose *ad hoc* prohibitions on production and consumption, for example by calling *l̠ikan* names over an item of technology such as a canoe (Warner 1937: 68, 130, 283, 287; Thomson 1949: 21, 42), and more recently such items as a truck or an electric cooker. Older men may also have benefited in that polygynous households were advantaged economically. The ability to get others to comply depended on the authority of those applying such prohibitions, as well as the legitimacy of the prohibitions themselves.

These aspects of the control of economy by religious means give some extra point to the episode in the Djang'kawu story referring to male appropriation of sacred objects and secret rituals (Chapter 1). Male control of *rangga* sacred objects and secret knowledge seemed to have arisen in the primordial relations among the *wangarr* ancestors. The Djang'kawu story presupposed that the sole right of men to have access to the secret ground where *rangga* sacred objects were displayed was so fundamental that even the sisters, creators and law-makers though they were, were bound by it: the men were able to retain the dilly-bag and sacred objects because the sisters were unable to go into the inside ground. The economic powers inherent in the right to deploy *rangga* sacred objects, together with connected names and designs, seemed to follow from this fundamental right from which even the ancestors were excluded.

The Changing Role of Leaders

The roles of male leaders changed during the 50 years following the establishment of the first missions. It is possible that their authority was greater before aggregation on missions; an older person living on his or her own land with a relatively small group could more easily assert the right to determine who had access to that land and its resources. Disputes among adults were less likely to arise and easier to resolve in small communities, by some protagonists moving to another camp, for example. However, feuding between people of different areas was endemic. Fierce warriors figured large among Yolngu legends about dead leaders. But prominent men emerged during the mission era as mediators between mission and Yolngu and, with the elimination of blood feud, as peacemakers.

The most prominent leader of Honeybee group emerged in the 1950s as a dominant man in the public life of Milingimbi Mission. He used structural advantage (as the oldest male of his group of siblings with the same mother) and consummate skill in the Yolngu domain as a springboard to a key role in community affairs, remaining Chairman of the elected Council from the late 1960s through the 1970s. He was the most dominant leader in ritual matters and a polygynist who was closely connected by marriage to many of the patrifilial groups represented in the community. The Honeybee leader represented himself not as a great warrior but as a peacemaker who, together with the missionaries, had ended blood-feuding in the region. He remarked to me:

> It was very bad in former times at the Mission. There were fights over women, children, and ceremonies. But it is quiet now because of me. If there was any trouble at the Mission, I was there. If there was trouble on the mainland, I was there. If there was trouble with Magpie Geese people, I was there. I wanted to make it a peaceful place for my children, peaceful like the white people. People settled on the Mission to get tobacco, not food—there was plenty of food in the bush. They settled also because of all the fighting.

The ability to control production, distribution, and consumption through religious means within a foraging economy became largely irrelevant in the context of settlement economies, through the decreasing use of the technology and food categories involved. Several women at Galiwin'ku did complain to me recently, however, that men had called sacred names over an electric cooking-stove, prohibiting its use, 'punishing' the women. Moreover, the widespread move out of townships to outstations may have restored the power to impose restrictions to some extent. One area of production of continued if transformed economic importance has been bark-painting (Williams 1976; H. Morphy 1983; Peterson 1988). Older men retained rights to determine who made certain of these, and the rights of groups have remained centrally relevant. Moreover, bark-painting for the market became an important context for the transmission of secret religious knowledge. Nevertheless, at Yirrkala, and more recently at Ramingining, women have gained the prerogative to 'look after' and to create painted designs on bark, especially with the decreased competence of male relatives through alcohol consumption (H. Morphy 1991: 212). More profoundly, conflict over new land-based resources, such as revenues from bauxite mining at Gove peninsula, has been conducted through claims to the ownership of country by members of patrifilial groups. These claims were based on ancestral connections, and expressed in terms of religious properties, joining the ancestral domain with the money economy.

Leaders of some groups have exercised their prerogatives in relation to mining companies, land councils, and other agencies. Settlement leaders attempted to legitimate their authority in relation to township affairs at Galiwin'ku through public display of sacred objects and through their control of the Christian Revival (Chapter 9). Not all male leaders of the larger groups have been closely involved in settlement affairs, however; some have remained aloof, leaving such matters to younger men and women who have had the benefit of schooling. Rather exceptionally, at Milingimbi a Yolngu woman whose group had the same *wangarr* ancestor as one of the indigenous groups, and who was married to a white community-adviser, emerged during the 1980s as the most powerful voice in the township administration.

This chapter has outlined the definition of groups, aspects of marriage and kinship relevant to religious practice and to the nexus of practices connected with religious secrecy and the power of older men, and some characteristics of leaders, who were kinds of kin. It has brought out a degree of indeterminacy, disagreement, and variation in the constitution of groups and leaders, the definition of kin, and marriage-practices; and it has suggested links between the control of religious knowledge and practice, marriage, and economic advantage. The next chapter looks at disputes over the group identity of land in the context of the ancestral constitution of country. It also asks under what conditions people could put their versions of reality into practice.

3

Country

Drawing a map of connections to country can give an impression of fixed, agreed areas and boundaries. Look at the way in which Aboriginal relations to land have commonly been represented—as apparently clearly defined sectors of coastline and coastal plain, bounded inland territories or enclaves (e.g. D. Turner 1974: 4; R. M. Berndt 1976: 148–50). Such maps all imply agreement about the definition of 'countries' and their boundaries. The mapping of Aboriginal land-tenure in the desert and semi-desert of the Northern Territory for the purpose of Aboriginal land-

Landing at Wa:l̲angur

claims has sometimes been more subtle in the suggestion that a group owns country defined by its focus on a key site rather than an encompassing boundary (e.g. Toohey 1982). Evidence given in some Aboriginal land-claims which have been lodged and heard since 1975 imply that while Aboriginal military technology and organization was not compatible with invasion and displacement of populations, competition for control of country was a motive for verbal conflicts and bloody fights. Of course, since Myers' (1986) description of Pintupi social life, reports of ambiguity of definition and contest over country should come as no surprise.

However, the ethnography of north-east Arnhem Land has given the impression of rather clearly defined groups with determinate estates. Yolngu groups were exclusive by comparison with Pintupi, and estates more clearly identified, but the greater clarity was a relative matter. Just as Yolngu people differed about ancestral events, the structure of groups, and patterns of marriage, so they contested the definition of country, as well as rights over it. The definition of country was not 'objective' but relative to a person's perspective, interest, and loyalties. People sometimes competed for control of country, each trying to make his or her account of its significance prevail.

Before settlement on missions and the institution of 'welfare colonialism' (Beckett 1987) the processes of contest in north-east Arnhem Land seem in the past to have been related to the dynamics of the growth and fission of patrifilial groups, and so ultimately to the adjustment of population to food resources and living-space (Peterson 1972). More recently, the appearance of new resources associated with the establishment of those missions, with development, and with the move to outstations, have led to conflicts over country.

Wa:nga—*Country*

The Yolngu word for place and country (in the sense of a person's country) was *wa:nga*. The use of the same word to mean the place where one sits and sleeps, a camp or township, and a named country, implies that camp and country were not fundamentally differentiated; there was no equivalent to the division between town and country. *Wa:nga* is one of a set of words for related concepts, incorporating rather similar sounds: *wangarr* ancestral being; *rangga* sacred objects placed in the country by the ancestors and which were the *ngaraka*, 'bones', of the ancestors; *nga:rra*, the ceremony ground at which the *rangga* were revealed to novices. All these things and beings implied links between people, living and dead, country, ancestors, and the ceremonies which followed them. The individual, group, and country were all identified with the bodies and *mali'* ('image', 'spirit') of the *wangarr* ancestors. In ceremony a *rangga* sacred object which represented the transformation of a *wangarr* ancestor or a part of an ancestor was its bone and flesh. The 'bone country' (*ngaraka wa:nga*) contained the transformed substance as well as the powers of the ancestors. The individual gained both his or her being and powers from the *wangarr*. At a person's death the spirit was believed to return to the waters on his or her country, the domain of the *wangarr* ancestor, and/or to a land of the dead over the sea, and/or to heaven, the spirit home of 'God *wangarr*'. But if people performed disinterment and reburial rites the body of the dead was reincorporated with the country and body of the ancestor in its manifestation as a hollow-log coffin. (See H. Morphy 1991: 221 for the relationship between paintings and country.)

From a Yolngu point of view, then, land was not mere dirt; land and waters consisted in part of the bodily substance of the *wangarr* ancestors. Some ancestors were still alive under the waters, and all were a source of ancestral power (*ma:rr*, *ganydjarr*). A certain rocky hill was the fat of Kangaroo; the subsoil beneath one of the outstations was the flesh of Shark, whose backbone was a nearby rock poking up out of the grass; Mangrove Log ancestor moved in an out with the tide in one of the estuaries.

The concept *l̠ikan*, literally 'elbow', denoted connections among places, ancestors, persons, and ceremonies. The word also signified junctions such as that between branch and tree, or a crescent form such as a bay or the new moon. Just as the elbow connects the upper and lower arms, or as the branches and roots are connected to the trunk of a tree, so *l̠ikan* names or *bundurr* called out at the climaxes of ceremonies, and geometric *l̠ikan* or *l̠ikanbuy* ('pertaining to the elbow') designs, painted on the body or sacred objects, implied connections between ancestor, place, group, and ceremony (see H. Morphy 1988: 250; 1991: 228). They made connections in a particularly focused way: *l̠ikan* names were at once the proper names of *wangarr* ancestors or things with ancestral significance, persons, and places; and the designs, instituted by ancestral action, condensed place, ancestor, group design, and sacred object into one multivalent form (see H. Morphy 1977: 228).

Places, small and large, belonged to patrifilial groups and moieties. Some Yolngu distinguished between 'big names' (*rapam*) which applied to a focal place and to a wide area around it, and 'small names' which denoted particular locations within that wide area. But, as Howard Morphy makes clear (1984: 26), it would be misleading to think of the country as divided by a hierarchy of areas marked by boundaries; rather, the names of particularly significant places were extended outward from foci, the degree of expansion depending on the context and on the knowledge of the speaker.

As a first approximation the relation between group and country can be treated as if places were clearly defined. People of each group (and I refer here to a proper name applied with the minimal degree of inclusiveness) held in their possession one or two extensive stretches of country, referred to by one or two big names. But most also had smaller areas encapsulated within other groups' larger countries, or forming small enclaves on the coast. These countries were often separated by distances of up to 60 kilometres. The spread of a group's places gave members control of a wide variety of types of terrain and waters, including inland and coastal areas, coastal waters, and in many cases parts of islands. As a rough guide (based on fragmentary information), and bearing in mind vagueness and disagreement over the group identity of some areas, a group's clusters of big-name countries, not including the smaller enclaves, varied in size from about 800 to some 10,000 hectares. The smaller areas were those in the rich, estuarine plains and the islands. Countries in the stony inland country were probably larger.

As we saw in the last chapter, some people agreed that their group was divided into subgroups, each of which held one country. In some of these cases each distinct subgroup was known by the name of the country as well as the group name. For example a subgroup of L̲iyagalawumirr people was the *Mirarrmin̲awuy mala*, the group pertaining to the place Mirarrmin̲a. Other groups divided responsibility for the places among their members but not along lineage lines. People of Crocodile group disagreed about how the subgroups were related to their countries, and this sort of disagreement may have been common. Even where a group was clearly divided, some would argue that in principle all members shared the countries equally.

This picture is a simplification, since in reality the definition of countries was not altogether agreed. Only in ecologically rich areas such as estuaries, coastal plains, swamps, mangroves, and the foreshore did people more or less unambiguously mark the places where two countries met, by features such as a rocky outcrop, a river, or the margin between salt-flat and grass-plain. Even in these instances members of groups whose countries were adjacent were likely to disagree about which mark was significant. But the moiety and group identity of these rich areas was, I think, less open to dispute than other areas. Yolngu associated these places with major *wangarr* ancestors and the important sacred objects and ceremonies associated with them. Ownership of such elements was, in my experience anyway, less subject to dispute.

In the tall open forest which stood in the broad interstices between the richer strips and parcels, and which formed an indefinite hinterland, the extent of a country was simply undefined until one came, say, to a creek which was clearly someone else's or led to someone else's place. Country was thus centred around areas rich in resources and religious significance, and this definition round a core was reflected in the way in which names were focused. But the ownership of forest was ambiguous in another way. We shall see that the *garma* (public) songs and ceremonies associated with places in the forest were widely shared among groups of the same moiety, and that people were likely to put their group at the top of the list, as owners, while the rest, they asserted, merely sang about the place. Thus north-east Arnhem Land was not neatly divided into moiety and patrigroup areas like a meandering chessboard (cf. Williams 1986: 77). Its social and religious geography, like all such geographies, was ambiguous and contested, clearly focused (if disputed) here, fuzzy and undefined there.

Ancestral Traces and Powers

Yolngu totemic geography was not just about identity and ownership but about ancestral powers. The activities of the *wangarr* left traces in the land and waters which the Yolngu referred to as the 'feet', 'footprints', or 'tracks' (*l̲uku*) of the ancestors. These traces took many forms: I have mentioned rocks and soil which were believed to be the transformations of ancestral bodies, and that people also believed that some *wangarr* ancestors remained beneath the sea or deep inside a lagoon. People interpreted other features of the landscape as marks made by an ancestor's action, such as a gully formed by Crocodile dragging itself along. Others were transformations of an object which a *wangarr* carried—a tree, for example, which was the digging-stick planted by an ancestral woman, or a rock which was once Dingo's bark canoe. There were additional kinds of ancestral traces. The designs which represented the places and traces of the *wangarr* ancestors, as well as the sacred objects which they placed in the waters and the songs and dances that 'followed' those they themselves performed, were all the 'feet/footprints' of the *wangarr*. Yolngu referred to these kinds of traces as 'foundation' in English, seeing them as the 'ground' (a similar metaphor) or basis of their mode of life.

There is consistent evidence that many Yolngu believed *wangarr* ancestors, ancestral substance, and some ancestral designs to be very powerful and in some cases dangerous. The power of *wangarr* ancestors was conceptualized in a way not unlike sorcery beliefs. The transformed bodily substance of ancestors, together with representations causally linked to them, were believed by many to be imbued with ancestral power which could be harnessed to enhance a person's own abilities and do harm to enemies. People believed that they could tap this power through ritual action.

Ancestral places were thought efficacious in another way. Through actions such as menstruating into the waters, *wangarr* ancestors established the conditions of continuing reproduction in the waters. One of the Yolngu theories of reproduction was that a child's 'image' (*mali'*) entered a woman from waters associated with *wangarr* ancestors—a lagoon, a creek, or the sea—when she became pregnant.[1] In a dream, reverie, or strange experience, a father 'found' the spirit of his child before it entered the mother to form the foetus. The following story told to me of spiritual

[1] A lagoon is *gapu gul̲un*, literally 'water-stomach', or in Shapiro's (1988) gloss, 'womb'. The body sign *gul̲un* denotes *waku*, the term denoting a woman's own child, or a man's or woman's sister's child.

conception is not untypical (see also Warner 1937: 21 and Thomson 1983: 9):

> The man's parents were sitting on the shore at M——, which is Dingo group's country, when they saw a strange jet of water come up from the place where the Dingo lay. Later his father saw a spirit in a dream, and his mother told his father that she was pregnant. Eventually, after consulting with the Stringybark people, for whom Dingo country was 'mother', it was agreed that he was 'of the water' of M——. The man was Honeybee by birth, but his own country was of the Dingo group.

The person was, in a sense, born of the *wangarr* ancestor and the waters, in Yolngu belief. However, these beliefs were not deterministic, for Hamilton's study of the neighbouring Gidjingali suggests that women had several alternative theories of the sufficient conditions for pregnancy (see Hamilton 1981*b*; Warner 1937: 21–5).

Kin Connections to Country

A person was related not only to their own country and country of conception, but by kin links to many other groups, places, ancestors, and *ma<u>d</u>ayin*. The terms *ba:pa* (father), *ma:ri'mu* (FF), and *ga:thu* (FFF) applied to one's own country. A person's mother's group, country, and *ma<u>d</u>ayin* was his or her *nga:<u>nd</u>i* ('mother') group, place, and sacra. Thus someone might point to a tree of a species which was a *ma<u>d</u>ayin* of his mother's group and say 'this tree is my mother'. The kin categories most frequently extended in this way were those of a person's own matrilineal kin, although I have also heard terms from the father's matriline used. Categories of relations derived from kin categories denoted a person's relationship to groups, countries, and *ma<u>d</u>ayin*. There were three varieties: two sets of expressions denoted categories of 'holders' of countries and *ma<u>d</u>ayin*; expressions in a third set signified a group in relation to a person (Table 3).

Members of a group 'held' (*ngayathama*) land in common, together with the associated rituals, sacred objects, songs, and designs. The suffix -*wa<u>t</u>angu*, which can be glossed approximately 'boss of', 'holder of', or 'owner of', denoted this primary relationship between persons, land, and sacra. People were *wa:nga-wa<u>t</u>angu* ('country-holders') in relation to their own group's country

TABLE 3. Kin Relations to Groups

Person in relation to a group		Group in relation to a person
Expressed as 'holder'	Expressed as group	
		ngapawanditj (one's) own group
ngaṉḏi-waṯangu mother-holder or *ngamiṉi-waṯangu* breast-holder	*waku mala* wC/ZC group	*yindi-pulu* ('large'-*pulu*) or *nga:ṉḏi-pulu* mother-[group]
		barrkimirri wife's or husband's group
ma:ri-waṯangu or *wayirri-waṯangu* MM-holders	*gutharra mala* wDC/ZDC group	*ma:ri-pulu* MM-[group]
waku-waṯangu MMM('wC/ZC')-holder	*nga:ndi mala* wD$\overline{DC}$ ('M') group	*waku-pulu* MMM-[group]
yapa-waṯangu sister-holder	*yapa mala* sister-group	*yapa-pulu* 'sister'-[group]

and *maḏayin-waṯangu* ('*maḏayin*-holders') of their sacra. An individual was 'mother-holder' of his or her mother's mother's group, country, and sacra; 'mother's mother-holder' in relation to the mother's group, and so on. The reciprocals of the kin-terms in the 'holder' expressions applied to the same categories of people: the 'mother-holders' (*ngaṉḏi-waṯangu*) were the 'woman's child/sister's child group' (*waku-mala*), and so on. The *-waṯangu* expressions appear to have had greater currency in the west of the Yolngu region, the *-mala* expressions greater use in the east (Nancy Williams pers. com.). The word 'group' in these expressions signifies a number of people, not a patrifilial group.

The third set of expressions denoted the relationship of a group, its country, and its *maḏayin* to a person. Thus a person's mother's group for which he or she was 'mother-holder', was that person's *ngaṉḏi-pulu* ('mother-*pulu*'), and the person was of the 'woman's child group' (*waku mala*). People in the east of the region also employed the terms *djunggayi* or *djunggayarr* to refer to a person in relation to 'mother' ceremonies for which he or she was responsible, terms which seemed to be increasing in currency.[2]

[2] According to H. Morphy the oldest living son of the oldest woman of the senior generation was the senior *djungayarr* (H. Morphy 1977: 76).

People could apply other kin terms to groups, their ancestors, and property. For example I have heard a man refer to one group and country as his *ga:thu* (FFF/FFFZ, FMM/FMMB), meaning probably his father's *ma:ri* (FMM).

Where, as often occurred, women of one group married men of more than one other group, each of the categories of 'holder' could be dispersed among several groups of the same moiety. Consequently not all members of a patrifilial group, especially a large one, were necessarily mother-holders and MM-holders in relation to the same groups.[3] Moreover, the network of claimants of rights in a group was open in two senses. First, there was no clear cut-off point in the line of matrilineal filiation, although rights got weaker with the distance of kin connection. Second, positions in the network were iterative. A person's or group's *waku* (wC/ZC) were required for certain ceremonies to take place, but if people for whom a group was directly 'mother' were unavailable, other, more distantly related *waku* could take their place, and might make equivalent claims to close *waku* in some circumstances.

These categories of 'holder' entailed various claims and obligations, and consistent with the kinship distance, a person's strongest connections were with his or her own group, mother's group, and mother's mother's group. For example, people claimed the right to be asked permission by others for access to their land, and claimed the right to the exclusive use of certain aspects of *ma̲dayin*. Mother-holders or *waku* were expected to help make the group's designs and sacred objects, could live at, and look after, its country, and took various roles in rituals. Mother's mother-holders or *gutharra* might be granted permission to make a group's designs, could perform its songs, and would claim to succeed to the land and sacra of their mothers' mothers' group in the event of members dying out. As Howard Morphy notes, the 'mother's mother' (*ma:ri*) group members had the power to withhold rights in women as wives' mothers in the case of breach of ritual rights (H. Morphy 1977: 84).

Relations of a person of group A to group B could also be extended to other groups with the same ancestor and ceremony as B. An individual might assert rights in the ceremonies of a group

[3] Compare H. Morphy (1977: 80), who argues that the relation of MM–wDC in the context of ceremony was group-centric and not individual.

with the same *ma<u>d</u>ayin* as his own, or maintain that he or she was mother-holder of a group with the same *ma<u>d</u>ayin* as his or her mother's group, and so on (see H. Morphy 1991: 63).

Kin Relations Between Groups, Places, and *ma<u>d</u>ayin*

Not only did an individual trace a kin relation to each group, its country, and its sacra, but groups were related to one another through kinship: one group was the 'mother' of another, which was its 'woman's child', and so on. The countries and *ma<u>d</u>ayin* of two such groups lay in a similar relation. Because groups were exogamous and patrifilial, these relations always followed those traced through women (Fig. 6) (see H. Morphy 1978). Accounts given to me by different people, which generalized matrilateral and affinal relations among people of the different groups, were quite consistent, although in some cases they conflicted with the general trend of marriages between groups.[4]

Yolngu men frequently expressed marriage as a matter of relations among groups, countries, and *ma<u>d</u>ayin*. Marriages ought to take place between people of geographically close countries (Shapiro 1981: 58). As well, men would say that marriage 'follows the *rangga*', that is, marriage followed the kin relations between sacred objects which were transformations of aspects of *wangarr* ancestors. One man put it this way to me:

> We get different groups and kinship by following the *rangga*—*ma:ri* people, places, and *ma<u>d</u>ayin*; *nga:<u>nd</u>i* people, places, and *ma<u>d</u>ayin*. *Ma:ri* married *ngathi* and made *nga:<u>nd</u>i* and *ngapipi*; *ngapipi* married *mukul* and made *galay*. (See also R. M. Berndt 1962: 75.)

That is, MM married MF/FMB and made M and MB; MB married MMBD/WM and made MBD/MMBDD. These kin relations among the *rangga*, the most sacred material manifestations of the *wangarr* ancestors, were those which ideally existed between groups. Kin relations and hence marriage relations between the *rangga* sacred objects of groups were supposed to set the precedent of present and future marriages (see Chapter 2).

The chapter so far has generalized about the terms in which

[4] The categories used by Dhuwal- and Dhuwala-speakers can be arranged to form a six-place structure, although some Galiwin'ku people generalized relations into a four-place 'semi-moiety' structure (see Chapter 2).

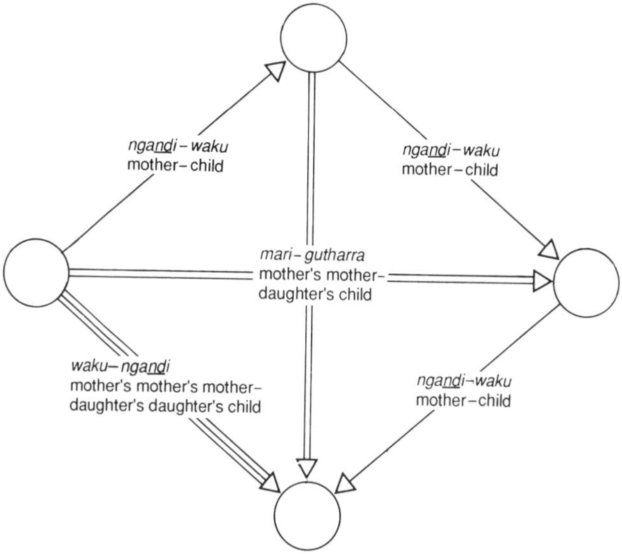

FIGURE 6. Kin Relations Between Groups

Yolngu constituted and expressed the identity and possession of country. Some variation has already emerged in the expressions used to denote 'holders' of country; particulars about the ancestral definition of country, and especially disputes over land, will reveal a variety of sometimes incompatible perspectives.

A Dhuwa Country and its mad̲ayin

Cloudy Water people formed a very large and growing group in the mid-1970s, a time of fast population-growth in the Aboriginal community more widely. At this time there were more than 200 people with this group identity. Their main country, Wa:l̲angur (this is a pseudonym), was centred on a mangrove-lined fan delta leading off a tidal creek between Langarra (Howard Island) and the mainland. The fan delta drained an area of grey muddy tidal flats, freshwater swamps, and grasslands giving way to clumps of vine jungle and extensive upland forests. The total area of this country was of the order of 5,000 hectares, although boundaries were indeterminate in the forest. On the mainland immediately to the west lay Stringybark county, also of the Dhuwa moiety, and to the east, a country of the Honeybee group (Yirritja moiety). For the most part Langarra was Yirritja, but Cloudy Water people had a small area there too, as did Stringybark people. Prominent also in the group's affairs was the country of Two Women people, mentioned in the last chapter, which lay at the south-west tip of Galiwin'ku, where many Cloudy Water people lived.

At the invitation of Wandhuma, a leader of the group, I visited the Cloudy Water country of Wa:l̲angur with my family twice during 1975, in May and August, travelling by dinghy and outboard from Milingimbi. On the first trip we were accompanied by D̲an.gi, a man of the Yirritja moiety who was married to Cloudy Water women. Wandhuma also made a point of bringing his son and a young man of Black-capped Tern group, for whom Cloudy Water group and country was *ma:ri* (mother's mother), and two other children who were Wandhuma's brothers' sons, so that they could see and learn about the place. Not all younger members of the group, or indeed the adults, many of whose parents and grandparents had settled at Milingimbi and Elcho Island Missions, had first-hand knowledge of their country, but people did take the opportunity to hand on knowledge where they could. On the second occasion, I went with Wandhuma and Buwat̲a, Wandhuma's 'son' (*ga:thu*). He was a man in his early sixties who, with his brother, led the Thunder group of the Dhuwa moiety, a close ally of Cloudy Water group, and whose country lay to the east of Wa:l̲angur. We travelled with Buwat̲a's youngest wife (a fifteen-year old), Wandhuma's wife Ngal̲indiwuy, and her daughter by her previous husband, Wandhuma's older brother. (We took the opportunity to hunt as well as tell stories; on the second trip Buwat̲a went in search of wallaby and honey, while the two young women went to get *dhukuray* (*Nerita lineata*) shellfish, just like the Djang'kawu.)

The country had two kinds of meaning: the traces and memories of more or less recent human occupation, and the traces and traditions of its *wangarr* ancestors. The story of the Djang'kawu sisters (Chapter 1) broadly shaped the ancestral significance of the country. But many of the brief explanations and snippets of story which illuminated this tree or that branch of a creek had not appeared in the telling of the larger narrative. They could be slotted into the bare outline of the sisters' visit to the country,

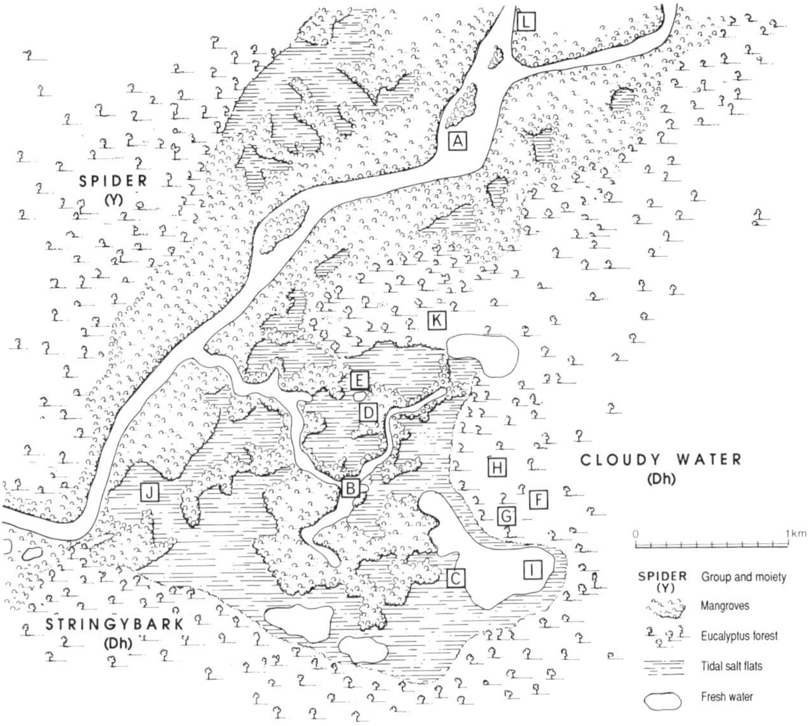

MAP 3. Wa:l̲angur

their creative acts, and their loss. Several significant places and features did not appear in the main myth told to me by men sitting in the camp at Milingimbi, because they were men's secrets, such as which particular *wangarr* made the Nga:rra ground, the use of which was restricted to men, and just how it was that the Djang'kawu procreated. The following summarizes ancestral explanations of country made on the two trips. (Letters in brackets refer to Map 3.)

Two islets in the creek were 'images' (*mali'*) of the two canoes of the Djang'kawu sisters, Dhal̲kurrngawuy and Bad̲at̲awuy; they were Canoe dreaming but were called 'Mistletoe Tree' (A).

Yolngu used to keep the landing, a muddy gap in the trees along a narrower side-creek which wound through the dense mangroves, clear (B). All the mangroves 'meant' (*mayali'*) the Ma:tjarra group (the string of groups created by the Djang'kawu sister, and celebrated in the Madayin or Nga:rra ceremony).

At Wa:langur the Djang'kawu hung up their Guwilirr dilly-bag/s full of sacra in a tree called 'Dilly-bag-hung-up' in a little island of mangroves (C), then went to collect shellfish which they speared in the water: *nonda* or *wa:lawun* (*Telescopium telescopium*) and *dhukuray* (*Nerita lineata*).

Then they created sacred waters in the mangroves which only old men could use, and waters for everyone to drink. They heard a bush-bird whistling in the mangroves and made another well at Wa—— (D), an outcrop of mangroves entered along a narrow path. Fresh water bubbled up in the mangroves to form a stream that disappeared underground after three metres or so. In former times people used to dig it out to make a well. The place where the spring bubbled up was the only well used by women and children; they were not allowed to go to the place where the water came out from the underground channel. All other fresh waters in the mangroves were 'sacred'. (Wandhuma used the English word.) Only 'big men' (men with grey hair) could use them. The Djang'kawu speared the spring waters with their Ganinyidi sticks.

At a mangrove thicket called Ga—— (E) was the 'number one sacred well', the one Wandhuma had sung about at the (washing purification) ceremony recently. A second well with fan palms was just for old men.

At the Nga:rra ground near the edge of the forest (F) the men hit the clapsticks and called the *marrawinydjun* chorus. The Djang'kawu heard them from the little lagoon near Dha—— (G), an area of paperbark and cabbage palm, fern-like *winikamu* vegetable food, and trees placed by the Djang'kawu. White clay could be found there for body-paint.

At Dha—— there stood a large well which people used to dig out. When a person died, his or her spirit returned to the water. The many trees alongside were the people: the paperbarks there 'meant' (*mayali'*) the Ma:tjarra group. A hole in the flowery scum on top of the water might be where the spirit of the leader who had recently died went into the water (see Chapter 7).

In the forest near Buwata's hut, which he had built some years earlier, were three trees which the Djang'kawu had planted: *wa:nduwa* (*Eucalyptus polycarpa*), *deti* (*Canarium australianum*), and a fallen *gaypal* (*Acacia auriculiformis*) (H).[5] From there the sisters walked south to collect larvae (*ga:murung*, longicorn and other species) which lived in the rush-corms in the lagoon (I).

Men crept from the Nga:rra ground (F) to steal the *malayin* from the sisters while they went to the mangroves to look for *nonda*, *dhukuray*, and *batjimurrungu* (*Casidula anguifera*) shellfish and put them in a dilly-bag.

[5] Species identifications are from Rudder (1977).

The two sisters returned and tried to get near the ceremony ground but heard '*binbir*', the old men had begun the ceremony. So they had to leave the dilly-bag and said, 'We'll go and work [*dja:ma*]; the men will make [*dja:ma*] the ceremony'.

The Djang'kawu made a well also at the adjacent Stringybark country to the east (J), which was *ma:ri* (MM/MMB) in relation to Wa:langur, for there was no water in the forest on Stringybark land. This was 'company' land for Cloudy Water and Stringybark countries. The ground was stony because it was Olive Python country. The Djang'kawu called out to Olive Python who 'went inside' (the water) at Stringybark country, and they made another well in the forest at Wa:langur.

An unusually tall cycad palm (*warraga*, it was 7 metres or more) was one of their Ganinyidi digging-sticks with which they made the springs; the palm was the image/spirit (*mali'*) of the digging-stick (K). The Djang'kawu planted that one after 'spearing' the springs, and carried another with them. The digging-stick image stood there now. (Note the tacit transformation.) It bore the name 'Mistletoe Tree' as a proper name, like the Canoe dreaming (A). The cycad palm was Yirritja moiety, but it was the 'idea' of the Djang'kawu (using the English word), implying again that it was odd for a Yirritja species to be a Dhuwa *madayin*. A Mistletoe tree they had planted had now fallen down, leaving only standing stones. In former times one man's father had broken one of the stones. Nearby was a stand of cycad palms whose nuts used to be eaten in the form of cakes during the Nga:rra ceremony.

An old 'warriors' track', as Wandhuma called it, implying perhaps that warriors formerly set out this way from Wa:langur to raid other camps, led through paperbark and palm forest to a landing to the north-east (approximately at A), connecting Wa:langur with the nearby islands. There the Djang'kawu (or Walawala, depending on the version of the myth) landed on their return from Galiwin'ku island. Here the canoe sank to become a rock and a sacred object.

It is instructive to consider the various kinds of memories and comments that such a visit provoked. Some were stories about the ancestral significance of particular places. Others were related ancestral stories, such as how the Djang'kawu gave various *madayin* to other groups and created many people. The visit provoked the men to recount and evaluate ancestral traditions of other groups with the same ancestors and whose country had similar significance. It was on one of these visits that Wandhuma protested that the story belonging to Sunrise people was a 'different story' and was not true; the Djang'kawu had no man, they alone followed the sun (Chapter 1). As well as the ancestral significance of this country,

Buwaṯa explained aspects of his own country, linked by stories about the Djang'kawu, which were similar to those at Cloudy Water country: how the sisters made spring waters (*milminydjarrk*) for everyone to drink; where the sisters' canoe sank (as it did also at Wa:ḻangur); where they urinated and menstruated, becoming 'sick', and stones stood there now, as well as at Thunder country and other places; where they went on to catch fish; and the place where the sisters heard the Yirritja moiety chorus from their Nga:rra ceremony (see Chapter 6). And he told me how it was that the sisters created the forebears of each group, without the aid of men, explaining details which women were not supposed to know.

Closely related as these stories were to ceremonies, Wandhuma outlined how the Nga:rra or Maḏayin ceremony was organized. In former times, he explained, his father and father's brother used the ceremony ground, then bathed at the end in the freshwater hole near the place where *ra:kay* lilies grew. Indeed, the old Nga:rra ground on the eastern side was the very one from which the ancestral men crept to steal the dilly-bag. The grassy camp, where people used to sleep, was the place where the Mistletoe Tree stood (the Riyawarra Tree erected for the public phases of the Maḏayin ceremony, see Chapter 6). In the thick bush was the 'sacred' place where the men danced. When people lived away from their country they recreated these places in ceremony. But wherever the ceremony ground was made, it represented the equivalent places of other groups.

The country was marked by traces of previous human and animal activity as well as ancestral traces. We remarked on foot-prints in the mud from our previous visit, as well as the track of a crocodile. Three sticks stuck into the mud at a camping-place marked the heights of three young men who had been camping there. Buwaṯa had built a small timber and corrugated-iron house some years earlier as the beginning of an outstation that had not then been completed, though it would be in later years. Further along the path was a well in the forest and some old gardens which Buwaṯa and his kin had made.

A major theme in song and ceremony was the succession of generations. The visit evoked memories of earlier events, such as how Wandhuma's father's brother and brothers used to visit the country when he was a young man, when he first saw the sacred well. Commenting on the importance of the young people coming to see the country, Wandhuma explained that the young man with us was currently making feather strings (*wana*, 'arms') used in the Madayin or Nga:rra ceremony, and was taking Wandhuma's place. The two old men recalled the names of past Thunder and Cloudy Water male leaders, their *ma:ri'mu* (fathers' fathers). This generational succession had its analogue in the trees, which Wandhuma equated with the Bungunbungun (or Ma:tjarra) groups, whom the Djang'kawu created (and who were celebrated in the Nga:rra ceremony). A visit to a place could cause a person to age, according to Yolngu theory: men attributed their greying hair to drinking water from a sacred spring.

Reflecting kinship as a dominant idiom of a person's relationship to place, Wandhuma recounted the kin relation of the country to the visitors and other close relatives: Wa:langur was Dan.gi's 'sister's husband' (*dhuway*, FZC) country and the 'mother' (*ngandi*) country of the Honeybee leader, Wandhuma's sister's husband.

Chapter 1 compared stories about ancestors of the same identity told by people of several groups. While adults of the same group subscribed to the same general body of myths, individuals were exposed to somewhat idiosyncratic family traditions, and had their own particular experiences which coloured their interpretations of country. Thus some years after my visit, the man I call here Wulakitj, a brother's son of the Cloudy Water leader, showed Robert Bos a rather different range of places at Wa:langur than Wandhuma had pointed out to me (Bos 1979*b*). Wulakitj explained how people used to strike a hollow tree with the proper name 'Mistletoe Tree' then listen to the sound of water bubbling up—it had been burnt in a bush fire; some women had drunk from one waterhole, causing one of them to become crippled; nobody was allowed to approach a very sacred hole on the north side of the lagoon where the Water Monitor *wangarr* ancestor lived. Wulakitj also claimed that his lineage looked after one part of Wa:langur while another lineage looked after the northern part.

Yirritja Connections

Characteristic of Cloudy Water country, indeed of the group's *madayin* in general, was its connections with just one mythological tradition, that of the Djang'kawu. By comparison the country of a

Yirritja moiety group which I visited was related to several public song-series and ceremonies as well as the Ma<u>d</u>ayin ceremony. Through each major *wangarr* ancestor and *mokuy* (ghost) the country and group was connected to a somewhat distinct set of Yirritja groups. The contrast with Wa:<u>l</u>angur is perhaps not as clear as I have implied, for if one traced each separate element in the Djang'kawu story of the Cloudy Water people and the associated songs—Mangrove Robin, Black Butcher Bird, Flying Fox, Black Cockatoo, Water Monitor, and so on—you would also find, I believe, that each was owned by a slightly different set of groups.

Two Windbreak countries lay on either side of the Glyde River estuary, separated by the country of the Mangrove group of the Dhuwa moiety. An outstation occupied mainly by people for whom the country was *ma:ri* (MM/MMB) and *waku* (MMM/MMMB) was situated in the eastern area. The group associated with the western country consisted of only one man and his child in the mid-1970s. It is possible that the two countries had been associated with a united group, and that the only adult member had simply lost control of the eastern area. On the other hand the two countries may have been defined as belonging to two rather distinct groups with the same *ba:purru* identity, for Warner (1937: 17, 50) records that the group associated with the easternmost place was nearly extinct, but he does not mention the western country. The remaining man had little control of his own rituals—Honeybee men both controlled Windbreak rituals and occupied the outstation on the eastern country.

<u>D</u>etjirimirri (a pseudonym), the western Windbreak country, consisted of several 'big-name' places. This country had been more regularly used and visited at the time than Wa:<u>l</u>angur, because of the outstation located by the main lagoon and because of its closer proximity to Milingimbi and Nangalala settlements. <u>D</u>etjirimirri was the name of a freshwater lagoon extended out to the immediate environs and to the country as a whole. Each 'small-name' area was associated with a particular set of *ma<u>d</u>ayin* and the related ceremony or ceremony elements. I visited <u>D</u>etjirimirri with Marrngu, a Honeybee man for whom the country was 'mother's mother'. He told me some of the stories about places in the presence of one man for whom the country was 'mother' and one whose group shared *ma<u>d</u>ayin* with Windbreak. The following summarizes what he told me that day. (The pronoun *ngayi*

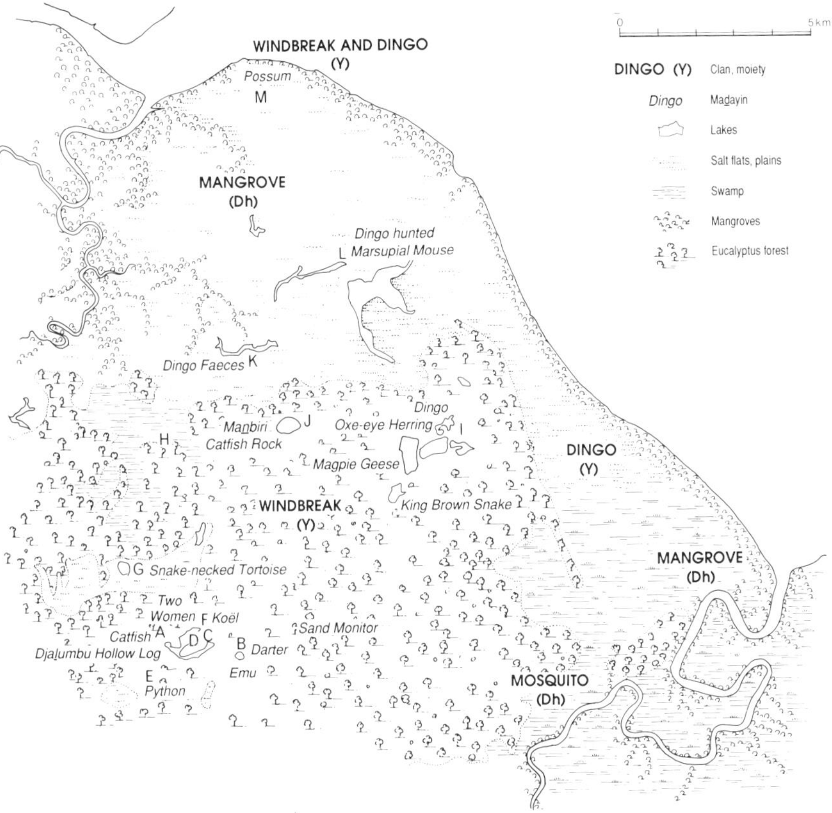

MAP 4. Detjirimirri

Notes

Catfish danced calling 'rrr, rrr, rrr', pulling Djalumbu Hollow Log along, and made the creek [A]. It was Bindangu Catfish, a freshwater fish, but the same as Manbiri Catfish, which joins fresh- and saltwater at Honeybee country where Hollow Log lies in the lake.

Darter came under the ground from M____, to come out at [B], looked back and said, 'Why has Emu closed up my place?' He went on to [C] and said, 'I will make the bullroarer and paint it.' He tied the bullroarer with grass (its string). The Windbreak Darter bird had painted himself as well, but he could not understand the language of the one from Honeybee country. He created a long lake. Black-headed Python was a major *madayin* of the Windbreak group, believed to lie at the creek to the south of the lake (E), as well as at the eastern Windbreak place. Emu came from Honeybee country and danced at B. Honeybee and Windbreak group had a common *ba:purru* name. Emu was a *madayin* of the 'forest' songs and dances.

Sand Monitor came from Rrimbitja and dug for yellow ochre near B. Under the camp (a mound by the lagoon) at Detjirimirri [D] the Two Women with the big breasts were living. They cooked lily bulbs in an oven made of termite mound.

Black-headed Python came from Tortoise country, crawled, and made a creek [E]. It is a snake but it means water. It said 'I am [the group *likan* name],' and sat on the south-east side.

At [F] Koël called in the jungle. He/she flew up and another came from the islands to the east. They met and each said he/she was [the group's *likan* name]. Then one went to the coastal country of a group to the west, and called out '*guwak*' [its onomatopoeic name]. Koël was welcomed there by his/her *ma:ri* (MM/MMB) who sang for him/her and said, 'We are of one *ba:purru*.' The Koël from the west usually comes here also, with possum-fur string for the circumcision ceremony, for the wake, and to send to *ma:ri* (MM/MMB) and *gutharra* (wDC/ZDC).

Snake-necked Tortoise with the broken hand lies at [G]. The male and the female danced from another place where they lie, and they lie here too.

Dingo travelled at [I], and Ox-eye Herring came from Tortoise country. Near there lies Magpie Goose, a *madayin* of Magpie Geese group. There are two small Dhuwa places nearby, one associated with King Brown Snake.

Manbiri Catfish danced round the rocks at [J]. Two Men made the Djalumbu Hollow Log, and Catfish danced. Djalumbu is at B country of Honeybee group and Spear-grass group country also.

Dingo of Windbreak group quarrelled at Dingo group's place with their Dingo. 'Where are you going?' Windbreak Dingo said. 'I had better go to the coast', said the Dingo *wangarr* of Dingo group. He/she went there and smelled the Whale meat. The Windbreak Dingo returned to his/her place looking back as he went. He/she made 'Dingo Faeces Lie' place where white clay now lies [K].

At [L] Dingo hunted for Marsupial Mouse, digging in the ground, but could not find the Mouse. Possum made [M] place. The country belongs both to Windbreak and to Dingo groups because they have the same 'mother' group.

---

and possessive pronoun *nhanngu* do not specify gender. Where unavoidable I have used he/she, and him/her.) (See notes to Map 4 for details of Marrngu's explanatory comments.)

Detjirimirri lagoon and its environs were the centre of Djalumbu Hollow Log ceremony, including the Log itself and Catfish *wangarr*, danced and painted on to the coffin (at A on Map 4). Nearby was a small lake (B) where Darter, associated with the Bullroarer dance for the Hollow Log ceremony, appeared in the myth. A little way to the south was the clearing which Emu made in the story (B). Other *madayin* belonged to the forest near Detjirimirri lake: Koël, Possum, Sand Monitor (B), and Murayana *Mokuy*. Under the camp by the lagoon at Detjirimirri lived the two Women with the big breasts; they cooked lily bulbs in an oven made of termite mound (D). Black-headed Python was a major *ma̠dayin* of the Windbreak group, believed to lie at the creek to the south of the lake (E), as well as at the eastern country of Windbreak group. To the north of D̠etjirimirri Snake-necked Tortoise *wangarr* 'lay inside' the area of swamp, wet-season lakes, and paperbark trees (G). This area also had an association with the Swamp songs, in which the Two Women, who lay beneath the camp-mound by D̠etjirimirri lake (on which the outstation camp stood) gathered lily bulbs in the song.[6]

In the north-east of the country lay a cluster of lakes rich in fish, geese, pelican, and other game, and in plant-life. One of the lakes gave its name to the whole area (I). Two important *wangarr* ancestors, Dingo and Ox-Eye Herring, related to sacred objects for the Ma̠dayin ceremony, belonged to this area, which was a former ceremonial centre for both moieties (I). Some parts of the area consisted of small Dhuwa-moiety places embedded in the Yirritja country. To the east of I lay Rock *ma̠dayin* in a lake where, according to Windbreak group tradition, Catfish *wangarr* swam round and round the Rock (J). Honeybee men performed the corresponding dance for the Hollow Log ceremony and in circumcision ceremonies. An area of tidal salt-flats was shared with another Yirritja-moiety group whose country lay to the north of D̠etjirimirri. There, according to the myth, the Dingo *wangarr* of the two groups quarrelled, and they agreed to go different ways, and the Windbreak Dingo returned to his own country (K). Dingo was the *wangarr* ancestor of the group, associated with the Ma̠dayin ceremony (See Thomson 1939*a*). He hunted for Mouse, digging into the ground, but could not find him (L). Possum made country belonging to both Windbreak and Dingo groups (M).

Each of these traditions was associated primarily with an ecologically discrete area as well as a particular ceremony or ceremony element: the Hollow Log with D̠etjirimirri lagoon, the Forest *ma̠dayin* with the area of forest nearby, Snake-necked Tortoise with the swamp, Dingo and Ox-eye Herring with the Ma̠dayin ceremony place, and so on.[7] The combination gave the group access to all the major types of Yirritja-moiety ceremony. More than half of the stories connected D̠etjirimirri to other Yirritja-moiety countries, both near and distant, by the journey of

[6] See R. M. Berndt (1948*a*: 19–20) for a Wan.gurri-group version of the songs, featuring the Moon, which is a *ma̠dayin* associated with a country to the west of D̠etjirimirri.

[7] See also Borsboom (1978*b*) on the ecological correlates of a Djinang group's *ma̠dayin*.

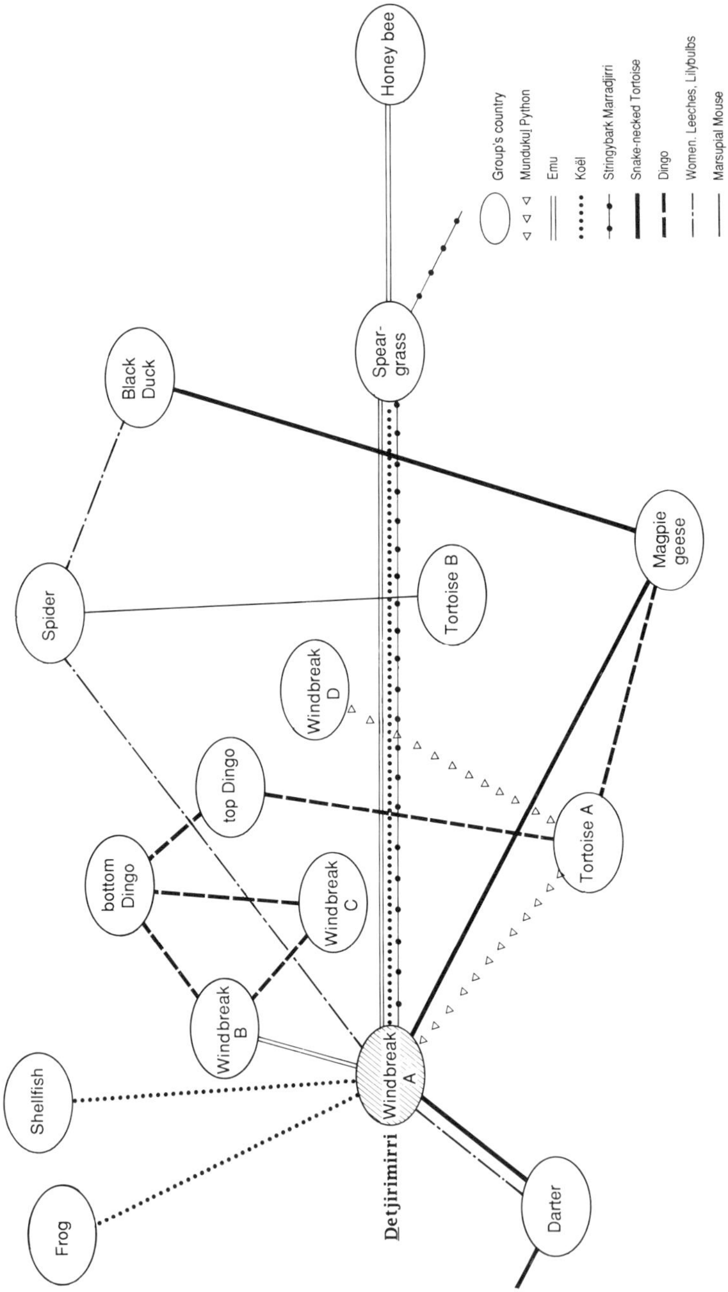

FIGURE 7. Links Through *mada<u>y</u>in* of <u>D</u>etjirimirri to Other Yirritja Countries

a being from another country, or to another country, or because two beings were 'the same' (Fig. 7). In these ways each group had a relationship to all of its neighbours of the same moiety through rather different sacra.

When people talked about the relations between groups and among countries they would say that they had the same or different tongues, songs, tunes, *mad̲ayin*, or ceremony. For example, Spear-grass and Rain people had the same Hollow Log ceremony, shared also with other groups, and Rain and Spider groups had the same Swamp songs (see Chapter 4). Some stories, however, explained the lack of a connection, such as the absence of an element from a collection in which it might be expected to occur. In the story about Dingo, for example, he or she hunted unsuccessfully for Marsupial Mouse. The latter is associated with similar areas belonging to other Yirritja groups.

These links joined groups into cross-cutting strings. The man who showed me D̲etjirimirri told me, 'This place is really of the Up-river people in part. You know Honeybee [B] country? Well, the two places are friends: both have Djal̲umbu Hollow Log, Darter and Emu—one tune, one law, one design.' The Upriver people consisted of Honeybee and Lightning Snake groups. People also said that Windbreak and Spear-grass people were both Spear-grass *ba:purru*. This common name came from sharing the Hollow Log ceremony (although not all other groups with the same Hollow Log ceremony were thereby Spear-grass). This connection implied potential claims in each other's land, and was one of the kinds of condition which enabled one group to control the ritual life of another. But the use of the word 'friend' was significant as a marker of sharing ceremony. Men organized the exchange ceremonies with the stated intention of creating 'friends' with distant people who would otherwise be dangerous.

The contrast between Wa:l̲angur and D̲etjirimirri exemplifies my point that there was no universal form of country. Around the common themes of ancestral creation and ancestral traces they were heterogeneous: some were replete with dangerous places, like the Bula country of the Jawoyn people (Keen and Merlan 1990); others, like Cloudy Water country, were associated with a unified tradition; at least one mapped the body of the Shark *wangarr* organ for organ on to country, as in eastern Gunwinggu religion, so that a representation of the ancestor as animal was at once a map of country (Franca Tamisari pers. com.; Taylor 1987). By the same token different groups possessed rather dissimilar total complements of countries, including one or more major areas and perhaps several smaller places as enclaves within or on the edge of other group's countries.

In spite of such heterogeneity, these descriptions of the significance of the two countries imply consensus about the identity and significance of each area. While Wandhuma and Wulakitj gave different emphases in their explanations of Wa:langur there was no real conflict in their accounting. Indeed, the kind of understanding of the significance of countries gained from visits with one or two people belonging to the group in question, or closely related to it, gives an impression of objectivity and absence of dissension. A different picture may emerge if one garners a wider range of opinion.

Conflicts over Succession

Eleven years after these visits to Wa:langur and Detjirimirri I had occasion to inquire into disagreements over land tenure and the distribution of royalty payments in an area in the east of the Yolngu region. This gave me a very different perspective on land tenure. My discussions with people showed that the ownership of forest country associated with public songs was in a sense indeterminate, while ecologically rich areas to which the 'big' *wangarr* ancestors and the major ceremonies belonged was more certain. People of contending groups tended to agree on the *madayin* significance of places, and about which groups had rights in the songs and ceremonies. Each ranked his or her own group at the top as the true owner, the others merely having the right to perform the songs. However, the story propounded by men of Cloudy Water about Walawala taking songs to his or her *mar:i* country and creating their spring waters (Chapter 1), which underwrote Cloudy Water claims to have succeeded to that country, shows that such disputes were not always confined to country related to public *manikay* songs.

People argued about at least four aspects of relations to land and waters: the extent of a group's country, the group identity of a particular place, the ranking of groups who possessed the songs and ceremonies about a place, and succession to a deceased or dying group's country. Some of these arguments reflected changing social and economic conditions. In the recent pre-colonial past, conflict among men over the control of country and its resources was probably linked mainly to competition over women and the control of ceremonies, for the non-exclusivity of rights in food resources of country and the flexibility of residence made it unlikely that there were great quarrels over access to those food resources unless major anomalies arose in the relations between the size of groups and country. Ownership of land did imply, ideally, the ability to exert some control over its use, and over the performance of associated ritual, both of which were major resources, especially for men. However, new resources such as the appearance of mission stations, and more recently a bauxite mine, introduced new reasons for competition, and brought about a disjunction between the definition and control of country and the value of these resources.

Succession to Country

In spite of the distribution of rights in land and related *ma̲dayin* among a set of cognatic kin, if a group's membership was destined to come to an end through having only female members and/or childless male members living, people of the same moiety took steps to 'look after' the land and *ma̲dayin*. Their descendants perhaps eventually took on the group's identity rather than re-identifying the country, for in spite of 60 years' writing about north-east Arnhem Land social life, there seems to be no documented case where the group identity of a country as a whole changed absolutely, although groups have extended their areas, eroding the countries of adjacent groups. I have heard Cloudy Water people refer to their *ma:ri* (MM/MMB) land, which they 'looked after', as their own, and Honeybee people assert that parts of Milingimbi island, where many of them lived and which was the *ma:ri* country of some of them, belonged to them. But the identity of these places was well known within a wide social network, and affirmed during ceremonies, so inhibiting radical re-identification.[8]

[8] Warner (1937: 17–18) and H. Morphy (1977: 81) have made similar observations, but Warner thought that the land would eventually be re-identified with the successors' group.

There were several possible grounds for a succession claim to the country of a dwindling or memberless group. The strongest lay with people whose mothers' mothers were of that group (the mother's mother-holders or *gutharra*) (see Williams 1986: 53 ff.). The country claimed was their *ma:ri* (MM) country. Among the cognatic network of kin they were the most closely related to the core group, as well as being of the same moiety. Claims were made on the basis of the kin connection of an individual to the deceased or dying group, although claims might also be made on the grounds of a relation between the groups as wholes as 'mother's mother' to 'woman's daughter's child'. Where there were no close *gutharra*, more distantly related *gutharra* might claim to look after the country.

People claimed to look after country on other grounds. According to Williams the joining of groups in myth as 'sister' groups, often with a name held in common, or in 'company' relationships, might provide the basis for the fusion of such groups. These 'subsidiary' rights could overlap those based on *ma:ri–gutharra* relationships (Williams 1986: 175; Peterson, Keen, and Sansom 1977: 1–2). A person might claim to have taken over his or her conception country which was their 'very own country' or country 'of the water'. The person's father and mother 'made' the person there and the person's *mali'* (spirit) came from waters of that group.[9] Secondary paternity could also provide the justification for taking on responsibility for the land and *ma<u>d</u>ayin* of a declining or deceased group. One Nangalala man asserted to me that after his father died his mother married one of the last male members of

[9] Men made the claims to conception country known to me, however one of these was also strongly asserted by the man's daughter: an area was their country because it was her father's country of conception. In another case a Dhuwa-moiety group of the Glyde River area was reduced to two women members and one mentally retarded male who was unlikely to marry. A man of another Dhuwa group who was 'of the water' of that group was said to be 'boss' of the *ma<u>d</u>ayin* although he did not sing the group's songs. This was because, men said, he was only 'of the water'. He was a prominent singer of his own group songs, but I imagine that he had not had the opportunity or desire to learn those of his conception group. The children and grandchildren of people who moved to Milingimbi mission earlier this century, and who were born in the vicinity of the settlement, were 'of the water' of groups of the same moiety whose country was in the vicinity of the mission station. This fact gave them grounds to claim the country as their own, although such a claim was in principle weaker than through group identity where such people (i.e. the *wa:nga-wa<u>t</u>angu*, 'land-holders') were still alive, or claims by the *ma:ri-wa<u>t</u>angu* (mother's mother holders or *gutharra*, wDC/ZDC), some of whom were residents.

another group. This man 'gave' his wife's son his ceremonies, the man claimed, so that he could perform them and make the sacred objects.

The fact that a number of people could lay claim to have succeeded to a deceased group's country and sacra for a number of reasons gave rise to conflict.

Disputes over Succession

Disputes over succession to land would arise for a number of reasons, and people might make claims on rather distinct grounds. More than one reason justifying a claim could be invoked simultaneously—that the country was a person's *ma:ri*, that one's group and the groups whose country was under claim were 'sister' group and had the same *ma<u>d</u>ayin*, and so on.

One area of difference was the degree of closeness in the *ma:ri–gutharra* (MM–wDC) relationship among those claiming to look after the country of another group. These differences might change over time, and such transformations could become quite complex. Suppose, for example, that the people first asserting custodianship were rather distant *gutharra* (wDC) of the group whose country they claimed to look after (the *ma:ri* group). Their claim might later be contested by people of another group who were close *gutharra* to the *ma:ri* group, but who were too young to assert a claim when the last male of the *ma:ri* group died, or not yet born. Meanwhile, a man of the original custodian group married a woman whose mother was of the *ma:ri* group so that his children would later be claimants to succeed to that same country.

The definition of country was not necessarily stable in such disputes. It was possible for a group of the same moiety to redefine country as its own, particularly forest country, while others identified that land as belonging to the *ma:ri* group. One strategy was to have one's cycad-nut bread and eat it too—the country in question was of the claimant's own group, and even if it were not, then that person had the authority to look after it as *gutharra* or as a member of a 'sister' group.

Re-identification of country could come about because of the association of stringybark forest with the Mokuy figures of public (*garma*) songs. These songs (together with the related designs, dances, etc.) were widely shared by groups of the same moiety (see Chapter 4). In one case of a conflict with which I became familiar, men and women widely agreed that five Yirritja groups shared the public songs associated with the disputed forest-area. However, older men and women of the groups with 'the same song' were inclined to say that their own group was the true owner of the forest country under dispute, while people of the other groups merely sang about it. However, a leader with close kin-links to claimants who were not of his own group supported their claim, unlike some other men and women of his group.

As I pointed out earlier, the group identity of country related to major *wangarr* ancestors and *rangga* sacred objects seems to have been harder to redefine, and to have been more definite than the hinterland forest associated with *garma* songs. The objects and designs were taken as evidence of the ownership of land by the group which held them.

Succession claims could be complicated by the existence of factions dividing claimant groups. Some people of a *ba:purru* group might make a succession claim to the country and sacra of another group, while others of their group contested the claim because of their kinship and *ma<u>d</u>ayin* links to claimants belonging to other groups of the same moiety. These conflicting loyalties could map on to distinctions among land-holding subgroups where these were recognized. Suppose that people (A) associated with one of a group's countries might make a succession claim to their *ma:ri* group. People (B) of the same group as (A) but with particular links to a distinct country might support claimants belonging to another group (C), stressing particular *ma<u>d</u>ayin* connections between their two countries not shared by (A), as well as close kin-links. Divisions and connections of these kinds would be reflected in patterns of residence, and were linked to factions drawn from networks of cognatic kin of both moieties. People of these networks would tend to support the claims of their close kin of the same and the opposite moiety.

People told 'warrant' stories in support of succession claims.[10] We saw earlier that Cloudy Water people justified their claim over their *ma:ri* country with the story about how the last members of the then deceased Two Women group asked Cloudy Water people to look after the country and *ma<u>d</u>ayin* (Chapter 1). Warrant

[10] For the notion of warrant story see Fogelin (1967) and Sansom (1980: 27).

stories of this kind put forward by contesting claimants bore a family resemblance:

> A of the dying group asked B to look after country *c*; B looks after *c* because B or B's father was born on/lived at/died at *c*; A and B were in a relation of *ma:ri–gutharra* (MM–wDC), were 'sister' groups, or had the same *mad̠ayin*.

Given competing claims to look after land, and incompatible accounts both of the extent of a group's country and the group identity of areas associated with *garma* songs, how did any version come to prevail? Certain matters were widely agreed. The general location of named places and their spiritual significance were widely known, although there were disagreements and variation in the depth of knowledge about details. Under many circumstances such matters could remain ambiguous. Where they had to be resolved, what counted was being able to act on one's version of ownership; to enact the prerogatives due to land-holder or *dja:gamirri* (one who 'looks after'), which include control of access to the land and use of the country, such as to establish an out-station, and performance of the related ceremonies. The size of the group was important for such abilities, but so was one's place of residence. Those who lived at or near a disputed country, who had the more detailed knowledge of it, and could muster the most support, were at an advantage. Sheer physical force, related to group size and demographic structure may also have been a factor. Under more recent conditions relations with external agencies may also have been crucial, such as representation on the Northern Land Council, a statutory body established to administer Aboriginal land under the Aboriginal Land Rights (Northern Territory) Act 1976, who might ratify an arrangement to control the distribution of royalties by one group rather than another, or determine traditional Aboriginal ownership of land under the Act.

The existence of new resources was obviously relevant to succession claims, and to disputes over the ownership of country more generally. In a long-term dispute over both the group identity of forested country being developed for bauxite mining on the Gove Peninsula, and control of the distribution of revenues from the mining, country that had previously been only of marginal value economically became economically central. It had become necessary to define boundaries where none had previously existed, and to specify ownership that was formerly imprecise.

Conclusions

While I think it unlikely that contestation over country was a new trend, the decreasing relevance of foraging together with competition over the control of new resources led to new patterns of conflict. Attachment to country through spirit conception and long residence as opposed to group identity probably increased after the more permanent movement to missions. Furthermore, the religious constitution of land depended on the transmission of knowledge from one generation to the next. Because of settlement a generation grew up seldom or never having seen their own country or that of closely related groups. In the mid-1970s people who were active in the religious life (i.e. the majority) did sing about their country, performed dances and made designs which referred to it, and made sacred objects connected with it. But many had the most intimate knowledge of the country in the vicinity of a mission settlement and were affiliated with it through spirit conception. Again, the outstation movement, which began to burgeon in the mid-1970s, brought a larger proportion of younger people into contact with their own places.

Control of country depended on residing on or near it, so that under settlement conditions many people became powerless in relation to country from which they were far removed, while leaders living on settlements on or near such country were able to assert control of the country and its resources. In two instances known to me, people of small and diminishing groups have had little influence within settlements established on their land.[11]

Principles of relations to country do not seem to have been radically transformed. The reason for this is not hard to discern. Even before the consolidation of the population on missions, there

[11] The history of Milingimbi shows early fights between local people and those migrating west, followed by alternating dominance of mission and township affairs first by people with the same *wangarr* ancestor as the group whose country it was, then by a group for whose leader the country was 'mother's mother'. This group's fortunes declined after the death of the leader, when people of the former group reasserted themselves. The male leader of one of the owning groups tried to exert some control by returning from Maningrida to Milingimbi in the early 1970s, but did not stay permanently.

was a distinction between group identity of country and residence. Detailed studies by Peterson and Shapiro in the late 1950s and early 1960s show that the residents of bush camps included a high proportion of people living on country not of their own group, a finding consistent with Warner's generalizations (Warner 1937: 19, 138–40; Shapiro 1973; Peterson 1972, 1978). Consolidation of the population on missions simply represented a large increase in that proportion. A more fundamental disjunction has appeared between the definition of country and interests in land in the Gove region. Thus land use changed in the half century to the mid-1970s in the context of settlement and outstation life, and the economic value of country was modified. However, with the widespread move to outstations since that period, patterns of use may have become more like those of the past.

This chapter has outlined the ancestral constitution of country, and relations of people to place through moiety, group identity and kinship. The descriptions of Cloudy Water and Windbreak countries illustrated something of the heterogeneity of significance given to places, as well as the dense web of connections formed in myth and ceremony among places and groups. It is not possible to describe that web of connections objectively from a transcendental point of view, because each individual and group constituted it in their own way, agreeing on some things, differing about others, and reshaping it in order to make and create relations or to claim resources.

Owning country was evidently closely connected with relations to ancestors and *maḏayin*, and it is to these matters that we turn in the next chapter.

4

The Same Song

The Right to Know

Like the 'word' of the Darwin fringe (Sansom 1980), Yolngu treated ceremonies and elements of ceremonies such as songs and designs for body-painting as durable objects that could be owned. These *ma̲dayin*, as Yolngu called anything connected with *wangarr* ancestors, were attributes of groups and of multiple, cross-cutting, and rather open strings of groups. People took sometimes strong measures to ensure the correctness of these shared forms, and also to maintain the differences that separated one group's *ma̲dayin* from another. But to write of a common form is an

Men dance a *bunggul* in the Morning Star ceremony

oversimplification: Yolngu also argued about the proper form of ceremony. They did so in part to establish or maintain differences, or conversely to deny differences between the *madayin* of two groups. Such an act often had an ulterior motive, for to assert that two religious forms were the same was to lodge a potential claim to resources associated with land and *madayin*. And, of course, objectified ritual entities were themselves resources for which men competed (Bern 1979).

Madayin were the kind of quasi-entities frequently referred to as 'totems' by anthropologists. One of the first things that I learned from a middle-aged man when first visiting north-east Arnhem Land was that Honeybee people had Wurrpan (Emu), Lurrpu (Sulphur Crested Cockatoo), and Guku (Honey, Honeybee), while Black-capped Tern people had *Mokuy* (Ghost) and Banumbirr (Morning Star). This could have been the beginning of a list like the one Warner compiled in which 'totems' are classified into 'high' and 'low' *rangga* (sacred objects), and *garma* ('public') totems (1937: 39–51). Not only was possession of these entities a kind of badge of group identity, but an individual could be identified with and by one of them, as we have seen. This point was brought home to me one day while with some men who were hunting on Mangrove country. Someone shouted, 'Hey, Dingo!' to attract the attention of a man of a group with Dingo as *wangarr* ancestor.

People took measures both to maintain and erase similarities and differences between ritual forms. A myth of the Honeybee group expresses this mode of action:

> Murayana *mokuy* went from Spear-grass-group estate to Rock group's country and found men of the Rock group making Djalumbu hollow-log coffin and the Garayurru exchange-pole. They called him over to see and he said,
>
> 'My pole is pointed, yours is blunt.' Murayana was covered with painted designs of flowers and plants, and he said,
>
> 'I am carrying all the designs to [my country], and so we have the same ceremony.' He went off laughing.

In another man's telling, Murayana was 'jealous' and angry, crying, 'Why have you used my pointed pole? Cut it! Yours is the blunt one!' And he called out his group *likan* names. The groups' poles were indeed distinct in just this way. Men sometimes stated these rights in terms of moiety as well as group: one man told me that only Yirritja men could wear the Rock painting for the Ma<u>d</u>ayin ceremony; if Dhuwa men used it they would be speared. Until they were finally eliminated by missionaries, police, and Yolngu mediators, violent feuds sometimes developed at times when people of one group accused another of using its design (Thomson 1939*b*: 2; Elkin, Berndt, and Berndt 1950: 62; H. Morphy 1977: 83).

The similarity of sacra was, naturally enough, a matter of judgement. On one occasion I asked the male leader of the Magpie Geese group whether his Dingo *wangarr*, the main ancestor of the group (in spite of its name), was the same as that of the Dingo group whose country lay further down the river from Magpie Geese land. His denial reflected, I believe, the political relations among the groups. Marriages connected Dingo group to groups of the Dhuwala dialect who were enemies of Magpie Geese. It is quite conceivable that some people judged the Dingo *wangarr* of the two groups to be 'the same'.

People did not just take steps to sustain differences among sacra, they also controlled the flow of information about them. Many were unwilling to impart information to me about groups to which they were not closely related, although this reluctance decreased with age and ritual knowledge. Standing with other onlookers at a ceremony, I would occasionally ask a neighbour the identity of the dance being performed. If the person was not closely related to the group whose dance it was, or if the person was young, the answer was often, 'I do not know' (*yaka ngarra marnggi*). People were also circumspect about giving information in which an older person had rights without that person's approval. When a mature man of the Cloudy Water group wanted to tell me about the Djang'kawu sisters, he told me that he was only partially knowledgeable (*ma:rr-gangga marnggi*) about the *ma<u>d</u>ayin*, whereas his 'father' (FB), the group leader, knew everything. He brought Wandhuma to visit me, and when the two men were together the younger man told me much more than the older, while often asking the latter leading questions for confirmation (Chapter 1). When the same man told me about a group design over which he claimed particular rights, we were alone.

From a Yolngu point of view ownership and knowledge of *ma<u>d</u>ayin* were inseparable from the belief in their ancestral origin. One of the first things anyone told me by way of explanation of what they were doing when singing *manikay* songs at a ceremony was that, 'We copy from before.' Another man said of ceremony that, 'It is from the beginning.' He made the latter remark to deny my suggestion that Yolngu rituals were like theatre. The term *dhin̠'thun* signifies following and decoding an animal's tracks as well as the 'footprints' (*l̠uku*) of the *wangarr* ancestors in the form of designs, songs, and ceremonies (Marika and Christie forthcoming). Individuals identified with the *wangarr* when painted and dancing as ancestors and in singing songs which ancestors sang, and at least some people believed that they could tap ancestral power through performing ceremonies. Thus ownership and differentiation of ceremonies and their elements was not merely a matter of semiotic relations among sets of groups and sets of symbols or of ideology (Lévi-Strauss 1964), but involved beliefs about inherent powers and shared bodily substance.

It seems reasonable to deduce from reported Yolngu statements about ancestral powers recorded at various times and places, that Yolngu thought of performances of songs and dances and the making of designs and objects not simply as representational or symbolic, but as potentially efficacious. As I suggested earlier, these beliefs had much in common with sorcery beliefs. These implied that a trace of a person or being (impression, track), an image of a person or being (object, painting, photograph, shadow), and a part of a person or being (hair, urine, faeces, sweat, nails) were causally connected to the person in such a way that action on the trace, image, or part would affect him or her. The flesh, fat, and blood of the dead were imbued with power which one who stole them could tap in order to enhance their power or harm others. These principles were common to beliefs about *wangarr* ancestors as well as sorcery, but acting in opposite directions. A person would work on the image, trace, or part of a person by heating, piercing, or singing, in order to harm them. Conversely, by making an image of a *wangarr* ancestor or some part or attribute of the ancestor, people believed that they could gain access to the ancestor's power to enhance their own ability and perhaps to redirect the dangerous power to harm enemies, or use the appropriate songs to heal the sick. Like bodies of the dead, the transformed bodily substance of *wangarr* ancestors was imbued with ancestral power according to Yolngu belief. Similarly, ceremonies had instrumental as well as communicative effects. Yolngu did not see them primarily as modes of communication with ancestral spirits but as modes of identification with the *wangarr* and access to ancestral power.

Because ceremonies and other *maḏayin* had their origin in the activities of *wangarr* ancestors and other beings, Yolngu denied that their meanings were the product of human thought, or that people were innovative. In Yolngu ideology their forms were the unchanging currency of inter-generational transmission, which people received rather than invented. An individual gained authentic knowledge from close kin and through revelation. Indeed, the primary senses of Yolngu concepts of thought and knowledge were passive. The word *marnggi*, 'to know' or 'to be knowledgeable', meant to have experienced something, even by accident, as well as to understand or know by personal cognitive effort.[1] Furthermore, the concept implied social competence and rights rather than simply the content of the mind (see Stephen Harris 1984: 44 ff.). People would say that a person 'knew' a ceremony even after experiencing it once, but a claim to be knowledgeable and to be able to impart that knowledge depended on such factors as group membership, gender, and age.

Men often claimed to have gained religious knowledge through some kind of trance or reverie in which spirits of dead ancestors revealed it to them, as the following illustrates.

> Ga:murung visited me at my house to talk about the Maḏayin ceremony to be held the following year, at which he would refurbish the sacred object. He said that he had been sitting in camp last year with his hand to his forehead and eyes, and had experienced something like a dream, although he was not asleep. He told his father about it who said, 'That is sacred [*dhuyu*], your *maḏayin*.' It was his *ma:ri'mu* (FF/FFB) who told him all about it in the dream.
>
> Discussing the possibility of filming the ceremony, he said that he wanted to tell the stories to me and to the people of Canberra. He was knowledgeable of everything but needed to get permission to tell others about the *maḏayin*. Then he qualified that to say that he did not know everything: there were some words that his father had yet to tell him.

[1] Similarly the word *guyanga*, 'think', means to worry and remember as well as to cogitate purposefully. See Christie (1984: 268).

Varieties of Ceremony

As re-enactments of ancestral events Yolngu ceremonies were not performed for audiences (although very often a group of onlookers did assemble at the climax) and they were not usually rehearsed. There were standards of adequate performance, but the point was not to achieve a flawless result. No one could command a performance, consequently their occurrence was usually unpredictable, and they took place when the essential participants were all willing to begin. A leader might say with great confidence that a ceremony would begin on the next day, and then in the morning delay it on the grounds that too many people had gone fishing.

Rom: Right Practice

A performance of a ceremony instantiated *rom*, which Yolngu translated into English as 'law' or 'culture'. The food that an animal or other creature customarily ate was its *rom*, but the latter included explicit rules, as well as the proper way of doing something such as a ceremony. I shall translate *rom* with the expression 'right practice' or 'proper practice'. The expression 'the way' would capture something of its religious connotations. *Ma<u>d</u>ayin* ('sacra'), as well as the explicit norms governing social life, had their origin in the *wangarr* ancestors in Yolngu belief. Thus one man contrasted *rom* from the ancestors with *yolngu rom*, 'people's *rom*', by which he meant sorcery.

A group's *rom* was encapsulated in its body of myths about *wangarr* and other beings. The Stringybark leader asked me if I was acquainted with his group's *rom*; when he learned that I was not, he told me the story of the Wagilak sisters, using a bark-painting to point out the main protagonists and places. Indeed, the *rom* of the Bible, taken to be a guide to the law of God which Balanda should follow, became increasingly relevant to Yolngu following the Christian Revival (see Chapter 8). Most of the performances of ceremonies described in later chapters were the *rom* of Cloudy Water people, who interpreted them with reference to stories about the Djang'kawu sisters recounted in Chapter 1.

A classification of Yolngu ceremonies into three genres—*garma* ('public'), regional, and the Nga:rra or Ma<u>d</u>ayin ceremony (Keen 1978)—only reflects Yolngu categories in a rough kind of way, for ceremonies did not group into neat, exclusive categories, but overlapped somewhat in style and function. Furthermore, Yolngu rituals did not form a neat 'system'. The distribution of types of ceremony was heterogeneous, local networks performed a ceremony of a given name in a particular style, the genres overlapped in function, style, and performance, and ceremonies did not fit into a neat hierarchy related to the life cycle. Ceremonies with the same name had different styles in different local networks; furthermore, each genre and type of ceremony was also current among neighbours of Yolngu. This heterogeneity reflected the contingent origins and adoption of distinct religious forms; indeed, Christian rituals have recently been adopted alongside other genres, and in mortuary ceremonies Yolngu now combine Christian elements with a *garma* framework, as well as elements from other types of ceremony (I. Slotte pers. com.).

Garma Ceremonies

The general term for 'ceremony' was *bunggul*, which could refer minimally to one man singing or to a complex sequence of songs and dances lasting for weeks. *Garma* ceremonies were the most clearly discrete type at Milingimbi in the mid-1970s (and see Warner 1937: 135–7). At their heart were songs (*manikay*) accompanied with clapsticks and drone-pipe or 'didgeridoo' (*yiḏaki*), with which were associated dances, painted designs, settings (especially sand-sculptures), and objects.[2] Every group possessed a suite of *garma* songs (with their dances etc.) on which members could draw in order to compile, in a huge variety of ways, a ceremony for a particular purpose. Not all groups had members who were competent to perform their *garma* ceremonies, however; lacking competent singers they had to ask men of other groups to lead. The purposes included the initiation of boys by circumcision, disposal of the dead and mourning, purification after a death or fight, exchange of a sacred object and of goods, greetings and partings, fighting, dispute settlement, and simple entertainment. Table 4 summarizes *garma* ceremonies, and out-

[2] The drone-pipe or 'didgeridoo' is a pipe made usually from a termite-hollowed branch about 5 cm. in diametre and some 1,500 cm. long. Players produce two tones with an interval of about a tenth, and a continuous note using circular breathing in which breath is taken in through the nose while the flow of air is maintained with the cheeks.

TABLE 4. *Garma* Ceremonies

Ceremony	Description
Circumcision initiation *dhapi* ('foreskin')	When a boy's father and other relatives agreed that it was the right time for a boy to be circumcised, they arranged for a ceremony to be performed. The boy was prepared on two or three successive occasions by painting a *likanbuy* design on his chest to the accompaniment of *manikay* and dances. On the final day the boy's foreskin was cut away, and the boy secluded with male relatives while the wound healed. This changed his status to that of 'bachelor'.
Songs for the sick	If a person was believed to be seriously ill, men of the sick person's own and related groups sang *manikay* or *bilma* songs, sitting with or near the person to keep vigil day and night. Sacred objects such as the group's *madayin* dilly-bag and *wana* cords might be hung by the person's bed.
Announcement of the death *mali'-buthuru-bitjun* ('listen to the spirit'), *ringgitjhirri*	When a person died, speaking the dead person's name or any words of a similar sound was immediately prohibited. People in the community were informed of the death using circumlocutions such as 'so and so's father'. The death was announced in other communities with a short ceremony called 'listening to the spirit', in which a male singer sang *ringgitjhirri* songs, while others present stood by him in a rough circle. At the end of the short ceremony women wailed, sang mourning-songs, and cut their scalps with sharp objects. Men sometimes wept.
Wake *gupa-djuy'yun*, *gupa-ya:n,guma* ('send on their way', 'say goodbye')	The corpse was wrapped in cotton cloth and blankets, and laid under a bough or bark shade in the camp. Men sang *manikay* songs. Women sang mourning-songs, wailed, and hit their heads with sharp objects. Men displayed aggression and desire to avenge the death. Close female relatives sat or lay by the body. Groups of relatives sang and danced to 'visit' the body. Men painted a *likanbuy* design on the chest of the corpse (in some cases) or on the coffin lid, and pulled out the hair to make into hair string and belt. Commonly a close male relative of the deceased accused someone of killing the person with sorcery, or gave reasons why no one need be accused. A post-mortem might be held to look for signs of a sorcerer's identity on the body. Those who handled the body purified themselves with ochre, smoke, and water (see below). The body was 'buried' into a vehicle to chorus and invocation of *likan* names, then driven to the cemetery for a short Christian burial.

TABLE 4. *Continued*

Ceremony	Description
Smoke purification *d̲ad̲a*, *man'tjarr* ('mangrove leaves')	After *manikay* songs a smoky fire was lit in a sand-sculpture to chase away the spirit and remove the smell. People were patted with smoking leaves. The place were death occurred, or the dead person's country, might be closed for some months.
'Rag' *yindi man'tjarr* ('big mangrove-leaves')	Small pieces of the dead person's clothing were burnt in a fire after *manikay* songs.
Washing purification *buku-l̲up* ('head-wash') *Wan'tjirr* ('leaf')	Men sang *manikay*, then relatives of the dead washed in a large sand-sculpture containing leaves and smeared themselves with red ochre.
Exhumation and presentation *bukubut̲* ('head-fly')	Men exhumed the body and cleaned the flesh off the bones, sometimes with songs and dances. This was followed by purification ceremonies. To songs and dances the men red-ochred the bones and gave them to female relatives to look after and carry. In the past the skull was often painted with a *l̲ikanbuy* design.
Hollow Log *d̲upun* ('hollow-log coffin') (Djal̲umbu, D̲aymirri, etc.)	During an elaborate ceremony with songs and dances, men crushed the bones and placed them in a decorated hollow log, placed upright and abandoned.
Exchange ceremony *marradjirri* ('ceremonial string' or 'pole') (Band̲arra, Ban̲umbirr, Djurrpun̲, Garayurru, etc.)	People sent a child's first 'gift' or a relative's hair to request an exchange ceremony. The requestees made a tall pole and string incorporating the hair or 'gift', and presented it in song and dance to the requesters at an agreed place. The recipients later incorporated the string into arm-bands and ceremonial dilly-bags. (The ceremonies could be incorporated into mortuary ceremonies.)
Dispute settlement *makarrat̲a*	After dances representing the main *mad̲ayin* of the groups involved, the kin of the deceased threw spears, first at the instigators ('pushers') of the killing and then at the killers. Then the offended groups of men thrust spears into the thighs of the killers. Old men urged the younger men to cool their feelings.

lines in a general way what was accomplished in each of them. As well as these categories of ceremonies to do with *mad̲ayin*, there were many varieties of children's songs, some of which introduced children to ancestral concepts (McGowan n.d.: 4).

The *garma* genre was not discrete, for Yolngu incorporated elements from one genre into ceremonies of another. For example, some Dhuwa-moiety groups incorporated *garma* songs into certain phases of one of the regional ceremonies called Djungguwan. Conversely, people could include passages from regional or the Madayin into a *garma* ceremony such as the wake. Furthermore, Yolngu joined with people of the Blyth River region whom they called 'Burarra' to perform *garma* ceremonies as well as other genres.

Regional Ceremonies

What I have classified as regional ceremonies belonged to Dhuwa-moiety groups, except for the more recently introduced Yabudurrwa, which was Yirritja moiety. The Dhuwa ceremonies were associated by many Yolngu with the Wagilak sisters, who, in the myths, were swallowed by a giant Python believed also to be a cause of tropical storms. People portrayed them as ceremonies 'for everybody', in spite of their Dhuwa ownership. These characteristics aside, the regional ceremonies were something of a mixed bag, and had a variety of functions: Mandayala was used for circumcision initiation; Gunapipi was a revelatory ceremony and a celebration of sexuality; Djungguwan associated circumcision of boys with the remembrance of the dead, and was also a revelatory ceremony; Ngulmarrk, which was no longer performed in the Milingimbi region in the 1970s, was a revelatory ceremony identified with the Wubarr of western Arnhem Land.

While they bore a family resemblance to one another, the regional ceremonies did not have a uniform style. In all except Djungguwan, songs were accompanied by paired boomerangs used as clapsticks, and men decorated their bodies with down or wild cotton, although this was done in a Djungguwan I saw in the west of the region. (Djungguwan incorporated *garma* songs and songs with large clapsticks.) And whereas three of the ceremonies had secret phases in which men (and in Gunapipi both men and women) revealed dances and objects to novices, in Mandayala, men merely decorated their bodies in seclusion and danced in public. The style of each particular ceremony varied across the region. I have already mentioned variation in the use of down in Djungguwan. As well, east-coast people included objects called Djuwany posts in performances of Djungguwan (H. Morphy 1977: 129) whereas these were not features of the ceremony which I saw put on in 1976 by Djinang people in the far west of the Yolngu region. We could say, perhaps, that the main thing the regional ceremonies had in common was their assimilation by Dhuwa-moiety groups to the story of the Wagilak sisters. This seems to be true for the central and eastern areas of north-east Arnhem Land, but some Djinang speakers associated the ceremonies with the Djang'kawu sisters (see Chapter 7).

Mandayala and Gunapipi were built around symbols which, until the adoption of Christian elements, were the most universal, namely giant snakes (*motj*) believed to lie in deep waters in the countries of many groups. But this symbolism was less important in Djungguwan, in which Sand Monitor (*djarrka*) was dominant. Dhuwal- and Dhuwala-speakers did conceive of the ceremonies as a set, and gave them quasi-kin relations: Djungguwan was the 'father's father' of Ngulmarrk, which was 'father' to Gunapipi and Mandayala.

These ascribed relationships may have reflected the later adoption of Ngulmarrk, which probably originated from the Wubarr of western Arnhem Land, of Gunapipi, which was introduced from the south and which was similar to ceremonies such as the Gadjari ceremony of the Warlpiri people and their neighbours of the semi-desert, and of Mandayala, which was also in a 'desert' style (in which decoration in wild cotton or down and tall head-dresses for the men were characteristic) and which was distributed south and west to the southern coast of the Gulf of Carpentaria.

In spite of being thought of as a family of ceremonies, the regional ceremonies were not evenly distributed. Gunapipi was not performed by groups of Arnhem and Buckingham Bays in the 1920s. According to Warner the ceremony had diffused north up Annie Creek and the Glyde River, a journey commemorated in myth (Warner 1937: 452–3). On the east coast it spread north from the Rose River to Caledon Bay, Blue Mud Bay, and near Yirrkala, where it was occasionally performed in the 1940s. The Gunapipi performed at Galiwin'ku in 1981 was apparently the first to be held there (R. M. Berndt 1951: 14–15; Bos 1988*a*: 127).[3]

[3] These writers contradict one another and are inconsistent with Warner, who reports (1937: 452) that Gunapipi was not performed by groups of Arnhem and Buckingham Bays. Yet Berndt says that the ceremony spread to Arnhem Bay from Milingimbi. Berndt also writes that the ceremony went also to Elcho Island from Milingimbi, yet according to Bos it was first performed there in 1981.

This pattern of diffusion was reflected in co-operation by people of different regions and language families in performances of the ceremonies. For example, Rembarrnga people regularly came north from the Roper River to participate in Gunapipi ceremonies, and Yolngu people reciprocated.

The Maḏayin Ceremony

Many men regarded the Maḏayin ceremony as the most important of the revelatory rites, associated as it was with a group's creators (see R. M. Berndt 1952: 14). Its main purpose was to reveal secret dances and sacred objects to young men, but people performed it for other reasons, such as to commemorate a dead leader, and the ceremony was also a re-enactment of ancestral creative acts.

It was like other revelatory ceremonies, with its secret men's ground placed some distance away from the camp, but it had a distinctive style. Men did not accompany dances in the Maḏayin ceremony with songs but with various calls and cries; songs performed in the evenings in the Dhuwa ceremony were accompanied by large clapsticks and no drone-pipe; and body designs were painted in ochres with orchid-juice fixative, often incorporating geometric *l̲ikanbuy* designs on the chest.

Maḏayin occurred much more rarely than Gunapipi in the Milingimbi region. Only a few groups of each moiety seem to have had the ability to initiate a performance, which they might do on behalf of a different group of the same moiety. It seems likely that it was those groups with rights to call *l̲ikan* names who could do so.[4] In spite of inequality in the control of the Maḏayin ceremony, probably all groups possessed *rangga* sacred objects, representing *wangarr* or their attributes, which the older men could reveal in the ceremony to younger men on the secret men's ground.

Its content as well as functions overlapped the other genres in that men could use dances from Maḏayin in *garma* ceremonies such as a funeral, and perform public songs during preparation

[4] It is my belief that Djungguwan and Maḏayin (or Nga:rra), which were quite similar in structure, were indigenous in the sense of being practised in the region longer than the others, and perhaps developed there. Certainly, western Arnhem Land mythology suggests that those people adopted Maḏayin from the Yolngu (R. M. and C. H. Berndt 1970: 122).

and in between the formal dances of Ma<u>d</u>ayin. All the ceremonies were intertwined in another way: they all referred to *wangarr* ancestors and other beings, and each cross-referred to content embedded in other ceremonies within a group's body of sacra.

Later chapters include examples of performances of ceremonies of the three genres. This chapter now explores how ceremonies and their elements were owned and shared.

Group Identity and Ownership of ma<u>d</u>ayin: *A Model*

Simple hierarchical models of 'clans', clan subgroups, phratries, and the like have the merit of being easily grasped; unfortunately they do not fit the ways in which Yolngu organized religious practice. The ownership of ceremonies and their elements was complex; but in order to try to make sense of the complex details about the ownership and sharing of the various types of ceremony, I will outline a general structure that seems to fit the varieties (Fig. 8). This owes much to Howard Morphy's (1991) analysis of the *ba:purru* concept, but modified by the approach to Yolngu groups outlined in Chapter 2.

In Chapter 2 a 'country' was taken as a basic unit for the understanding of group identity. As Chapter 3 showed, each

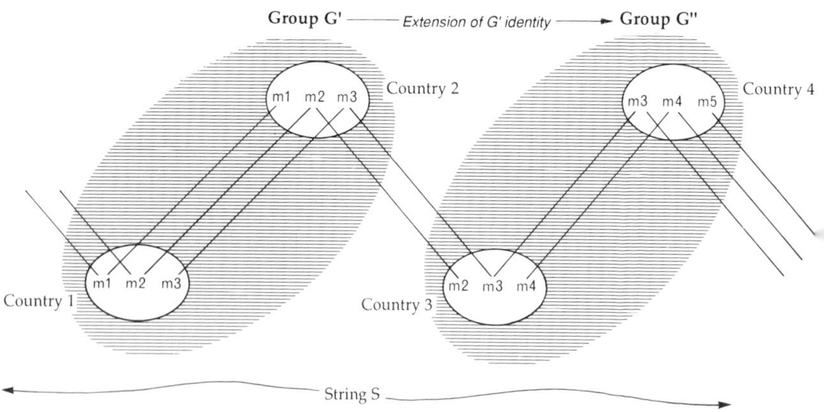

FIGURE 8. A Model of *ba:purru* Identity and Connections

group possessed a body of myths, types of songs, designs, and sacred objects, and programmes for ceremonies, all related to their country. These *maḏayin* were shared with other groups (m1–m5 in Fig. 8). A country had links to many others of the same moiety through ancestral travels and shared *maḏayin* (sacra). Different groups of *maḏayin* linked the country to distinct strings of other countries and groups.

Two aspects of ownership and sharing of ceremonies can be distinguished: the structure of the 'string' of links between countries and the groups holding the countries; and the assertion of group identity through the possession of places and sacra. Relations along a string had the quality of connection, of being 'the same', and the quality of individuality, of being 'different'. This relation of 'same but different' was manifested in various ways: transformations among a polythetic group, similarity of overall forms which were differentiated by details, a common name which referred to different manifestations, a common genus represented by different species, a shared clapstick-pattern but different tunes, and so on. *Ba:purru* identity implied a common *wangarr* ancestor, and hence shared sacra, so that groups with places along a string of countries were 'one *ba:purru*' with a proper name, but their common identity was qualified by local features.

The least divisible identity was that of a group holding one big-name country (which might be ambiguously defined and potentially divisible), but as we saw in Chapter 2, a least inclusive *ba:purru* name was often associated with several countries (again, potentially divisible), each with rather distinct links with other countries of the same moiety. People of one group were inclined to extend their particular name to other groups of the same string of groups, or to groups with the same tongue, but in some cases people of different groups agreed on a more inclusive proper name. Thus, some *ba:purru* names applied to local points on a string, and clusters of such points, but were potentially extendable to others of the same string. Other names applied to all points along a string. In the politics of group identity, people might assert control of other countries along a string, and so stress the common identity of that country and their own. Conversely, people associated with a particular country might assert their distinct identity, choosing a *ba:purru* name that distinguished them from former consociates.

The patterns of ownership of the three genres of ceremony are variants on these themes.

The Same Song: Sharing garma *Ceremonies*

It is convenient to think of the basic building-block of *garma* ceremonies as a complex of related stories, songs, dances, designs, and objects. Each such complex was built around a series of a dozen or so songs which evoked journeys around, and associations with, country. Many Yolngu classified *manikay* song-series (used for *garma* ceremonies and the Ma<u>d</u>ayin ceremony) according to topographical associations, or referred to them by the name of one of the constituent songs or the general character of the series. Thus there were forest, plains, salt-flat, beach, and sea song-series. By way of example, one of the Dhuwa-moiety forest series was named after the focal topic of Stringybark (*ga<u>d</u>ayka*), and one of the Dhuwa sea-series after Black-capped Tern species (*djarrak*). People tended to group these series into more inclusive sea (*gapupuy monukpuy*, 'pertaining to salt water') and land (*<u>d</u>iltjipuy*, 'pertaining to the forest/bush') categories of series (see Warner 1937: 418). Songs and song-series, together with their related dances, objects, and designs, were group attributes or 'properties', so that one could talk about a Lightning Snake (group) song, or a Stringybark song, while the possessive case showed a person's relation to a song, as in *ngarraku manikay*, 'my song'. Men sang their group's songs using drone-pipe and clapsticks, whereas women 'cried' (*nga:thi*) songs with words similar to those of the men, mainly during funerals (see McGowan n.d.: 2).

In language similar to spoken language, but with a simpler syntax and incorporating special song-words, the songs described the activities of spirit-beings, including *mokuy* ghosts and *wangarr* ancestors, usually interpretable as a journey through specific places and as a passage of time from sunrise to sunset.[5] The songs also referred to the seasons, and to the relationship of the life cycle of species to meteorological phenomena such as the winds and onset of the rains. Dances mimicked the beings' actions and journeys,

[5] The Yolngu classified the language of a song according to dialect-group and tongue. For Anbarra people of the Blyth River, however, many *manikay* words were in a foreign tongue, some from Yolngu dialects.

sand-sculptures represented places associated with them, artefacts reproduced objects which the spirit-beings used, and painted designs referred to the beings themselves, in one transformation or another, or to places, things, or creatures associated with them. The discursive structure of songs and stories diverged: whereas myths recounted the actions of the beings at particular places, songs projected images of the actions, attributes, and locations of things and creatures identified with spirit-beings, using simple clauses and lists of names and place-names. (There is an example of a *garma* song below.)

People evaluated songs, tunes, dances, designs, objects, and ceremonies of any two groups of the same moiety as the same (*balanya bili*) or 'one' (*wanggany*), close (*galki*), or separate (*ga:na*). The similarity of two groups' song-series was often qualified: the forest songs of Stringybark and Black-capped Tern groups had 'the same' *mokuy* ('ghost'), but each group's *mokuy* had a different name: one was called Wanydjuk (Stick-insect), the other Wurray, but they were 'the same'. Myths like those in Chapter 1 explained a relationship of identity by the journey of a spirit-being or a chain of interconnected journeys, or the close identity of two beings. Each group belonged in this way to several distinct but overlapping and open *ba:purru* or *mala* strings of groups.

The song-series belonging to several groups with 'the same song' formed a polythetic group (Table 5). Any two groups said to have 'the same' song-series had about half their song topics in common, but only two or three song topics were in common to a number of groups with 'the same song'. Moreover, a group's song-series might resemble two groups' song-series each of which was *not* regarded as the same as the other. Some groups with 'the same' song had the same tune or clapstick rhythms. Men explained that songs of the Fly group, for example, had the same clapstick rhythm as those of the Black-capped Tern group. However, the subjects of their songs also resembled those of the Stringybark people. As with designs, each group's total constellation of songs was unique.

Song-series as entities had a certain fluidity, but the programmes of ceremonies with proper names were a little more concrete, and the pattern of sharing was consequently simpler. The countries of groups which shared the named Hollow Log ceremonies such as Djal̲umbu and D̲aymirri were widely separated in space; such

TABLE 5. Songs Shared by Morning Star Group with Two Other Dhuwa-moiety Groups

	Group		
	Morning Star	Sun	Black-capped Tern
Sea series			
Freshwater Snake	x		
Rope	x	x	x
Seaweed	x		x
Flying Fish	x		
Bark Canoe	x		
Seawater	x		
Plover	x		
Forest series			
Cuckoo	x		
Sugar Squirrel	x		
Stringybark Tree	x	x	x
Longnose Honeybee	x	x	x
Wurray *mokuy*	x	x	x
Fire	x	x	x
Whimbrel	x		
Gecko	x		
Frogmouth	x		

groups had similar song-series, although not necessarily the same tune or the same tongue. The Dhuwa Hollow Log and exchange ceremonies linked groups with rather different song-series, but all the song-series were about the Morning Star. The exchange ceremonies were shared by even more widely separated groups than in the case of Hollow Log ceremonies.

Another important element of *garma* ceremonies as well as of the Maḏayin ceremony was the invocation of *ḻikan* names of *wangarr* ancestors and other sacra, and accompanying chorus, performed at climaxes of the ceremonies. As with songs, strings of groups of each moiety were united by having a common form of chorus, but only some groups of each moiety had the right to lead the invocation. The Honeybee group (Yirritja moiety) was the main one with rights in Yirritja invocations at Milingimbi; the Cloudy Water group called the Dhuwa invocation. Several older men of these groups were the 'power-men', *ḏalkarramirri* or *djirrikaymirr yolngu* ('power-having person'), who called out the names.

Groups with *madayin* in common were frequently said to have exchanged names, or people explained that one group made a gift of a name to another group. (I use real names of groups here to make the point clear.) The Guyula people of the Dhuwa moiety, for instance, were also called Liya-wulma-mirr, 'thundercloud people', and Liya-dhaliny-mirr. It was said that the group gave the name 'Guyula' to the Djapu group in exchange for the name 'Djarrangguykuy', which referred to the Djarrak (Black-capped Tern) *wangarr*. 'Djarrangguykuy' was also the name of the Djambarrpuyngu of Flinders Peninsula (Schebeck 1968: 73; Williams 1986: 41).

What was it, though, that was owned and shared? It would be a formidable theoretical task to try to specify just what made a performance a song of Honeybee group rather than Black Duck. But briefly, whether they were shared or not, equivalent to those of other groups or separate, certain attributes of a group's songs were relatively invariant; these included the contour of a tune, clapstick rhythm, topic, the repertoire of song words and phrases on which singers drew, and proper names. However, the ability of individuals to innovate and improvise implies the generation of new forms.

Yolngu emphasized tradition and the need to follow (*dhudakthun*) both immediate ancestral and *wangarr* precedent, yet ceremonies were inherently improvisatory and contextually determined (see Chapter 7; H. Morphy 1984). Moreover, Yolngu effected a nice compromise between individual autonomy and group co-operation, and between group identity and inter-group collaboration. Co-performers of *garma* songs beat clapsticks in unison and roughly followed the same melodic outline or a variant of it, yet each improvised from an encyclopedia of appropriate expressions. Members of different groups who shared identical melodies and clapstick rhythms would sit and sing together the part they shared, diverging for other series, or sing snatches with each other's tunes. Or singers of one language would pause while a singer of a different tongue would insert part of his own equivalent series. Here similarity of song amounted to a common name (translated from one language to another) or equivalent names (such as varieties of catfish) together with the kinds of things sung.

Putting Together a Performance

Thus most Yolngu *garma* ceremonies were rather flexible frameworks in which participants could slot sundry bits to realize the ritual in its most basic form, or greatly elaborate it according to circumstances. The number of people who came and the complexity of the occasion depended very much on the renown and resources of the focal person or group. People's choices of song, dance, and design depended upon many factors, including these: the relationships between co-operating groups; the identity of the initiand in a circumcision (or the dead person in mortuary ceremonies); the intention of marking the prestige of the organizer (or of the deceased in the case of a wake); wanting to indicate affiliation to group and country; wanting to lay claim to another group's *maḏayin* and country; and matching the symbolic content of the performance to its purpose (see Chapter 5; and cf. H. Morphy 1984).

Men improvised an individual item (a song, dance, design, or object) around a framework of clapstick rhythm, group tune, design matrix, and conventional representation, drawing on a repertoire of figures. For example, singers of *garma* songs did not reproduce set texts but improvised from a stock of proper names and phrases. The evidence suggests that the contents of the stock differed from performer to performer. Here, for example, are phrases from two verses of the song of Sulphur-crested Cockatoo, rendered by two men of different Honeybee lineages, each sung at the beginning of a different session:

Singer A

Ḏan.gi Gunbuma Ḏan.gi	Sulphur-crested Cockatoo
Ḏan.gi Gunbuma Ḏan.gi	[names]
nga:thi marrtji Ḻurrpu Marrbuma	crying, going, Cockatoo
miṯil miṯil	calling, calling
ganydjarr nganya Gunanganydjulili	[fly] powerfully towards
Milawuluwululili, Gunanganydju	Gunanganydju, Milawuluwulu
Gunbuma Ḏan.gi	Sulphur-crested Cockatoo
nga:thi marrtji Ḻurrpu	crying, going, Cockatoo
djikiḏingdhun	calling out
Gunbuma Ḏanggi	Sulphur-crested Cockatoo
nga:thi marrtji Ḻurrpu	crying, going, Cockatoo
yingirri Lunggurrmay Mirrarray	North Wind blows his/her feathers

Singer B

ngupil ngupil ngupil	calling, calling
Marrbuma Ḏanggi	Sulphur-crested Cockatoo
Marrayaḻyaḻ	breeze from the forest
Marrayaḻyaḻ	[makes him/her happy]
djikiḏingdhun	calling out
Galumaluma	happy Cockatoo
Gunbuma Ḏan.gi	Sulphur-crested Cockatoo
Marrayaḻyaḻ	breeze from the forest
Balthawunngura	in the large paperbark tree at Balthawun
Butju bumara Djalathang	South Wind blows on him
Lunggurrma Ngirndirngindir	North Wind blows

The verses have the same theme: Cockatoo(s) fly and call, the wind blows on his/her/their feathers and make him/her/them happy, but the theme is somewhat differently expressed. It is as if each singer drew on a body of acceptable names and phrases in his own way to tell a similar story. Still, the impression that a person could do just what he or she liked would be quite wrong, for participants monitored one another's actions and might take steps to correct them. Obviously a man of age and status was more likely to intervene in what someone else was doing than a younger or less confident person.

Ownership of the Maḏayin Ceremony

Like Gunapipi, Djungguwan, and Nguḻmarrk, the Maḏayin ceremony was a revelatory rite, although it had many other purposes and effects. The content of the ceremony emphasized group identity as well as the connections between groups of the same moiety—connections through women necessary for reproduction. The putative origin of these connections lay in the journeys of major *wangarr*. Again, the pattern of ownership of the ceremony and its elements linked each group to several strings of groups according to various criteria. Yet, as with *garma* ceremonies, each group retained its particular identity both in the total configuration of its attributes and in the detailed specification of each thing it owned. But there were differences in patterns of ownership between the two moieties. Dhuwa people seemed to be more unified than Yirritja.

According to Yolngu stories certain *wangarr* created the Maḏayin ceremony itself. In one story of the Honeybee group, Quail and Honeybee (or 'Sugarbag') made the first Yirritja Maḏayin ceremony:

Long ago five men including Quail and Honeybee walked at night, made a fire, and carried the fire along a track to make the Maḏayin ceremony, just as people do now, copying them. Quail is boss for Fire, and made the Fire *rangga*.

Honeybee painted his body with yellow ochre and, carrying the (named) Dilly-bag, went to the sea, mixed saltwater with freshwater, and called the *likan* names in the invocation.

Honeybee and Quail heard Murayana Ghost's drone-pipe, so they went to Spear-grass-group country and copied it. Murayana told them that the drone-pipe was his alone, not theirs.

Niwuḏa or *birrkuda* is a large, yellowish, stinging bee which builds nests in the tops of trees and makes orange-coloured pollen-balls (Rudder 1977: 58). This was the identity of the 'big' Honeybee-group *wangarr*, who made the group's first *rangga* sacred objects and was its ancestor. *Rangga* sacred objects represented Honeybee's nest, the entrance to the hive, and related objects, each referred to simply as 'Honeybee'. The Honeybee ancestor was connected in particular to the group's coastal-plains country, but myths described Honeybee's journeys from that place to another country belonging to the group, and from other groups' countries to Honeybee places. This was merely a Honeybee perspective; other groups of the Yirritja moiety attributed the origin of the Maḏayin ceremony to other beings, especially La:nytjung and Barama (H. Morphy 1990).

The Dhuwa Maḏayin ceremony organized by Cloudy Water men followed the story of the Djang'kawu sisters (Chapter 1, Chapter 7). Here, the string of groups with rights in the ceremony were linked by a long ancestral journey. Groups with country along the coast on the path of the Djang'kawu sisters owned the ceremony (although people did not altogether agree about the course of the journey, as Chapter 1 showed). Dhuwa groups whose main countries did not lie on this track had small places embedded in the countries of other groups which were on the Djang'kawu track, so giving them rights in the ceremony. However, Dhuwa-moiety men outside the Yolngu region attributed the ceremony to other beings (Elkin 1972: 86). People might affirm or deny links through ancestors and disagree about their relationship. Partly as a consequence of constantly shifting kin-relations, particular connections were continually transformed through the exchange of names and ritual forms, sometimes in order to construct quasi-kin relations of amity.

Performances of the Maḏayin ceremony followed a common general form, but the participating groups placed particular elements which they owned, such as dances, into the relevant slots in the ceremony. The elements most relevant to this discussion were *biḻma* songs (without drone-pipe accompaniment), coarse designs (*la:rr*) or 'colours', the invocation and chorus, *ḻikanbuy* designs, the *rangga* sacred objects, and certain dances.

From the Honeybee leader's perspective, Yirritja-moiety groups were divided into distinct sets in the ownership of *biḻma* songs. He explained to me that his people owned no *biḻma* songs; when they performed the Maḏayin ceremony they had to ask Crocodile, Hook-spear, or another group to sing for them. (They did possess the right to call the invocation, however.)

Songs

From the point of view of the Cloudy Water group, people agreed that all the Dhuwa groups in the region possessed the Djang'kawu *biḻma* songs which 'followed the sun' like the sisters' journey, but that the Sunrise group had a different tune. In this case the string of groups with 'the same song' was divided in different ways, by the particular set of songs they performed, and by the tune. Another differentiating marker was the total complement of kinds of song. The Cloudy Water group thought of themselves as unique in not possessing *manikay* songs with didgeridoo. They used *biḻma* both for *garma* ('public') ceremonies and in the Maḏayin ceremony. Cloudy Water men explained to me that because they had no *manikay*, they had to ask others to perform the fire purification (a *garma* ceremony) for them.

Conversely, Cloudy Water men considered themselves to be 'boss' (*bunggawa*) of the Dhuwa Maḏayin ceremony because they owned the *biḻma* songs. They shared this status with other groups such as Thunder and Mangrove, but the Morning Star people whose country was a distant island, together with the closely related Two Women group with country at Galiwin'ku, were 'workers' (*dja:mamirr*) for Cloudy Water with respect to the ceremony.[6] An episode of the Cloudy Water group's Djang'kawu story explained the origin of this relationship (see Chapter 1). Standing at Wa:langur, the Djang'kawu, or someone called Walawala, heard his (her or their) *ma:ri* (MM/MMB) singing across the water at Galiwin'ku. They crossed over:

> Walawala told his *ma:ri* (MM/MMB) to sing with a slow clapstick-beat. The singers of Morning Star and Two Women groups said, 'Very well *gutharra* [wDC/ZDC], you are bosses of the Madayin ceremony, we will work.'

A 'worker' group had to ask a 'boss' group to lead the performance when they wanted the ceremony performed. The Cloudy Water men sang, called the invocation, and led the dancing. It is not clear why these two groups were in this relationship, but it may reflect the fact that the Morning Star and the closely related Two Women groups did not claim Djang'kawu as ancestors but were linked by the journey of two different female *wangarr*.

Colours

I mentioned earlier that Honeybee men as 'power-men' possessed the right to lead the Yirritja invocations among Milingimbi groups, and Cloudy Water men could lead the Dhuwa invocations. But the Honeybee leader's status as 'power-man' went beyond mere specialization: he was widely regarded as the 'boss' of the Yirritja Madayin ceremony at Milingimbi and the nearby mainland communities. If Barramundi or any other Yirritja moiety group wanted the Madayin ceremony, they would have to ask him, even those with quite different main *wangarr* and *rangga*. This man seemed to have effective control over whether the Yirritja Madayin ceremony was performed at Milingimbi, not just nominal rights. But when he died in the early 1980s his group did not maintain its dominance.

Coarse painted designs used in the Yirritja Madayin ceremony represented a variety of *wangarr* ancestors. The Honeybee leader explained rights in the Yirritja designs solely in terms of the pigments used, and was characteristically assertive in placing his own group at the head:

[6] Both *bunggawa* and *dja:ma* are words of Macassan origin, and Yolngu used the English words 'boss' and 'leader' as synonyms.

I would say, 'You Whale group have both [colours]; you Magpie Geese have white; Barramundi, white; Lightning Snake, white and yellow; Water Snake, yellow; Hook-spear, yellow; Honeybee group, yellow; Koël group, white; Rock group, white.' Yellow has the name Galangarr, and Honeybee group is the leader.

We shall see later that these divisions reflect the sharing of rather distinct *madayin*.

The Dhuwa moiety had less diversified rights in colours, reflecting the more unified Dhuwa traditions in respect of the Madayin ceremony, although this unity may have been more apparent than real, reflecting the perspective of the Cloudy Water leader who gave me this information. He explained that all Dhuwa groups had Kingfisher colours, consisting of horizontal stripes of red ochre (*miku*) or haematite (*ratjpa*)—which is a purplish ochre with a silky sheen—yellow ochre (*buthalak*), and white clay; and one invocation chorus, the *marrawinydjun*. Within this overall uniformity there was variation: Sunrise group had only white clay, because it was of the salt water; Thunder and Cloudy Water groups had red and yellow ochre, and they had white clay because they were of the salt-flats and plains; Two Women group had haematite, yellow ochre, and white clay. The unity reflected the linking Djang'kawu story and *ba:purru* identity as Ma:tjarra or Bungunbungun; the differences reflected associations with other *wangarr* such as Shark. We shall also see that these groups interpreted the Madayin ceremony rather differently in spite of joining to perform it (Chapter 6).

Likanbuy Designs

The distribution of *likanbuy* designs (*miny'tji*, 'design'), employed in *garma* ceremonies, as well as the Madayin, made further discriminations among the groupings formed by the colours of 'coarse' designs, as Howard Morphy's analysis shows (H. Morphy 1988: 255–6). Yirritja-moiety groups employed designs based on the diamond, pointed oval, triangle, and zigzag, while Dhuwa groups used combinations of squares, curved and straight parallel lines, and circles. Each type of pattern was associated primarily with a particular *wangarr* ancestor, and each linked the owning group to others with the same *wangarr* ancestor and design, in the same way as elements of *garma* ceremonies. For example, the diamond design, associated primarily with Honeybee ancestor, belonged to at least six groups, including Honeybee, Lightning Snake, Crocodile, Water Snake, and Rock; while triangle design, associated with Whale Ancestor, belonged to six groups or more, including Lightning Snake, Crocodile, Whale, and Dingo.

Each group had its own particular myths about these beings and interpreted the similar forms differently. For example, the Hookspear group associated the diamond design with Freshwater Turtle as well as Honeybee, but the pattern signified Crocodile Ancestor and Fire to the Crocodile and Water Snake groups, while for another branch of Crocodile group it signified Sand Goanna. From the perspective of the Dingo and Whale groups the triangle design signified Cloud (H. Morphy 1988: 255–6). We will find a similar relation of shared form to disjunct content in the interpretation of dance sequences in the Ma<u>d</u>ayin and Gunapipi ceremonies (Chapter 6, Chapter 8).

Within the common general form the designs belonging to distinct groups were distinguished in their details. Although Morphy found that members of 'clans' publicly acknowledged joint ownership of all the designs and associated countries, 'subgroup' designs were distinct in some respects (H. Morphy 1988: 262–3). Each design was not associated with a group's territory as a whole but with a particular part of it, related to a particular *wangarr* ancestor.

Thus, to take the examples in the previous paragraph, Lightning Snake group had the diamond design of Honeybee and Fire, associated especially with one inland country, and linking it with Honeybee and other groups. It possessed also the triangle design, associated with its island country, linking it with Whale, Dingo, and other groups, but not Honeybee (H. Morphy 1988: 255). This pattern of shared designs mirrors connections of <u>D</u>etjirimirri to other countries through myth (Chapter 3).

Like the *rangga* sacred objects which they depicted, then, ancestral designs linked places of distinct groups, but the set of designs possessed by a group was unique, and each differed in detail from those of other groups (H. Morphy 1988: 266). This feature does not provide a criterion for a 'clan' level as opposed to clan subgroup, however, because the sacra associated with one country and its associated group was similarly unique, and so was the total complement of a more inclusive group.

The distribution (where recorded) of the features of coarse-

TABLE 6. Relations Between Ancestor, Colour, Design, and Chorus

Ancestor	Colour	Design	Chorus
Koël	white	open diamond	?
Rock	white	open diamond	?
Mangrove	white	open diamond	?
Dingo	white	triangle	*minggathun* *murrmurryun*
Cloud	?	triangle	?
Whale	white	triangle	*minggathun*
Honeybee	yellow	diamond	*mum'mumdhun*
Crocodile	yellow	diamond	*mum'mumdhun*
Fire	yellow	diamond	*mum'mumdhun*
Yellow-ochre	yellow	diamond	?

design colour, group design, chorus, and major *wangarr* ancestor, as reported by different writers, closely correspond. Groups with yellow ochre (*buthalak*) had Honeybee as *wangarr*, the Digging Stick *rangga*, and the use of ironwood for the body of the sacred object. The groups with white clay were associated with a variety of other *wangarr* ancestors, particularly those to do with the sea. These include Rock, Barramundi, Mangrove-wood, Whale, and Black Duck, as well as Dingo and Long-necked Turtle (Thomson n.d.; H. Morphy 1988; Groger-Wurm 1973). Table 6 summarizes these relations.

Sacred Objects

In spite of these commonalities, each group possessed a distinct set of *rangga* sacred objects related to its *wangarr* ancestors, each representing some aspect of the ancestor. Myths described the journeys of the major ancestors across the land and waters, leaving *rangga* sacred objects in the ground or water at some specific place at each group's countries. People explained possession of related *rangga* by the journey of the same being connecting the two groups. Wandhuma, the Cloudy Water leader, explained to me that the Djang'kawu gave each Dhuwa group somewhat different sacred objects, but all joined for the Ma<u>d</u>ayin ceremony (see H. Morphy 1990), once more reflecting differences within connections.

As with designs which represented the sacred objects, and like elements of the *garma* ceremonies, a group's complement of *rangga* connected it with several other strings of groups, and each group's complement of *rangga* and associated ancestors was unique. A group also maintained its distinctiveness through the particular features of its own *rangga*. Groups with 'the same' *wangarr* ancestor might have *rangga* each of which represented a different aspect of the being, or distinguished by a detail of design such as width or contour, as with *likanbuy* designs. As well, two groups with several *rangga* in common put different values on them: the 'big' *rangga* of one group could be the 'small' one of another. For example, Lightning Snake and Honeybee groups (who together formed the Up-country people), shared several sacred objects, including those connected with Honeybee. But whereas Honeybee *rangga* was the most important from a Honeybee group perspective, Lightning Snake people put Black-headed Python in this position. The most important *rangga* represented the creator of the group. The various *rangga* which a group owned could also be assigned to distinct subgroups, where these existed, and to individual men to 'look after', with the leader or leaders holding the most important.

As H. Morphy remarks (1977: 87), people might extend the ownership of a design by making a gift, following a close mythological relationship between two groups, and sharing of some *rangga* and group designs came about through gift-exchange. I was told that the Shellfish group had no sacred objects until given them by another group with the same ancestor.

Where there were a number of mature or old men of the same group-identity, responsibility for a group's various *rangga* was shared among them, like the associated *likanbuy* designs. Wandhuma claimed to hold two 'big' *rangga* of the Cloudy Water group, a third was held by him and his 'sons' of one lineage, while people of another lineage held a 'small' *rangga*. This differentiation by *rangga* and *likanbuy* designs associates divisions of a group with different places in the group's country, although, given the prevalence of disagreements over group structure as well as responsibility for sacra, I hesitate to propose an unequivocal association between group divisions and places.

Owning Regional Ceremonies

The regional ceremonies had more of a set pattern than the *garma* ceremonies and the Madayin. They did not break down to the same degree into elements which could be recombined in many different ways. Despite their formal ownership by groups of the Dhuwa moiety, people of most groups of both moieties participated fully in the regional ceremonies, and there were no closed congregations. People from a wide area attended performances, especially of the Gunapipi in the west of the Yolngu region.

Among some Yolngu groups, myths which explained the origin and ownership of the regional ceremonies were variations on a long epic about two women, the Wagilak sisters, similar in some ways to the Djang'kawu myth (see Chapter 5). Like this myth, the Wagilak story ends with a catastrophe which causes the women to lose the ceremonies and the men to take them. In both genres of ceremony, people understood the dances as re-enactments of the events which gave rise to them, and the songs as those which the *wangarr* ancestors sang.

Yolngu did not entirely agree about the ownership of regional ceremonies. Some people said that the Stringybark group 'looked after' or possessed all the regional ceremonies although they were 'for everybody'. The Stringybark-group leader told me that the ceremonies were his 'dreaming' and that he was the 'boss' (or *bunggawa*) for them, qualifying this assertion by saying that there were many bosses. Indeed, leaders of certain other Dhuwa-moiety groups could initiate and organize a performance, although people sought the agreement of the Stringybark-group leader.

Sacred Objects

The common link between owning groups was not so much the Wagilak story as the possession of certain objects associated with the regional ceremonies—the Drone-pipe (object and being) associated with the Djungguwan ceremony, and the Poles associated with Gunapipi. These types of object were conceived of as sacred transformations of *wangarr* ancestors residing in their country but not necessarily related to the mythological journey of the Wagilak. The Drone-pipe *ma<u>d</u>ayin* counted as the 'shade' or 'image' (*mali'*) of King Brown Snake or Olive Python, for each Dhuwa group had one or the other associated with its country. It is likely that each one of the various Snakes and Pythons in the story was related to the Drone-pipe of a particular group and country. Gidjingali people, however, associated the objects with Hollow Log man (Clunies-Ross and Hiatt 1977: 136). Secret proper names denoted both the objects and the *wangarr* of which they were metamorphoses. (They were 'big names' but men would say them in a stage-whisper which anyone could hear.)

People differed also about whether some groups did possess the relevant sacred object or not. According to the Stringybark leader, only some groups of the western Yolngu area possessed Drone-pipe: Stringybark, Blossom, Drone-pipe, and Cabbage Palm groups did so, but Mosquito, Black-capped Tern, and unspecified Burarra groups did not. One Mosquito leader privately claimed to me, however, that Drone-pipe did lie in his country.

Many Dhuwa groups which did not possess Drone-pipe or the Pole being did hold other *ma<u>d</u>ayin* relevant to the regional ceremonies, some of which were equivalent in meaning, and as well as the giant reptiles, people believed that other *ma<u>d</u>ayin* associated with the regional ceremonies stood or lay at the countries of Dhuwa-moiety groups. These included Fish-trap, the Forked Upright and Rails common to the structure of a dwelling and fish-trap, and Cabbage Palm, associated in form with the reptiles and in function with house posts, for cabbage-palm leaves were used for the construction of shelters (and the trees are good to sit under). All these *ma<u>d</u>ayin* corresponded to elements of the regional ceremonies. Fish-trap was equivalent in the traditions of certain groups to the Djapanbani Trestle used in the final stage of the Gunapipi ceremony; both were formed from two forked uprights and a rail, like the stringybark house of the Wagilak sisters. People equated Cabbage Palm with the Giant Snakes and the Gunapipi Poles. In addition to Giant Snakes lying in their waters, some groups which possessed the regional ceremonies had Two Women as *wangarr* which were not identified as the Wagilak. We will encounter one such group in the next chapter, who interpreted the regional ceremonies by referring to the Djang'kawu sisters.

Subsection Designs

Whereas the regional ceremonies had programmes that were somewhat less variable than *garma* or Ma<u>d</u>ayin ceremonies, some of their elements could be recombined. These were designs made with blood and wild cotton and the related dances, as well as variants in the style of songs. The former were assigned to subsections (Table 7).

Each of the two Snakes, represented in the ceremonies in the

TABLE 7. Subsection Designs

Moiety	Subsection	Design/Dance
Yirritja	Ngarritj Ngarritjan	Kangaroo
	Gadjak Gutjan	Chicken hawk, Gu_d_utjurrk lizard
	Banga_d_i Banga_d_itjan	Emu
	Bulany Bulanydjan	'Cooked' Kangaroo
Dhuwa	Balang Bilinydjan	White-breasted Sea Eagle
	Wa:mut Wa:mutjan	Blackbreasted Buzzard, Blacknosed Kangaroo
	Burralang Galiyan	Agile Wallaby, Rock Wallaby
	Gamarrang Gamandjan	Wedgetailed Eagle

form of Poles and engravings on the walls of a special pit, belonged to four subsections, two of each moiety. The Djapanbani trestle of the Gunapipi belonged to two subsections of each moiety (see Chapter 8). Songs, wild-cotton designs, and related dances for the Djungguwan ceremony were also subsection-specific. Men chose wild-cotton designs to wear and the related dances to perform according to their relationship with a group *ma_d_ayin* as well as by subsection, so that the complement of designs and dances varied at each performance according to the mix of participants' groups.

Thus the regional ceremonies transcended differences between local *ma_d_ayin*, and they also united the moieties. Novices of both moieties were 'swallowed' in the Gunapipi and Man_d_ayala. The Giant Snakes in Gunapipi were 'meaningful' (*mayali'mirri*, 'meaning-having') and relevant to both Yirritja and Dhuwa people not only through their making thunderstorms and swallowing the novices, but because people of Yirritja groups as well as Dhuwa believed that such Snakes lay in deep waters on their country. This universalism was extended even further with the growth of Yolngu controlled Christianity, seen as uniting Yolngu and *balanda* (light-skinned people). (See Chapter 8.)

Openness to Interpretation

The contrast between co-operation in the enactment of ceremonies, together with variation in the interpretation of a ceremony's significance, corresponds to the contrast between commonalities among a string of groups and differences in detail. Shared forms of ceremony implied that people could co-operate to bring a ritual into being, but in doing so people of each group enacted their own particular variant and projected their own specific beliefs on to the common matrix, interpreting shared performances of common forms in terms of variants of similar ancestral events or even radically different myths. Even participants belonging to the same group might give rather different explanations of a dance or design. Commonality of form was not straightforward, however, and might come down to a type of object or a proper name used in discourse as if it referred to a spirit-being. What kind of being the name referred to, and to whom the object related, was open to varying interpretations.

Warner made evident one way in which episodes of Yolngu myths corresponded to segments of major ceremonies. Ritual and myth, he argued, were but two parts of one social phenomenon. The Wagilak myth was fundamental to, and organically related to, a certain group of Yolngu rituals which expressed the oral rites (Warner 1937: 245). All but one section of the Djungguwan ceremony dramatized the myth-symbol of the Snake swallowing the Wagilak sisters (1937: 372), whereas Gunapipi brought in the reproductive cycle and the symbols of animals running away from the unclean women in the Wagilak story (1937: 376). The mapping of myth on to rite was not uniform, however, for each ritual expressed a distinct set of the episodes of the Wagilak myth, although these overlapped (1937: 377).

Stanner's ethnography of Murinbata religion (Stanner 1963) took possible relations between story and ritual further: some myths appeared to have no associated ritual while some rites had no corresponding myth. Murinbata circumcision and burial rites were not associated with myths in the same way as the bullroarer rite, although they had a certain 'support' from the general system of beliefs (1963: 107). In spite of this lacuna, Stanner was able to detect structural parallels between mythless rites and a myth of the Snake Woman which had no apparent connection with any rite (1963: 81 ff., 131).

It became evident to me that different groups in north-east Arnhem Land, and indeed individuals of the one group, interpreted a performance in which they were involved with reference to distinct myths. They were able to come together and co-operate in an enactment without subscribing to a common orthodoxy. Chapters 5 to 8 provide examples of various interpretations given to elements of *garma* ceremonies and performances of Ma_dayin and Gunapipi. Dhuwa people explained Ma_dayin partly with reference to Djang'kawu stories with the variety seen in Chapter 1 (Chapter 6), whereas participants in Gunapipi mapped myths about Wagilak, Djang'kawu, and Nagorrko on to episodes of the ceremony (Chapter 8). It was the ambiguity of names, designs, objects, and songs which enabled groups to map their own interpretations on to shared forms. People of different groups interpreted painted designs with a similar form in different ways, as we have seen.

The Stringybark leader and his sister's husband explained the distribution of ownership of the regional ceremonies to me in terms of the journeys of the Drone-pipe and Pole *wangarr* to various countries, using the proper names to denote beings. These proper names were unambiguous in referring to objects made in the ceremonies, but ambiguous when referring to *wangarr* ancestors. From the perspective of one group the names were the esoteric proper names of Olive Python, while from that of another group they signified King Brown Snake, and of another, Hollow Log Man. Groups with different Giant Snakes and other *ma_dayin* could agree that Drone-pipe was a *ma_dayin* and the transformation of a being of the same name. By refraining from specifying their own beliefs about its origin when in the presence of members of groups with different beliefs, and by refraining from discussing to what *particular* being the names were equivalent, groups from a very wide area could agree that they all had the same *ma_dayin* and ceremony. They could explain the diversity by saying that the being was transformed into each of these forms, or gave them to the groups. Indeed, as manifestations of the ancestors, transformations such as these, united by a common form, name, and origin, were seen as intrinsically related, having an essential identity, not a conventional one. The names, then, were semiotically similar to group designs where the ambiguity of a form allowed each group to stress a different meaning. (The polysemy of proper names as well as geometric designs, and the distinction between outward forms and inner content, were also bases of secret religious knowledge, as we see in Chapter 7.)

Not only were common forms given different interpretations, but the forms too were disputed. Some Yolngu ceremonies had more of the character of compilations than others, although all were collages to some extent. Among the mortuary rites, the wakes were compilations of songs and dances of sundry visiting groups to 'visit' and 'say goodbye' to the dead. Gunapipi and the Ma<u>d</u>ayin ceremony, like Djunggwuan and Ma<u>nd</u>ayala, were more fixed in overall structure, and had a framework of more or less invariant acts, together with slots in which the various participating groups could do their own thing. People negotiated through public and private discussion what the course of a ceremony should be and when its phases should take place, but it was over the proper way to enact this overall framework that major disagreements occurred.

Not unnaturally, people of a local network who interacted with the greatest frequency and density, concurred most about proper forms of set rituals like Gunapipi and Ma<u>nd</u>ayala. Thus one found local varieties and styles in different regions, and the greater likelihood of arguments over forms when people of one region visited another. The mediating strategy was (or should be) if not silence then at least a certain delicacy and obliquity when referring to the significance of forms in the presence of those affiliated to incommensurable traditions.

New Forms

In spite of an ideology of the ancestral origin of sacra and the denial of the creativity of the individual as an agent, Yolngu religious practice was continuously innovative, though in some groups more than others. For example, different groups with 'the same' *garma* songs used for public ceremonies interpreted similar items differently, and some attributed obviously new meanings to old words. The Magpie Geese group had rights in the Macassan song-series which described the Macassan *wangarr* and the resources which they brought to Arnhem Land, such as tobacco, knives, alcohol, and rice. But the men said that the 'inside' meaning of the words described missionaries doing such things as unloading sacks of flour at the wharf. In this way the songs were reinterpreted to give them an immediate significance. I asked a leader of a Honeybee lineage about this interpretation. He said that it was wrong, for the songs were 'from the beginning'.[7] This case contrasts the conservatism of the Honeybee men with the more innovatory attitude of Magpie Geese people. It is, of course, quite conceivable that the songs about Macassans arose out of reinterpretation of songs about other beings.

A similar process has been occurring with children's songs which referred to *ma<u>d</u>ayin*. McGowan (n.d.: 14) reports that a motor car replaced the Mangrove Log in a song performed with the same tune. The bark of a mangrove log held the body as a coffin and a hollow mangrove-log served as a coffin for the bones. Similarly the car has become the hearse. Like the image of bark being stripped from the log as it drifted with the current, so the tyre being stripped from the wheel as it was driven along the river was analogous with the flesh that rots and decomposes, leaving only the bones.

Some other changes had occurred and were occurring in the mid- to late 1970s. First, painting was transformed and elaborated as bark-paintings, both for the market and as a vehicle for the transmission of religious knowledge. Second, on larger settlements the Hollow Log ceremony was removed from the liturgy to be replaced by a longer, more elaborate wake and burial on settlements. Third, various indigenous rituals were discontinued, such as Ngulmarrk in the west of the region, while people adopted others such as Yabu<u>d</u>urrwa. Fourth was the adoption and development of forms of Christian ritual. Since the indigenization of the Church in the late 1970s, Yolngu have included more and more Christian elements in mortuary ceremonies.[8]

New forms of music, especially rock-and-roll and film shows (and later videos), gradually replaced *manikay* as evening entertainment, and my impression is that the performance of major revelatory ceremonies declined at Milingimbi after the Second World War, especially after the move of numbers of people to new mainland settlements, which became the centre for these ceremonies. Circumcision and mortuary ceremonies continued unabated. It is possible that a smaller proportion of younger men

[7] I thank Dr J. Reser for the information about Magpie Geese songs.
[8] John Rudder pers. com.

and women directly participated in ceremonies in the mid-1970s than in the past, but it is remarkable just how robust interest in these rites has been. In spite of 50 years of the Methodist (later Uniting Church) mission, their forms were largely unchanged, and they were still regularly performed. It was not just a minority of the population who participated, although some of the younger men and women who would formerly have been active in the religious life took up roles in education and settlement administration and, in some cases, in Church and Land Council organizations as well as Northern Territory politics.

Conclusions

This chapter has explored the way in which Yolngu people took measures both to maintain and erase similarities and differences between ritual forms. It has described the 'forms' which groups owned and which connected them in multiple, cross-cutting strings, both through shared languages of practice and co-operation in performance. People guarded differences in contexts where they wanted to assert autonomy, minimized them when asserting control over the sacra and country of another group or trying to gain access to other resources. Yolngu were generally unwilling to impart information about groups to which they were not closely related, although this reluctance decreased with age and ritual knowledge. But to write of common forms would be an oversimplification: Yolngu also argued over the proper shape of religious practice.

Yolngu ceremonies were rather flexible frameworks into which participants could slot sundry bits in order to realize a ritual in its most basic shape, to greatly elaborate it, to accommodate different affiliations and perspectives, or to incorporate new elements. The purposes of different genres and types of Yolngu ceremony overlapped, and although the categories of ceremony were general throughout the region, and indeed beyond, groups did not possess a uniform suite of ceremonies, for the various types were not uniformly performed across the region. Furthermore, ceremonies of the same name had different styles in different places, belonging to local networks.

Those with the same group identity 'held' ritual forms and country in common, although factions in the dispute over succession to country described in Chapter 3 cross-cut group identity. Moreover, relatedness through distinct attributes such as language and elements of the three genres of ceremony seldom overlapped, for the Yolngu maximized options for relations.

Part II of the book examines religious secrecy separating men from women and old from young as well as one group from another. But before moving on it may be useful to discuss some of the implications of Part I.

The form given to groups, the character of country, a group's complement of ceremonies, and local styles were heterogeneous. People sometimes profoundly disagreed over group structure, leadership, identity of country, and the meaning of a ceremony or *maḏayin* element. Perspectives on ancestral events and the interpretation of ceremonies were relative to a network centred on a group or subgroup, the locus of the greatest degree of orthodoxy, and to the individual. More profoundly perhaps, ambiguity and indeterminacy lay at heart of the construction of ancestral worlds, so that typologies of beings were found wanting.

An inclusive hierarchy of types of group was inapplicable for a different reason: the metaphors around which Yolngu constructed groups were not those embedded in notions of clan and corporation. Yolngu constituted groups through relations to ancestors, ceremony, and country conceptualized in terms of metaphors drawn from plants and the body. They extended identity out from a centre and joined groups in strings following the logic of ancestral journey. Groups were not the bounded entities or fictive property-owning persons of corporation theory.

The variety and relativity of perspectives were made possible by shared, though ambiguous, negotiated and hence open languages and frameworks—of kinds and elements of ceremony, assumptions about powers of ancestors, beliefs about ritual action, and so on. Many of these frameworks extended like long fingers into the rather different social worlds of neighbouring peoples; others could be interpreted or revised to accommodate different ways of doing things. But there was no firm boundary between a Yolngu 'system' and other Aboriginal systems. Forms of practice were distributed in a mosaic pattern, and speakers of Yolngu languages joined with others in common forms of practice.

The anchoring forms of *maḏayin* were just that—'forms' or *shapes* of objects; shapes of sounds sung and exclaimed; painted and sculpted shapes on bodies, objects, and ground; shapes of dance movement. Yolngu referred to the relationship between shape and significance as *mayali'*, 'meaning'.[9] They and their neighbours thus subscribed to a common illusion, that a shape was inalienably connected with its significance, its ancestral origin and powers, and its human practitioners. Illusory because each saw a distinct inalienable web of connections in the same or similar forms. But people *subscribed* to the illusion in a process of 'misrecognition' (Bourdieu 1977: 168–9), a pervasive failure, or rather refusal, to make motives and conditions explicit.

[9] Stanner commented on shared shapes in ceremony whose significance was not made explicit by the Murinbata. Such forms, sequences among which seemed to have a form congruent with the general phases of the rite, were conventional signs which designated and indicated 'mysterious first things of the long ago, the mythical past', Stanner suggests (1963: 63). The important insight in this set of conjectures is that a form need have no particular interpretation to be significant in a more general way. That is not to be signifier in relation to signified, or symbol in relation to interpretant. In this case the Murinbata took the dances to be patterns laid down at the beginning of things and that must be followed, not the creation of people (Keen 1986*b*: 40).

PART II

Root and Branch: Secrecy, Age, and Gender

The image of the tree appeared earlier as a symbol of the relations of one group to many others and of the distinct identity of the group. The same image will serve to make the contrast between secret and public knowledge: a Milingimbi woman complained to a visitor recently that whereas Yolngu were teaching white people about their way of life, the white people were keeping knowledge to themselves, hidden like the roots of a tree. They should make the knowledge available to Yolngu, like the tree's branches. The roots beneath the ground represent hidden knowledge, roots exposed are knowledge revealed.[1]

[1] My thanks to Jo Erskine for this story.

The chapters of Part II are centred around descriptions of four ceremonies: a circumcision, Ma<u>d</u>ayin, a wake, and a washing purification ceremony. They consider the control of knowledge as competence, and what has often been discussed under the rubric of 'secrecy', treated as having to do with secret societies and collectives (Herdt 1990). Seen in this light Yolngu religious secrecy was about the separation of men and women and the authority of older and more knowledgeable people over younger. Together with revelatory rites and the encoding of secret knowledge in publicly disseminated forms, male initiation was integral to this process of separation and control. However, we will find an interplay between the control of access to religious knowledge on the basis of age and gender, and the guarding of religious forms and the transmission of information between groups and group-centred social networks.

Part II begins with the initiation of boys (Chapter 5), moving on to the revelation of secret religious knowledge to young people, especially young men (Chapter 6), and finally (Chapter 7) the encoding of secret religious significance within religious forms enacted in public.

5

Initiation and Gender

The division of esoteric from exoteric religious knowledge had to do with relations between old and young and between men and women, as well as between different groups. Initiation and revelation have to be taken together as a mode of practice separating boys from women, and as phases in the growing competence

Two Yirritja men dance 'Dog' at a Mandayala circumcision ceremony

especially of men. In the course of a Yolngu person's life the former division began formally with a boy's circumcision initiation, which marked the divergence of a boy's life from that of a girl, and the beginning of his admission into men's secret religious knowledge. This chapter will describe a *garma* circumcision ceremony in some detail. The description will serve as a vehicle to exemplify some of the themes of group identity and connectedness discussed in Part I, before the chapter goes on to consider the implications of the ceremony for gender separation and as a technique of control.

Initiation and Puberty Rites

The ritualization of a girl's puberty was not the occasion for a public ceremony.[2] Warner records that when a girl had her first menses the mother and other older women would paint her, place her inside a dwelling, which was not a special structure, and leave her there. She was supposed to remain in one place and use digging-sticks as crutches when she moved. This action represented the two Ancestral Sisters (such as the Djang'kawu) who walked with the aid of two digging-sticks, the older of whom was menstruating. When first menses were over, the girl could cohabit with her husband, although sexual intercourse sometimes took place before first menses (Warner 1937: 75–6, 98). I was informed that the seclusion of girls at first menses had been discontinued by the mid-1970s. A boy's circumcision (*dhapi*, 'foreskin'), on the other hand, was the occasion for a major public ritual which marked his passage from the category of uninitiated 'children' (*djamarrkuḻi*, sing. *yothu*), symbolically assimilated to that of females (*miyalk*), to that of 'bachelor' (*yawirriny'*, sing. *gaḏaku*), which was aligned with other categories of initiated males. These two broad divisions were fundamental to the control of religious knowledge (see H. Morphy 1977: 254).

[2] Yolngu male circumcision ceremonies are appropriately labelled 'initiation' ceremonies, since they marked the transition from childhood to bachelorhood, were mandatory, and usually involved more than one initiate. 'Puberty rites', on the other hand, are specific to individuals according to that typology (see Y. A. Cohen 1964; Allen 1967; La Fontaine 1985: 14). Maddock classifies Aboriginal ceremonies into initiatory and fertility ceremonies. Initiatory ceremonies have at their essence a conception of personal life as progression by stages along a path known in advance (1982: 125).

People could choose from a variety of genres for the purpose of initiation: they could use songs and dances of the *garma* ('public') genre for circumcision, or either of two regional ceremonies, or commission ceremonies such as a *malgarri* from neighbouring Rembarrnga people. Probably all Yolngu groups possessed the ritual resources to put on a circumcision ceremony of the *garma* genre, for they required only *manikay* or *bilma* songs and the related dances and designs. Not all groups had the requisite competence, however. In such cases members had to ask another group to perform on their behalf. The performers might incorporate some of the commissioning group's songs and *likanbuy* designs into the ceremony.

Men could put on a Mandayala or Djungguwan ceremony of the regional genre if their group had rights in these, or commission an owner to organize one on their behalf. Most Yolngu groups related the two circumcision ceremonies of the regional genre to stories about the Wagilak Sisters. These were Mandayala, a public ceremony, and Djungguwan, a revelatory ceremony with extensive esoteric dances for men. Initiands were of either moiety, usually between the ages of about 8 and 15 years. (Djungguwan was also a mortuary ceremony, performed as a memorial to a dead person, or in conjunction with a mortuary ceremony such as the *bukubut*, in which disinterred bones of the dead were distributed to close relatives.)

A Honeybee-Group Circumcision

The *garma* circumcision initiation ceremony had a relatively simple outline. I describe here a Honeybee group circumcision ceremony.

The Honeybee group of the Yirritja moiety held circumcision ceremonies at Milingimbi in three successive years during the years 1974–6, for the group was large and prolific. In the late dry season of 1976 Wurrpan, a leader of the group in his early seventies, organized the initiation of two of his own sons, his brother's son, and a boy of the Spider group whose father had died, and whom Honeybee people 'looked after'.[3] Members of the Spider group, which was a *ma:ri*

[3] Formerly older relatives sent (or were supposed to send) each candidate on a journey to visit distant relatives before the initiation ceremony. Although it did happen before a Mandayala performed on the nearby mainland in 1976, this did not occur at Milingimbi in the mid-1970s.

(MM/MMB) group of the Honeybee people, did not control their own ceremonies, for they were said to have no singers, the previous generation of men having died without teaching the younger men; Honeybee people 'looked after' their ceremonies.

The circumcision took place in Wurrpa<u>n</u>'s camp at Milingimbi over the course of two weekends in the dry season. The first painting-up of the initiands coincided with the final stage of a joint circumcision-ceremony put on by two Dhuwa-moiety groups, Black-capped Tern and Cloudy Water, closely related by marriage to the Honeybee group. This joint ceremony itself began on the final day of a Lorikeet-group circumcision, a group closely related to some of the Cloudy Water people. The three ceremonies thus formed a series involving four groups as the main participants, which had taken over a year to arrange. (Related groups frequently co-operated in this way.) The main participating groups in this case were closely related by kinship. Lorikeet group was *ma:ri* (MM/MMB) to Cloudy Water, and Cloudy Water was 'mother' to Honeybee. The latter group was 'mother' to Black-capped Tern, so that Cloudy Water was *ma:ri* (MM/MMB) to Black-capped Tern. These two groups usually co-operated to perform ceremonies, for they were 'con-joint' (*dha:manapanmirri*).

The main parts of *garma* circumcision ceremonies generally took place during the day. For several days before the main part of the ceremony began, men prepared the way for the main ceremony by singing *manikay* songs in the camp, especially during the late afternoon and evening. Beginning in the morning, several men would sit in a group under a shady tree in the camp to sing, and women sometimes danced some distance away, while other men, sitting with the singers, decorated the initiands with body-paint and *wa<u>n</u>a* cords. The daily programme was usually repeated two or three times; at Milingimbi it was scheduled for successive weekends in order to fit in with the week and work regime imposed by the missionaries.

The Honeybee ceremony was similar in form on all four days. On each day of the ceremony, men performed Honeybee- and Spider-group songs, seated in a small group on blankets under the tamarind trees in Wurrpa<u>n</u>'s camp near the beach. Other men who sat with the singers painted the initiands with red ochre and attached Yirritja-moiety *wa<u>n</u>a* cords, of possum fur on a hair-string base, to their arm bracelets. On all days but the first, the men danced with the boys to the cutting-place, the 'place of blood', simply a designated area of the clearing where the women later prepared a hay mattress for the circumcision operation, some 50 metres from where they made preparations. On the final day three men circumcised the boys.

As in the Mandayala and Djungguwan regional ceremonies, throughout the ceremony women relatives of the initiands circled round the group of men and initiands seated on the ground, calling in high voices, insulting the men, and snatching food, money, and tobacco from people sitting in the camp.

The songs, dances, and body-painting increased in elaboration from day to day. On the first day the Honeybee men simply sang while the sisters' sons (*waku*) and sisters' husbands (*dhuway*) of group members (i.e. the 'mother-holders' of the group) smeared the initiands, who were wearing cotton loin-cloths, with red ochre all over their bodies and attached the possum-fur *wana* cords, appropriate to the Yirritja moiety, to the boys' arms. Wurrpan then took a major role in the final stages of the Cloudy Water/Black-capped Tern (Dhuwa moiety) circumcision.

A week later, on the second day of the ceremony, the men sang Freshwater and Inland Saltwater songs which belonged particularly to the Hollow Log ceremony: Grass, Freshwater, Eel-tailed Catfish, Sulphur-crested Cockatoo, Wild Plum, and Red Goshawk. Inland Saltwater songs referred to lakes and tidal creeks which were either fresh or salt according to the state of the tide and the wet-season rains. Eel-tailed catfish (Manbiri), which tolerates both salt and fresh water, featured in these song-series.

While the Honeybee men sang, their *waku* and *dhuway*, who were of the Black-capped Tern and Cloudy Water groups, decorated the boys, and women circled around. Meanwhile, four mature men had their chests painted with Rock *likanbuy* designs relating to a Windbreak-group place on the mainland adjacent to Milingimbi. (Windbreak country is described in Chapter 3.) Thirty-three men prepared for the dances by painting Manbiri Catfish on each other's chests and thighs. All these designs were 'coarse' (*la:rr*) in style, and unrestricted.

When everybody was ready, the men danced in peregrinatory fashion to the 'place of blood', where a rock had been placed in the sand. There they danced up to the rock thrusting spears at it, as 'Catfish spearing the Rock', referring to a place on the adjacent mainland belonging to Windbreak group.

Next, the men executed various dances to *manikay* songs of the Forest series in an area by the 'place of blood'—Sulphur-crested Cockatoo, Goshawk, Murayana Ghost, Cyclone, Emu, and Water. They performed in a group facing the singers, who sat at the edge of the dance area. Women danced standing at the edge of the area, lifting their feet to the clapstick rhythm and making hand and arm movements appropriate to the topic. The songs and dances finished in the early evening.

On the next day, the third day of the ceremony, the singers sang through the Forest song-series, finishing with the Freshwater song, then the men danced 'Eel-tailed Catfish eaten by Shark'. Shark (*bul'manydji*) was a Dhuwa-moiety *wangarr* ancestor of the Black-capped Tern groups, whereas Manbiri Catfish was Yirritja. Black-capped Tern men danced as Shark, while the Honeybee men danced carrying a long string decorated with feathers representing the Catfish 'children' eaten by Shark. Many of the paintings on the Honeybee men represented half-eaten Catfish, related to their own country as well as Windbreak-group country on the mainland near Milingimbi, which was probably the conception country of some of them.

In the dance two senior Black-capped Tern men wore *l̲ikanbuy* designs of a Rock at the inland forest-country of the Honeybee group, the Rock which Catfish spears in the dance, and of Catfish Eggs. These two men danced up to the rock set in the ground by the 'place of blood', thrust their spears at it, and stared fiercely, identifying with the fierce Catfish *wangarr*.

During the course of the ceremony on these two days the singers and dancers travelled, figuratively, east from Windbreak-group country on the mainland near Milingimbi to the initiands' own group countries.

On the final day, the programme was repeated, but elaborated with men's dances, sometimes incorporating dances from the Mad̲ayin genre.

A week later, seated at the same place, the men painted the initiands' chests with *l̲ikanbuy* designs as they lay face up on blankets, to make them 'sacred' (*dhuyu*). The designs depicted the *rangga* sacred objects: each boy wore a design of his own group and *rangga*. Wurrpan̲ commented to me that the painted designs were intended to 'say goodbye' (the English expression used by the Yolngu to name the wake, and a gloss of *gupa-ngamathama*, 'make good') because the boys' penises would be hurt.

The initiands who were Wurrpan̲'s sons and brother's sons, all of the Honeybee group, wore designs representing Honeybee nest together with Jungle Fowl footprints, and Honeybee Nest with the Digging-Stick sacred object. The boy of Spider group wore a design of the Rock *wangarr* which united the coastal Sandfly groups, including Barramundi, War, Fish, Spider, and Black Duck peoples. He wore his own group's version of the design.

While some men went on with the painting the singers sang Swamp songs of the Up-river group (Honeybee and Lightning Snake groups), freshwater songs with the Spider group's tune, and Up-river-group

Forest songs, ending with Sunset. They went on to sing Frog, with the Fish-group tune, and Fire, ready for the Fire dance.

The dances on this occasion were from the public aspect of the Maḏayin ceremony. First, the men of both moieties danced Quail, a Yirritja *maḏayin* associated with Fire (quails run out of the grass when the grass-plains are burnt off), and then Fire, to the 'place of blood'. These dances were from the Maḏayin ceremony. During the Fire dance Wurrpaṉ called the *ḻikan* names of his group *wangarr*: Quail, Fire, and Honeybee.

At the 'place of blood' men circumcised the boys using razor blades (formerly a stone flake), as the boys' affines held them down, and a group of young men stood around them, forming a circular wall, ululating loudly to hide initiands' cries and oaths. Initiands incurred no shame if they showed emotion. The circumciser was a 'doctor' known to possess the skill, and did not have to be in any particular kin-relation to the initiand. During the circumcision Wurrpaṉ invoked the sacred names meaning 'Paperbark Tree falls over'. As was usual, the father of the initiands avoided this phase of the ceremony, sitting some way off.

Bachelors then carried the boys to a place away from the houses and shelters, where they applied soothing leaves to their wounds, and where they were placed under restrictions not to touch the ground with their hands, but to use small sticks when moving, and not to drink for twenty-four hours. Sometimes older men 'tricked' the boys at this stage by telling them that all the water in the universe was salty. The boys were *moḏa'*—circumcised but not yet healed.

When the wounds were healed several days later, the boys moved to a bachelors' house. At the completion of the ceremony the initiands took on the social status of 'young men' or 'bachelors' (*yawirriny'*, sing. *gurrmul*), and were no longer 'children' (*djamarrkuli*) or young boys (*gaḏaku*). As I said, rules of avoidance between brother and sister and between a boy and his *mukul rumaru* (MMBD) and *mumalkur* (MMMBD) came into effect before circumcision.

The prohibitions on drinking could be lifted using a small 'water' ceremony in which the initiates were again 'tricked'. An elder would solemnly get the boys to repeat 'sacred' words after him such as '*bing'ping*', which means literally a sucking noise with the lips, and give them water to drink from a bamboo container. A fire purification lifted remaining restrictions.

As we saw above (and in Chapter 4), a group's songs and other elements provided a stock upon which its members could draw in an infinitely varied way. Within the limits of what they could perform (or what they could get away with) people made choices about what elements to use, for a variety of reasons. It would not be possible to analyse here the symbolic significance of all the songs, designs, and dances of the ceremony described, even if they were known to me. First I will look back to the issue of group identity and connectedness discussed in the first part of the book, then go on to consider age and gender relations as aspects of participation as well as symbolism.

Group Identity and Relations

The ceremony marked the *ba:purru* identity of each initiand. Each wore a *ḻikanbuy* design depicting his own *rangga* sacred object and country, even though one group organized another group's ceremony. Both *ḻikan* names and *ḻikanbuy* designs represented in different ways the boy's country, *rangga* sacred objects, and *wangarr* ancestors. Indeed, the body-decoration of the circumcision initiands made them similar to the *rangga*, painted as they were with red ochre and *ḻikanbuy* designs, and adorned with *waṉa* cords. The *ḻikan* names which the leader called in a Honeybee-group circumcision were of the Paperbark Tree, the public and 'outside' counterpart of the *rangga*, falling into the waterhole, an image with manifold possible connotations and analogical equivalences, including the *wangarr* and *rangga* lying beneath the waters, and the incorporation of an individual in the domain of the *wangarr* through initiation or death. Thomson (1939*b*: 2) described the 'making good' and invocation as 'rites of aggregation'; each was an 'affirmation of the bond' between members of the group and the 'shade' of the *wangarr*. Certainly the acts put the initiand in contact with a manifestation of the *wangarr* in relation to the group country, and he was dressed in a similar way to the *rangga*. But the Honeybee leader interpreted the act as a form of ritual death: the painting-up was to 'say goodbye', an expression used to describe the purpose of a mortuary ceremony. Moreover, in being decorated like the *rangga*, which represented the bone and flesh of the *wangarr*, the initiand was treated in the same way as a corpse (or coffin) in mortuary ceremonies (see Chapter 7). The image of the falling Paperbark Tree also expressed this ritual death.

Many ceremonies associated new group-members and new life with the spirits of the dead. People used the same songs, designs, and dances of the *garma* genre for circumcision, mortuary, and other purposes, and the two functions were embedded in the Djungguwan ceremony (See H. Morphy 1984). The dances which the Honeybee leaders selected from the Ma<u>d</u>ayin ceremony for the circumcision ceremony described above combined expressions of fertility with the return of spirits of the dead. Such conceptions gave an extra dimension to the continuity of a group through identifying the dead with the yet to be born, a continuity expressed in the *ba:purru* concept, which also implies connectedness between groups through a common *wangarr* ancestor as well as relations between the living and the dead, and continuity of the population through marriage (Chapter 2).

Connections

Performances of *garma* circumcision ceremonies expressed relations among groups as well as group identity. In the *garma* circumcision ceremonies the songs, designs, and dances were those primarily of the initiands' groups. However, ceremonies always implied connections with other groups—of the same moiety in the linking of the Honeybee circumcision initiands' distinct groups and in the Rock dance, and groups of the opposite moiety in the Shark and Catfish dance. Congeries of groups of the same moiety owned *garma* songs and their components, as we saw in the last chapter, and members of various groups often sang together. Many of a group's songs and dances referred specifically to a group of the opposite moiety related through marriage, again as in the example of the Shark and Catfish dance, and men and women of both moieties participated together. Furthermore, people frequently co-ordinated the circumcision ceremonies of closely related groups in order to run consecutively or to overlap.

People's choices of song, dance, and design depended upon the relationships between co-operating groups and the identity of the initiand (or dead person in mortuary ceremonies). The Honeybee ceremony described above moved in dance and design from near Milingimbi to the country of one of the initiands. Then on the last day the singers created a journey from one initiand's group-country to another, through Spear-grass, War, Fish, and Honeybee countries, linking them. They were able to do so because the Honeybee group possessed several song-series belonging also to various other groups, and because the component songs referred to *ma<u>d</u>ayin* relating primarily, although not exclusively, to the countries of several groups in the way outlined in the last chapter (see also H. Morphy 1984).

The Honeybee men sang Forest songs of the Spear-grass group, whose country lay on the middle reaches of the Woolen River, and which was *ma:ri* (MM/MMB) to Honeybee group. The Freshwater songs used in the ceremony related especially to the Spider group, its country called 'Place of the all-in fight', and a nearby swamp, both at Langarra. Honeybee men used the Spear-grass-group tunes for these, for Spear-grass and Spider groups had the 'same song'. Within the Forest song-series, Pigeon and Kangaroo songs belonged to Fish-group country; they were that group's *wangarr* ancestors. Honeybee and Fire songs belonged to one Honeybee-group country (its coastal plains country), the Macassan songs to two Honeybee countries (its island country and forest place), and the Hollow Log songs to the coastal plains and forest countries. The Sea series was associated with Lightning Snake group's island country. (These places were separated by countries belonging to other groups.)

Through these associations the Honeybee singers traced a conceptual journey in the following way:

> Spear-grass group's swamp songs with Spear-grass tune, related to Spear-grass country;
>
> Spider-group Freshwater songs, with Spider tune, relating to Spider-group country (that of one of the initiands);
>
> Fish-group Frog song, with Fish-group tune, relating to Fish-group country;
>
> Honeybee-group Fire song, with Up-river-group tune, relating to its coastal plains country;

The final dances in the ceremony were of the two big *wangarr*, Fire and Honeybee, which belong to Honeybee country also at its coastal plains place.

The dance representing Shark *wangarr* of Black-capped Tern groups eating Catfish of Honeybee group, appears to have been a metaphor of the relations between these groups, who were in a 'mother–child' (*nga<u>nd</u>i–waku*) relation. That is, most Honeybee young women were married to, or 'eaten' as Yolngu would say, by Black-capped Tern men, and most Black-capped Tern men married Honeybee women. As well, the eating of Catfish was an appropriate metaphor of circumcision, especially since, as we see in the symbolism of the Maḏayin ceremony (Chapter 6), certain fish connoted conception spirits and reproductive functions. But the connection between Honeybee and Spider groups also had a political dimension.

Looking After

Group identity was iterated in the ceremony, but leadership was not confined to the group; one group and its leader or leaders sometimes controlled the religious affairs of another, expressed by the word *dja:ga*, 'look after'. People 'looked after' their own and other people's children, the countries of absent groups, and the countries and ceremonies of memberless groups and groups without knowledgeable leaders (cf. Myers 1980, 1986). During the 1970s the large Honeybee group, whose leader was styled 'the headman' of the mission by the missionaries, dominated the religious life of three other Yirritja-moiety groups in the context of circumcision and other *garma* ceremonies, including that of one of the initiands.

Wurrpan̲, the Honeybee-group leader claimed that he looked after the Spider-group *maḏayin*. These people lived mainly at Milingimbi, several spending time also at Maningrida, Langarra, and Galiwin'ku. None of the men was over the age of about 35, although there were several older women of the group. Wurrpan̲ sang a few Spider-group songs among his own group songs in the circumcision ceremony described above, as one of the initiands was of the Spider group. He told me that there were many of these songs that he never sang, and indeed I never heard any young Spider men sing, and so they seemed never to have the opportunity to learn the songs of their own group. Wurrpan̲ had rights in Spider-group *garma* ceremonies as well as those of Lightning Snake and Barramundi groups. This was because Barramundi was his actual mother's mother's group, Lightning Group joined with Honeybee as Up-country people, and the Spider group was in a more remote *ma:ri* (MM/MMB) relation to him.

Honeybee men also dominated the religious life of Lightning Snake people at Milingimbi during that period. Along with the Fish group, the Honeybee and Lightning Snake groups were both Up-country people, a name also given to their tongue. In kinship terms the two groups lay in a relation of *ma:ri–gutharra* (MM–

wDC/ZDC); they had some distinct *wangarr* ancestors and others in common; and they shared many songs, public ceremonies, and types of *rangga* sacred object.

From the ethnographic writings of Warner, Thomson, and Webb, it seems that the Lightning Snake group had been independent and possibly even the stronger of the two before mid-century (Warner 1937: 44; Webb 1933: 409; Thomson unpublished field-notes). By the mid-1970s the relative position of the two groups was asymmetrical. The Lightning Snake men conducted no ceremonies of their own, and they had no power-men, so that they were largely dependent on the Honeybee group to conduct their ceremonies, although they did paint their own group designs, for example as bark-paintings. Lightning Snake men explained to me that their fathers had died young without teaching them the songs; the Honeybee leader (Wurrpa<u>n</u>'s older brother) was also their leader; their country was his place 'of the water' (i.e. of spirit conception). If they wanted a Lightning Snake sand-sculpture or ceremony they had to ask the Honeybee leader, who was the 'boss' of their country, and he customarily agreed. The Lightning Snake group did have two male leaders; the older man lived at Elcho Island, and reputedly did not know the group's songs, while the younger man, a man of about 38 who lived at Milingimbi, was at that time something of an outsider, partly because of his relative youth, but also because he had been closely involved with the missionaries and had contracted a Christian marriage.

The Honeybee-group leader, and other male members, asserted that only he and two of his brothers were knowledgeable of the Lightning Snake Sea songs, which were 'truly' Up-river-group songs. (They attributed ownership of the Forest song-series that they used to the Spear-grass group.) Wurrpa<u>n</u> told me that they never performed them because the young people did not know them. I witnessed only one performance of the Sea songs, and that was by the leader of one Honeybee lineage who lived at Nangalala, where no members of the Lightning Snake group lived. It is possible that the Honeybee-group leaders tried to ensure the continued incompetence of most males of the Lightning Snake group by not performing the songs regularly and in their presence. Yolngu people learned by participating in performances over a long period, gradually taking a more and more prominent role as they got older, and as older people became less active or died. Whether intended or not, the effect seemed to be to reproduce incompetence.

The Lightning Snake group was not entirely dependent upon the Honeybee leaders, however. In 1976 a Lightning Snake man and his young Honeybee-group *ma:ri* (MMBSS) asked Hook-spear people of a remote outstation to perform a circumcision ceremony for their sons. Members of three other Yirritja groups (Black Duck, Whale, and Crocodile), who were from settlements and outstations to the east, joined with the Hook-spear to perform the Sea songs and dances, which the Lightning Snake also possessed but did not sing. Honeybee-group members joined in for the invocation and chorus. The Lightning Snake men danced with the Hook-spear men, whereas they usually danced with Honeybee men at Milingimbi.[4] This was the position in 1976. By 1990, however, Wurrpan̠ and his older brother had been dead for a decade; the Honeybee and Black-capped Tern groups no longer had control of settlement affairs; and younger Lightning Snake men had revived their songs, incorporating songs of the Black Duck, Hook-spear, and Whale groups.[5]

Thus relations of age and competence had implications for group relations and autonomy. Another dimension of social relations enacted and symbolized in the ceremony was that between men and women, and these were major dimensions of roles in the ceremony, but as the last chapter would lead us to expect, kin relations to individuals and between groups in part determined participation in the ceremony.

Participation in the Ceremony

Who participated in a ceremony like the one described here, and what they did, depended on a variety of factors, including

[4] Some groups retained the right to call the invocation and chorus, yet did not dominate the religious lives of other groups. This conclusion is supported by Warner's and Thomson's data, for the Yirritja Mad̠ayin described by Warner (1937: 356–70) was of the Black Duck and Whale groups, judging by the dances recorded by Warner. At that time the dominant leader at the mission was of the Black Duck. The ceremony which Thomson (1939*a*) observed was organized by a leader of the Dingo group, who was dominant on the mainland, and who had the reputation of being very fierce. His group was linked by marriage to the then more prominent Stringybark group.

[5] I am grateful to Franca Tamisari for this up-date.

patrifilial group-identity, kin relations to key participants, gender, age, and competence. Those involved in an affair of this kind comprised a focused network rather than corporate groups. Just who went depended on the position in the social fabric of one or more key individuals—the initiand, the deceased, the main organizer, and so on—as well as on situational factors such as disputes and sickness. Some key roles and responsibilities inhered in specific kin-relations, especially matrifilial relations. Then, a matrifilial relation such as *waku* (wC/ZC) to a central figure in the ceremony, such as the initiand, was expanded out to a kin relation between groups. The appropriate general model, then, is the kindred, modified by relations through common *ba:purru* identity, kin relations among groups, age, and gender. Yolngu described the kindred network with the English phrase, 'all the families', and the group relations as 'all the *ba:purru*'. Other aspects of participation which should be stressed are the separation of male and female roles, and the growth of competence with age.

Most of the participants in the Honeybee circumcision ceremony described above were members of the initiands' groups, and *waku* (wC/ZC) of these groups. Members of three other closely related groups—Lightning Snake, Dilly-bag, and Thunder—also came in some force. The senior men of the Honeybee initiands' own *waku* (wC/ZC) group took the most prominent roles in singing, dancing, and painting, so contributing to an initiation for a person of their 'mother' group. The male *waku* helped to paint the boys up, to play the drone-pipe for the *manikay* songs, and to perform their own and their 'mother' dances. The leading dancer in the Rock dance, who wore the *l̲ikanbuy* designs, was senior *waku* or 'mother-holder' for the Honeybee group. Ga:murung, the son of Wandhuma, the Cloudy Water leader, and who was a *waku* of the Spider initiand, painted the boy's chest with a Spider *l̲ikanbuy* design. The Honeybee leader, Wurrpan̲, was Ga:murung's *waku* (in this case, father's sister's husband). He taught his young *ngapipi* how to paint the design, which was Wurrpan̲'s 'mother'.

Those who performed dances of Black-capped Tern and Honeybee groups included men of other groups, each of whom took part because the dances were 'the same' as the person's own group's dances, or because they were their 'mother', *ma:ri* (MM/MMB) or *waku* (MMM) dances.

Gender roles in *garma* ceremonies were quite distinct, consistent with the purpose of the ceremony in 'making a young man'. Men sang *manikay*, painted designs on bodies and objects, performed the circumcision operation, and called the invocation. Men and boys danced in the arena in front of singers, whereas women and girls danced on the spot near the singers and at the side of the dance arena, or took part with men and boys in peregrinatory dances such as Kingfisher, danced as a short journey, pausing from time to time to perform some action (see Chapters 6 and 7). Men did not dance on the spot like women, but I have seen one woman take it upon herself to perform solo dances usually performed by men. Women did not participate in singing *manikay* or in painting *l̲ikanbuy* designs, and did not join the group of male painters of an initiand or male dancers. In the circumcision they kept away from the painting and the cutting-place, but as we saw, some women ritually attacked the men and had licence to take other people's goods.

Differences in age and knowledge shaped people's roles too. Young bachelors had a limited role in *garma* ceremonies: they were expected to dance and form a screen around circumcision initiands. The bachelors and young married men of some groups (especially those of Galiwin'ku), took a prominent part in singing, but young Milingimbi men sang quietly when they were with the older men, chins down on to their chests, coming into their own only late at night when the old men had retired. They would be 'taking the place' of the older people as they grew in age and competence, 'becoming large'.

Mature men deferred to the 'big men', their grey-headed elders, and took something of an 'executive' role requiring both authority and energy. They seldom acted as 'power-men' in the *garma* ceremonies or Ma̲dayin, but one or two at Milingimbi were singers of some renown, and took a prominent role in the technical aspects of performance, leading the singing, and directing a dance by providing the guiding clapstick rhythm. They and some active old men danced solo at circumcisions or wakes, which young men seldom did. Older men with ritual competence directed or helped execute the *l̲ikanbuy* designs, painted on the initiands' or dancers' bodies (or on a coffin in mortuary ceremonies). 'Direction' consisted of keeping an eye on the way a painting was going and suggesting changes when appropriate, and being asked for final ratification of a completed effort. Still, I did once witness an older man being corrected by a young man who said that his painting was misaligned. The older man rubbed it off and started again.

The older Honeybee, Black-capped Tern, and Cloudy Water men led the singing. Wurrpan̲, the Honeybee leader, was also the 'power-man' (*d̲alkarramirri*) for the circumcision, calling the *l̲ikan* names. The older Honeybee, Black-capped Tern, and Cloudy Water men, and one Lightning Snake man (of Up-country group along with Honeybee people and Fish group) directed or executed the *l̲ikanbuy* designs, as well as the dances.

Gender Separation

There was clearly a strong relationship between Yolngu circumcision initiation, the constitution of patrifilial groups, and the separation of males and females, for men took boys into the category of circumcised males at initiation, while songs, dances, and designs iterated their group identity.

Themes of detachment from women and attachment to the male category were prevalent in circumcision ceremonies of both genres. Women ritually protested and attacked the men in *garma* circumcisions and the regional varieties. Man̲d̲ayala dramatically enacted this separation: the initiands were dressed as women with breast girdles, and men teasingly captured the boys in an episode in which they ran towards the adult women. Stringybark people and their relatives interpreted Man̲d̲ayala as an enactment of the central episode in the Wagilak myth in which the Olive Python (in the Stringybark-group version) made a flood, encircled the women's camp as they sang to try to stop the rain, and swallowed and regurgitated the women. In the ceremony the line of dancing men was the Snake 'swallowing' the initiands. This can be interpreted as a symbol of separation from women and incorporation into male categories (Hiatt 1975: 155), although other interpretations have been made.[6] The similarities of prohibitions during first menses and after circumcision also identified circumcision initiands with females, and their blood-loss identified them in the Man̲d̲ayala with the Wagilak Sisters. Rules came into force after initiation which separated boys from the hearth, the domain of women in Yolngu discourse.

[6] For varieties of anthropological interpretation see Warner (1937: 387); Roheim (1925, 1945); Bettelheim (1954); Hiatt (1971: 88); Taylor (1990); and Alland (1988).

Garma circumcision ceremonies evinced images similar to the Snake swallowing the initiands, but they related more to identification with the *wangarr* ancestor and to group identity. For example, a Black-capped Tern circumcision which I witnessed included the Black-capped Tern (*djarrak*) dance, in which the boys were encircled by a string attached to a carving of the bird's head. A string of Dhuwa-moiety groups shared the Black-capped Tern *wangarr*, of the sea songs and dances. Honeybee men often danced in the circumcision as adult Emus looking after the chicks (the initiands) huddled in a nest. (It is male emus that incubate the eggs.) The implication seems to be that the boys entered the domain of adult men, and in painting the boys with *likanbuy* designs, men put them in a close relationship with images of *rangga* sacred objects in a way not available to girls and women until death. Women, however, were identified with *wangarr* ancestors and more obliquely with *rangga* sacred objects, through the designs they wore in public aspects of ceremonies, and identification as female ancestors in dance.

Thus at circumcision boys began the journey into the men's domain, a journey very different from that typical for girls and women.

Marked gender-differentiation occurred early in Yolngu childhood. After a prolonged infancy, in which babies experienced intense social and physical contact with adults and older children, boys gradually joined the play-group of other boys, but often sat with their fathers at public ceremonies, and later associated with other older men, returning to the hearth generally only to eat and sleep. Adults asserted little control of the boy's group. Girls, however, remained more closely linked to their mothers, learned to cook, helped with young children and with collecting firewood, attended rituals, and learned women's dance movements. These aspects of socialization were very similar at Maningrida among Gidjingali speakers and Yolngu (Hamilton 1981*b*: 17, 83, 104; Stephen Harris 1984: 21–7).[7]

The life cycle of boys and girls diverged most at about puberty.

[7] In some households a mother would keep her eldest daughter away from school in order to help with a baby sibling, food preparation, and cleaning. A young wife in a polygynous household was often the same age as a daughter of a senior wife, and would obey her older co-wives and take little responsibility. (I am grateful to Sue Harris for this information.)

Men seized boys for circumcision initiation, after which the boys moved to a bachelors' camp, although they visited their families to eat and socialize. At semiotic and normative levels, then, boys were detached from women and incorporated into the category of adult males as bachelors. Boys had to act with restraint in the presence of their sisters, women in the 'mother-in-law' category, and the mothers of these women. They gradually took on male roles which in the past had been primarily those of hunters and fighters. Boys were not only admitted to competence in *garma* ceremonies, as were girls, but to exclusive men's rites, to men's secret knowledge, and to related practices such as painting (Hamilton 1981*a*: 18; Warner 1937: 77).

Girls used to begin to cohabit with husbands at about the age of 13, a few years after boys of the same age were circumcised, although this stage was delayed by a few years among those attending school. A girl's circumstances changed as she took on a formal relationship with a man, perhaps an asymmetrical one, and with co-wives. However, girls could expect to remain with their parents for a period even after marriage, while the husband performed bride-service, living with his wife's parents and providing them with game and/or money and other gifts. Education in women's ceremonies and other aspects of women's roles continued without the marked break that circumcision made in a boy's life. Girls learned public songs and ceremonies, and girls of some groups learned the secret songs and dances of Gunapipi.

Yolngu bachelors (*yawirriny'*) had few responsibilities and a good deal of personal autonomy outside ritual contexts, and were extremely mobile. They could make demands on mothers, sisters, and other relatives for food and money, but contributed little in return (cf. Cowlishaw 1978). Their main obligations were to make gifts to potential in-laws in the hope of receiving a promised wife, and to learn the religious traditions.

Schooling had not appreciably changed this pattern in the mid-1970s. According to genealogies collected by Mrs Sue Harris and myself, among the twenty mothers who were born after 1950, the average age at the birth of the first child was 15.6 years. Sixteen of the twenty bore their first child when under sixteen years. Because they were in the middle of their reproductive lives, this average was two to three years lower than senior cohorts (Lalley 1988: 28). Nevertheless, girls were beginning to live with husbands two or three years later than in earlier years, and girls were often reluctant to go to promised husbands, especially where these were men much older than themselves. Thus before and even after schooling was intensified from the 1950s at Milingimbi, there was a rapid transition from childhood to marriage for girls, and to bachelorhood for boys.

Through participation in the ceremonies from an early age, young men and women were inducted into a way of life which required their conduct to be more constrained, and in which, at least in the context of ceremonies, they came under the direct authority of their elders. For young men especially, initiation was not just part of the process of separation and incorporation into the categories of older males; it was also a technique of socialization in a broader process of control, and an aspect of the religious *dispositif*.

Initiation and Control

It is possible, indeed likely, that Yolngu circumcision ceremonies, together with revelatory rites, were important instruments of socialization, and included techniques which had a profound effect on the personalities of initiands. The effects which Yolngu men attributed to such rites was to make difficult or troublesome young men 'quiet', as well as to make them knowledgeable. They described revelatory rites as a kind of 'Aboriginal school'.

Myers (1980: 207) asserts that Pintubi ordeals produced, over time, 'real personality changes'. During initiation and subsequent rites, elders secluded Pintubi boys, instructed them, and subjected them to physical ordeals, including tooth evulsion, nose-piercing, circumcision, subincision, fire ordeals, fingernail pulling, beatings, threats, and hunger. Yolngu initiation had a rather less drastic character than the Pintubi variety. I do not have direct evidence of personality change, however it may be that the regime of initiation and revelatory rites produced particular effects on the behavioural dispositions of individuals in given contexts. Perhaps the combination of pain and other privations inflicted by close relatives as a condition of fellowship, especially in conjunction with other rituals which reinforced their effects as 'rites of intensification', tended to subdue egotistical desires and dispose boys to associate with male peers and older men rather than with mothers (Chapple and Coon 1947; P. Spencer 1970: 147).[8]

There is strong negative evidence supporting this general hypothesis. Murinbata elders complained to Stanner about 'a great change of social personality among boys and young men' after the suppression of the major revelatory rite by mission authorities, as well as the irrelevance of circumcision initiation due to infant circumcision in hospital. These changes 'had produced a generation as much at odds with their own elders as with the missionaries' (Stanner 1963: 150).

We can see this control as serving the interests of the community as a whole in guarding against anarchic behaviour on the part of young males. However, the whole system of religious knowledge, including initiation, served the particular interests of older Yolngu men; without that religious backing young men were likely to refuse to support the interests of older men which entailed costs to themselves, in particular the late marriage which polygyny entailed.[9] It is certain that the ceremonies brought about normative

[8] A number of theories have been proposed about the relation of initiation to socialization, the conditions in which initiation rites are found, and their effects. Certain varieties of Freudian theory are concerned with gender identity and the prevention of incest as well as parental authority (Whiting, Kluckhohn, and Anthony 1958; Norbeck, Walker, and Cohen 1962; Paige and Paige 1981: 164–5). A second theory, applied to certain societies of the Papua New Guinea highlands, describes the use of teaching, rhetoric, and rapid behavioural conditioning in initiation to effect 'radical resocialization' in the form of abrupt maternal detachment, separation, and individuation, and 'routinized masculinization' (Burton and Whiting 1961; Herdt 1982). Harrington has demonstrated a cross-cultural correlation between circumcision and strong differentiation in the socialization of the sexes, whereas superincision tends to be associated with convergence in the treatment of the sexes in childhood (Harrington 1968). A second body of theories about the socialization effects of initiation has to do with the enhancement of relations of authority and solidarity. One theory, drawing in part on theories of so-called brainwashing, holds that the experience of a 'traumatic' event, or an induced 'destabilized state', 'critical period', or 'peak experience', enhances suggestibility, but resultant changes in attitude or beliefs remain after the cause of stress is removed. (See Y. A. Cohen 1964; Herdt 1982; Kiefer and Cowan 1979; Morinis 1985; Sargant 1957; P. Spencer 1965, 1970.) A second theory stresses the experience of submission to coercion (P. Spencer 1970: 144). A third theory suggests that the experience especially of pain brings about attachment between initiands and initiators, and among co-initiates (Y. A. Cohen 1964; Schlegel and Barry 1980; Morinis 1985).

[9] A Darwin Aboriginal fringe-camp studied by Sansom was not embedded in an environment in which young men had to participate in initiation and revelatory rites. Here, young men were able to prevent an older man of the camp from marrying a young promised girl (Sansom 1980: 254–6).

effects. Permanent mutilation marked the changed status of initiands, who had to adhere to recently imposed norms of avoidance as well as dietary restrictions. As bachelors they slept apart from women and girls, and were admitted, when invited, into more secret aspects of ceremonies.

This chapter has considered the separation of boys from women through circumcision initiation and its symbols, but continuing the theme of group identity and differentiation of Part I. It has suggested a link between gender separation and age-related authority created in ritual. The next chapter describes a performance of the Ma<u>d</u>ayin ceremony, a revelatory rite which continued the process in which older men revealed secret knowledge to younger men and which separated men and women, as secret ground was distant from the hearth.

6

Revelation

Revelatory Rites

After circumcision initiation a young man was eligible to be shown access to esoteric aspects of Yolngu revelatory rites, in which older people revealed secret dances, songs, designs, and objects to younger people. These ceremonies did not mark a formal change in status except for the gain of a title indicating knowledge of the ceremony. Revelatory rites were classified as 'of the *ma<u>d</u>ayin* side', meaning that they included secret aspects in which men revealed sacred objects. They included the Ma<u>d</u>ayin (or Nga:rra) ceremony and the regional ceremonies—Gunapipi, Djungguwan, Ngulmarrk, and Yabu<u>d</u>urrwa. Of these, only Ma<u>d</u>ayin and Djungguwan had been regularly performed throughout the region in the mid-

Dhuwa women dance 'Kingfisher'

1970s. These rites all contrasted a men's secret dance-ground, where sacred objects and dances were displayed, with the public camp, where men and women danced or where, in Gunapipi, women danced to the exclusion of men. Mandayala could marginally be included as a revelatory rite, for the men decorated their bodies in secret, but apart from esoteric interpretations of songs and designs which were features also of public ceremonies, there was no more to the secret aspect of Mandayala than that.

Except for Djungguwan, in which the young circumcision initiands themselves saw the secret dances and objects before the operation, circumcision initiation was a prerequisite for young men to enter the secret aspects of all the esoteric rites. However, the rites did not form a graded hierarchy which young men or women should move through in a definite order, as Howard Morphy (1991: 84) points out. Yolngu did not perform the ceremonies at strict intervals, nor did any set group perform each one; rather they occurred *ad hoc*. Group membership or close kin-relationship to a group determined in part what ceremonies a person participated in.

Nevertheless, the revelatory ceremonies continued the series of steps towards full adult responsibility for a young man. In order to begin living with a promised wife, a man was supposed to have participated in secret aspects of Madayin, and young men had to become familiar with the equivalents in ceremonies of major species before they could hunt them. It may be that Yolngu men, like the eastern Gunwinjgu, had been able to participate fully in hunting only after ritual grading, for the latter were not allowed access to hunting-weapons until familiar with the Gunapipi ceremony (four to five performances), although they could use fishing-equipment (Peterson pers. com.; Altman 1987).

Secrecy in revelatory rites was created through spatial separation. People guarded and avoided boundaries, creating barriers to participation, sight, and hearing, depending on the distance of separation. What was actually done was secret rather than the meaning of a code, and in Gunapipi and Djungguwan an important contrast was drawn between public versions of what occurred in the ritual and what novices discovered. Consequently men enjoined young participants in the secret aspects of ceremonies, under threat of ensorcellment or violence, not to reveal in words what they had seen and done. Through the use of special sounds, people maintained a limited degree of communication between inhabitants of the exclusive and the public domains in Gunapipi or between men's and women's exclusive domains, but adults propagated false information about the origin of these sounds.

The metaphor of 'inside/outside' marked this separation, but Yolngu also distinguished levels of esoteric knowledge with other pairs of antonyms:

djinaga	inside	——	*warrangul*	outside
maḏayin	sacred	——	*garma*	public, in the camp
dhuyu	taboo, sacred	——	*yarangu*	ordinary, profane

The metaphor 'inside/outside' was more literal in two cases. A men's 'inside' ceremony ground, which incorporated a secret tree, stood away from the 'outside' of the 'public' camp, where a public tree stood. The space of the 'inside' ground was divided again between the dance ground and the 'inside' space of a bough-shade housing the sacred objects until men brought them out in dance. In another more literal application, men often stored secret objects 'inside' the ground or secreted them under deep waters. 'Inside/outside' was more metaphorical in the case of painted, sculpted, danced, and sung forms, which were subject to various levels of exegesis. Here, 'outside' could apply to a form which men were willing to perform or to display in a public forum, or to an interpretation which they were willing to give, but with the potential for a more profound exegesis. It does not seem to have been the case that there was a fixed 'outside' story or interpretation; rather, interpretations lay in a relation of esoteric to exoteric along a continuum, and these relations could change (H. Morphy 1991: 78–82).[1]

The Maḏayin Ceremony: Secret Ground and Hearth

This chapter describes a performance of one kind of revelatory rite, the Maḏayin or Nga:rra ceremony. Along with the other revelatory rites, Maḏayin had a focal place within a wider body of procedures centred on the control of religious knowledge. The

[1] Morphy also points out that the 'inside' name or meaning of something was not always secret; it might be the name of a snake as ancestor rather than the common name, or the underlying motive for an action as against the ostensible one.

Madayin ceremony as a type had a currency far beyond the region in which Yolngu languages were spoken: among Gunwinggu people living during the 1950s and 1960s at Oenpelli some 200 kilometres west of Milingimbi (R. M. and C. H. Berndt 1970: 135–8), and still further west among 'Kulunglutchi' and Ga:gudju of the East Alligator River at the beginning of the century (B. Spencer 1914: 150–2).[2] People living in the area in between, including eastern Gunwinjgu people and Gidjingali speakers, also had this category of ceremony in their repertoire (Taylor 1987: 146; Hiatt 1965: 67–8). Two hundred and thirty kilometres to the south of Milingimbi, Rembarrnga people living at Mainoru commissioned Yolngu men to help them with a performance of Madayin in 1949 (Elkin 1972: 47–97).

Thus, like the other revelatory rites (Gunapipi, Djungguwan, Ngulmarrk or Wubarr, and Yabudurrwa), Madayin was prevalent over a wide region as a ceremony type. As a type it had certain diacritical features, but the flexibility of the ceremony's form enabled people to co-operate in its enactment in interpenetrating networks, and to sustain its legitimacy as an ancestral re-enactment, each group interpreting the significance of the ceremony in its own way. In this way people could sustain the kind of group identity and differentiation described in Part I, while co-operating with each other. In the performance recorded in this chapter a string of groups from Gove Peninsula to Cape Stewart was represented, centring on groups represented in the community where the organizers lived. All groups of the same moiety were conceived as the offspring of a common creator and as kin in relations of mother's mother to daughter's child (*ma:ri–gutharra*). It seems to have been common practice to perform the ceremonies of the two moieties together (Elkin 1972; R. M. and C. H. Berndt 1970), extending its participation to a scope similar to that of the regional ceremonies.

The Madayin ceremony had the explicit purpose of revealing secret dances and objects to young men, and to give them the opportunity of taking on roles of greater responsibility. Two men commented to me that the aim of the ceremony was to discipline young men, continuing the process of control begun with initiation.

[2] The Ga:gudju also possessed objects similar to Yolngu *rangga* which B. Spencer (1914: 189) refers to as 'muraian' (equivalent to *madayin*) objects.

People performed the ceremony for other reasons as well—to say goodbye to someone near death, or to commemorate a dead leader. One of its themes, then, was the passage of responsibility from one generation to another who were 'taking the place' of the former. Involvement in the secret part of the ceremony was an important step in a young man's religious life, which began with circumcision initiation. In this way the lives of young men and young women were very distinct. Other themes in the ceremony refer us back to myth and country—the creative acts of the *wangarr* ancestors, conditions for continued reproduction, and relations of reproduction among groups of the same moiety.[3]

Variation in the content of the ceremony reflected distinctions among the moieties and groups. However, the fundamental and pervasive division was that of women from men, who exercised greater control over access to hidden performance and significance. The spatial organization of the ceremony divided the domain of the hearth and of women from that of men and the secret 'inside' ground. By means of the threat of force and ensorcellment men guarded the threshold to the secret ground where they kept highly valued sacred objects and performed secret dances. Admission to this space and to the roles played within it was formal and incremental. Cloudy Water people organized the performance described here.

The Cloudy Water Madayin

The Context

The Cloudy Water group had performed a Madayin ceremony at Milingimbi in 1970, and joined the Sunrise group for a performance at Yirrkala, 220 kilometres to the east, in 1972 (Dunlop 1984). The Cloudy Water organizers timed the ceremony to begin at the very end of the dry season, when the thunder-clouds were beginning to build up, but before the onset of the north-west monsoon.

During 1975 people of the D lineage of Cloudy Water group (see Chapter 2) living at Milingimbi and Langarra were planning a Madayin

[3] See Keen (1991) for a detailed analysis of aspects of symbolism in the Madayin ceremony.

ceremony, but the group leader died at Galiwin'ku, some 70 kilometres east of Milingimbi Mission. According to some men, the leadership passed to the (reluctant) shoulders of Wandhuma, now the oldest man of the group, and the ceremony was postponed. Members of the group did not all agree, however, that Wandhuma was the man to replace the former leader. His main support, or pressure, came from his own lineage (D).

The deceased leader, who had lived at Galiwin'ku, was a member of L lineage. Most of B lineage lived at that settlement also, whereas some people of G and D lineages lived permanently at Milingimbi, and others moved between Galiwin'ku and Milingimbi. As the oldest man of the group and sole male representative of his genealogical generation, Wandhuma considered himself to be the new leader, although he characterized his sister Ma:tjarra as the 'true leader', for she was the first-born of the group. He was insecure in the role, comparing himself unfavourably with Buwaṯa, a leader of the Thunder group. He knew the songs well enough, he told me, but Buwaṯa and the deceased leader were really big men. One or two men of Wandhuma's lineage, as well as people of B lineage, thought that Ngaṯili, a man of B lineage, should be leader. However, Ngaṯili admitted to me that Wandhuma was now the leader and 'boss' of the *rangga*, although this admission may have been due to his knowledge that I was working with Wandhuma.

At the end of the following year Wandhuma and his 'sons' of his lineage, including his actual son, decided to put on a ceremony to 'say goodbye' to his older sister Ma:tjarra, who was thought to be terminally ill. Neither the dead leader's sons nor people of his lineage (L) seemed to be involved in the decision to begin the ceremony, and none were present at the beginning. Ngaṯili, another potential leader, was not living at Milingimbi at the time, and indeed did not attend the ceremony at all. The formal consent of Wurrpaṉ, Wandhuma's sister's husband and senior *waku* (wC/ZC) of the Cloudy Water group, was necessary, and he may have been instrumental in making the decision to hold the ceremony. Wurrpaṉ was at that time the most influential man at Milingimbi.

Ma:tjarra was the oldest wife of Wurrpaṉ. She witnessed the public parts of the ceremony, which took place at her camp. The course of the ceremony was linked to her health: the older men said that if she recovered from her illness they would go ahead and finish the ceremony; if not, they would cut it short. The ceremony was also to be performed so that Wandhuma's son, a man of about 22, could refurbish the *rangga* for the first time. He had been urging his father to hold the ceremony for two years or more.

Before the ceremony could begin, it was necessary to have enough *wana* cords, worn hanging from just above the elbows by some men and women dancers, and tied also to the *madayin* Dilly-bag of the organizing group, and to the *rangga*. Two years before this performance began, a Cloudy Water man told me that all the men of the group must make the cords, but that the planned performance was held up because some people had been slow. Hair and banyan-fibre string, as well as plenty of red-collared lorikeet feathers, were needed to make the Dhuwa-moiety cords, and to wind round the *rangga*. Women would make the string, and men made the *wana* cords (see also Warner 1937: 342). It was also necessary to prepare the ceremony ground, and to refurbish the *rangga*.

The men prepared a rectangular 'inside' dance-arena among trees on the sand dunes above the beach, several hundred metres along the beach, to the north of the public dance-arena in the camp of Wandhuma's sister and her husband Wurrpan, and built a brushwood shade at one end. Men also erected a Riyawarra Tree, a small forked tree stripped of leaves, twigs, and smaller branches, in Wandhuma's sister's camp, because her husband and her children were the *waku* (wC/ZC) for this particular performance. Her children were *waku* for Cloudy Water group through her, and her husband was *waku* through his mother (her father's sister). The ceremony had taken place four years previously at the same inside ground, but on that occasion the Riyawarra Tree stood near the house to the north of Wurrpan's camp, belonging to a Black Duck man who was the *waku* for that ceremony. Who took the role of senior *waku* on a particular occasion thus depended on a personal relationship to the main organizer.

The *rangga* to be refurbished on the inside ground and shown to some of the men on this occasion represented the older Djang'kawu sister, a 'mature woman' (*gungman*) one of whose names was also one of Wandhuma's sister's names. It had to be dug up from under the inside ground where it had been buried at the end of the previous performance, and then refurbished. It was repainted with haematite, a purplish ochre with a silky sheen, mixed with water-and-glue fixative, at a small Dhuwa-moiety country on the island. A few days before the dancing began, the men brought the *rangga* to the inside ground, and placed it in the brushwood shade. Each day Wandhuma, his son, and later one of his *waku* (ZS), would sit at the inside ground, preparing the *rangga* by winding it around with banyan string and red-collared lorikeet feathers.

According to Wandhuma, the time taken to refurbish it would determine the length of the ceremony, although in fact the men worked on it for only a few hours each morning for the first week or two of the ceremony. Only the one *rangga* was displayed on this occasion, although several *rangga* had been displayed at each of the Madayin ceremonies described by Warner (1937: 354–5), R. M. Berndt (1952: 17), and Elkin (1972: 80).

The Ceremony as a Journey

The course of the ceremony represented a journey from the country of one group to that of another. Each day at the Riyawarra tree women danced as the Djang'kawu sisters, while the men danced Little Kingfisher from the inside ground and 'chased' the women from the tree. The *djirrikaymirr* ('power-man'), usually Wandhuma, invoked the *likan* names of one Dhuwa country, changing the country every two days or so to include the majority of Yolngu Dhuwa groups, then the men danced various species of fish and men, and women danced as Little Kingfisher 'playing' at the tree. In this way the ceremony traced a journey from west to east across the northern part of north-east Arnhem Land from, and then back west along the coast (Map 5). The Djang'kawu sisters followed the Sun in myth, but the notional journey of the ceremony initially went the other way, then followed the Cloudy Water version of the sisters' journey from Sunrise-group country in the east to Cloudy Water country at Langarra and the adjacent mainland, passing through the countries of the closely related Black-capped Tern and Thunder groups. The inside dances also traced a journey with reference to other, related *wangarr* ancestors.

Wandhuma explained to me that the Dhuwa groups were in a *ma:ri–gutharra* (MM–wDC/ZDC) relationship in the context of

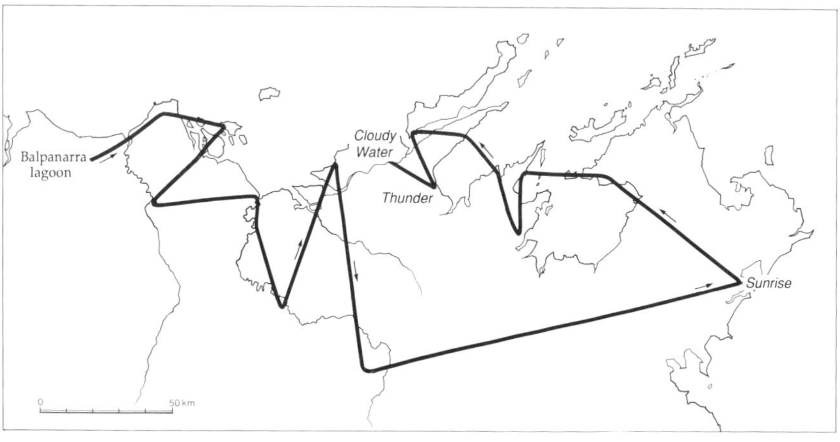

MAP 5. The Journey of the Cloudy Water Madayin Ceremony

the ceremony, although some were in fact in a 'sibling' relationship. The ceremony united them: 'Whatever *ba:purru*,' he said, 'all are together. Whatever *ma:ri* group—this *ma:ri* group, that *ma:ri* group, the other *ma:ri* group—all are together.' All groups with the Djang'kawu as common creator bore the *ba:purru* name Ma:tjarra or Bungunbungun, although each group's *likan* names, as well as its public names, distinguished it from others of the same *ba:purru*.

The Course of the Ceremony

Once the preparations were complete, the ceremony lasted for thirty-three days. I arrived from Nangalala outstation community, where we had been based for several months, on the fourth day of dancing. On three separate days through the course of the ceremony no dancing took place.

The first phase consisted of daily dances at the inside ground and at the Riyawarra Tree, with a similar form each day.

Days 1–18 During the afternoon of each day, the young male dancers of the Dhuwa moiety gathered under the shady trees near the inside dance-ground, painted their bodies with vertical stripes in red and yellow ochre and white clay (the 'inside' meaning of which should be known only to men), then began to dance in the late afternoon, unaccompanied by singing.

Women and children stayed right away from the inside ground. Men watched out for children or others who approached too close, and shouted out to them to move away if they strayed along the beach towards the ground. An owner or a *waku* (wC/ZC, Yirritja moiety) would escort an authorized newcomer to the ground, approaching slowly until welcomed by the men sitting near the shade. A man who wished to dance but who was not of the Cloudy Water group might say on arriving, 'Is it all right if I paint up?'. Men who had not already experienced the ceremony did not come unless invited. About a dozen Dhuwa men painted up and danced each day, but I cannot describe the inside dances in detail here.[4]

Each day, after the inside dances, all the Dhuwa men, together with male *waku*, lined up with spears and spear-throwers, then marched out from the inside ground, tapping their spears with their spear-throwers as a Yirritja man called in a falsetto voice, '*garrarra*' and the dancers replied '*yay! rrrr*'.

[4] For details of the dances see Keen (1978: 256 ff.), and cf. Warner (1937).

At the Riyawarra Tree several Dhuwa women, mainly Cloudy Water group, stood with their children (their *waku*, Yirritja moiety) and their brothers' children (their *ga:thu*, Dhuwa). Some women were smeared on exposed skin with red ochre, wore *wa<u>n</u>a* cords, and carried a dilly-bag to represent the Djang'kawu sisters with their dilly-bag and *ma<u>d</u>ayin* objects which they carried. Many other members of the community stood and sat around in a rough circle to watch, the numbers varying from day to day.

As the men approached the tree the women also called '*garrarra*', then the men ran up to the tree growling '*mmmm*'. The women ran to one side and lined up, as the men formed a circle around the tree. This sequence recalled the actions of the men in the myth, as they stole the dilly-bag hung in the Mistletoe Tree (Djota), and the loss of the dilly-bag with its contents on the part of the Djang'kawu sisters. According to Cloudy Water interpretations, the sound also connoted the incoming tide. In the related song, Black Butcher Bird calls as the tide comes in, and Mangrove Robin calls at the ebb.

Wandhuma, the Cloudy Water leader, climbed the tree, and, as the other men called the *marrawinydjun* chorus, called the *<u>l</u>ikan* names of the country and group of the day. Then he climbed down and called the invocation again as other men called the chorus.

Then Dhuwa men danced Spangled Perch, accompanied by sounds from the chorus of Dhuwa and Yirritja men standing by the Riyawarra tree. The chorus called '*rrrrrrr, rrr-rrr, rrr-rrr, waaaa* . . .', then the two pairs of dancers started to dance from the tree, each pair dancing in a circle in opposite directions, crossing over, dancing around again, crossing over, and then dancing forward, away from the tree to kneel down, as the chorus ran after them from the tree and surrounded them to the call '*goooo, yay*!'. Striped Butterfish, the next dance, was similar, except that each dancer held a spear-thrower behind his back, lying along the spine.

The Dhuwa women and men then danced Kingfisher in a group by the tree, accompanied by a Yirritja man playing a drone-pipe with the proper name of Gu<u>l</u>uwurru. Each person danced on the spot with a movement combining flapping the elbows with a Charleston-like step, provoking a great deal of laughter from both performers and onlookers.

In a later part of this phase men introduced novices to the inside ground, and performed new nocturnal episodes.

Introduction of Novices The dances by the Riyawarra Tree usually finished at about sunset, and the ceremony resumed on the following morning with refurbishing the *rangga* and painting-up. On two evenings, however, the Cloudy Water men sang *bi<u>l</u>ma* ('clapstick') songs in Wurrpa<u>n</u>'s camp near the Riyawarra Tree, following in song the journey of the Djang'kawu from Sunrise-group country in the east to their own; and on the evenings of days 11 and 14 a distinct sequence was performed called 'feeding the *ma<u>l</u>agatj*'.

On day 12 the older men began to introduce novices to the inside ground from time to time. Throughout the ceremony the novices had to stand facing the shade with Yirritja men, their backs to any dances that were new to them. The older men would put under-arm sweat on the boys' eyes, turn them around, tell them the identity of the dance, and enjoin them not to eat the animal or bird represented by the dance until released from the prohibition. The dancers would stare wide-eyed at the novices, provoking giggles and wry comments such as 'Oh very fierce!' from the older men.

Feeding the ma<u>l</u>agatj On the evenings of days 11 and 14 the Cloudy Water mature men told women and children in the camps that spirits of Cloudy Water and Honeybee group's human ancestors would visit the settlement and the ceremony. Each evening a rhythmical banging noise originating on the salt plain behind the camps along the beach resonated through the settlement. The men told the women and children that the spirits were making this noise by stamping on the ground with one foot. While this was going on Yirritja men who were *waku* for the ceremony walked through the camps collecting flour, tobacco, sugar, etc. 'to feed the *wangarr* and the spirits', and later took the food to the inside ground.

The day after the first 'visit' of spirits of the dead, some of the children of the mission were very excited about 'a great monster which came out of the sea'. Women had a different view. At least one or two mature women were eagerly looking forward to the spirits' arrival. One Cloudy Water man told a Thunder woman of about 50, who was one of the two who would dance as the Djang'kawu sisters on the last day of the ceremony, that the ceremony would last a long time because the spirits of the dead had to come from many places. Moreover, the spirit of a particular dead relative had made *wa<u>n</u>a* cords for her to wear. She asked the men for some food to be put close to the camp so that she might hear their voices, for only men were supposed to see the spirits. She explained to the mission linguist that the spirits of her dead 'fathers' were like barramundi fish in appearance, with the characteristic nose and tail, because they lived in the group waterhole, but they could take off this apparel and appear as men.[5]

Another man told his women relatives that a deceased man of the group wanted to see Wandhuma's sister, for whom the ceremony was being performed. She should therefore sit with a sheet over her so that the spirit would speak to her. I asked Wandhuma whether the old women

[5] I am grateful to Beulah Lowe for this information.

knew what the men did in the ceremony. Wandhuma replied that they were ignorant; his sister and another old woman listen and say, 'That's so and so'; women say, 'That is my *ma:ri* or *ba:pa* or *ngapipi*'. He did not know if they really came, but in telling the women those things, the men followed *rom* ('right practice', 'law'). The Djang'kawu said they should do that. I also asked two of the other senior men why they told the women these things, and they replied that it was to make them 'believe' (*ma:rr-yuwalkthirr*) the *rom*.[6]

Days 19–30 displayed some major variations in the dances, as the ceremony visited countries of groups represented among the participants.

Variations Men added King Brown Snake to the set of public dances as the ceremony reached the countries of Black-capped Tern groups with this *wangarr* on its symbolic journey. The dance appeared similar to the Fish dances, except that the dancers carried spears bent in their hands like bows to represent the Snake.

The dancers were still predominantly Cloudy Water men, with two men of the Digging-stick group, but on day 21, when the ceremony was 'visiting' Digging-stick country, two senior Black-capped Tern men attended the inside dances, and one of them took part in the dancing. During this period two Mangrove-group men also joined the ceremony. Few members of other Dhuwa groups came until day 24, when the ceremony reached Black-capped Tern country. On that day ten men of this group, including two of its leaders, two young novices, and several *waku* of the group, joined in. These *waku* were Lightning Snake group. Six more Black-capped Tern men joined the dances at the Riyawarra Tree.

On day 28 the ceremony reached a peninsular Black-capped Tern country associated both with the Djang'kawu sisters and with Shark *wangarr*, who travelled from Sunrise-group country on the east coast of Arnhem Land, according to the myths. In one version of the myth which I recorded, Shark met the Djang'kawu sisters at that place. The men danced Shark, wearing the appropriate body-painting. Three Black-capped Tern men took part in the dances at the inside ground, and four novices for whom Shark was 'mother' were introduced into the ceremony at this point. They were *waku* of the Black-capped Tern group which holds the country associated with Shark.

The dances on days 29–30 were the same as on days 1–18, except that Lesser Salmon Catfish was added to the dances at the Riyawarra Tree.

[6] R. M. Berndt (1951: 55) reports a similar explanation by men: 'we have to trick them'.

On day 31, more dances were added to the daily sequence in the day and evening, some of which echoed the nocturnal dances.

The Final Three Days During the morning, four Cloudy Water men painted themselves with the Little Kingfisher (*Djirridiḏi*) design, prepared long red-ochred sticks, attached *waṉa* cords to hang from their upper arms, and wore their individual *maḏayin* dilly-bags behind their heads. The Little Kingfisher design consisted of alternating red ochre, yellow ochre, and white horizontal stripes all down the body.

Early in the afternoon two of the mature men danced Little Kingfisher from the camp belonging to one of the Cloudy Water mature men, through the seaside camps, and towards the inside ground. In this dance the performers walked along in file, stopped to face each other, speared the ground with the sticks, knelt face to face, and levered the sticks reciprocally back and forth. (When several people perform the dance they form a circle.) They performed this sequence at fourteen places within the two seaside camps. Meanwhile two male *waku* (wC/ZC) collected food, tobacco, and money from the women in the camps, and took it to the inside ground for the men. As the dancers approached them, some women 'cried' as in a mortuary ceremony. A man explained to me that they did so 'because it is their Law', implying, I think, that they mourned for the loss of the *maḏayin* on the part of the Djang'kawu. This dance represented a diurnal parallel to the 'feeding the *maḻagatj* sequence'.

The country visited that day in dance was of the Thunder group, whose leader, Buwaṯa, had arrived from Galiwin'ku the previous day. He, together with six other old and mature men, four of the Cloudy Water group, and one each of the Sun group and Black-capped Tern group, painted their torsos with a distinctive design. The white torso and head was divided down the middle and across the collar bones with a red stripe bordered with yellow ochre. The secret meaning, supposedly known only to men, was a major game-bird *wangarr*, and women knew it as a variety of Little Kingfisher design. The other dancers painted themselves as before, while Wandhuma sang. It was the first day of the concluding sequence. Buwaṯa commented several times to those present at the inside ground that Cloudy Water group and Thunder group had the same Little Kingfisher design (*miny'tji*), they were not separate. The Honeybee leader remarked to the young men to whom the new dance was revealed, 'This is our *rom*, Dhuwa and Yirritja; it is the foundation [*ḻuku*, 'footprint']'. Another dance was also added to the sequence.

At the Riyawarra Tree the men danced only Striped Butterfish, each dancer balancing a dilly-bag on one side of his head, followed by the Little Kingfisher dance.

On day 32 the ceremony reached main Cloudy-Water-group country. The themes of thunder, lightning, and fire were introduced, foreshadowing the final day's dance, which depicted the creative acts of the Djang'kawu.

On this day the men decorated the *rangga* with feathered *wana* cords, standing it upright inside the bough shade at the southern end of the dance ground. The senior male dancers prepared themselves in the shade of the trees. Two men shaved, put red ochre over their faces, and stuck short *wana* cords on their chins as 'beards'. Again two novices were turned around to see the dance. At the Riyawarra Tree men again added the Lesser Salmon Catfish dance to the sequence of Spangled Perch and Striped Butterfish, followed again by Little Kingfisher.

Up to this point the only formal evening activities consisted of singing on two, or possibly three, evenings, and 'feeding the *malagatj*'. On the evening of day 32, however, the men and women performed a sequence of both 'inside' and 'outside' dances which differed from the daily dances. The themes were Thunder and Lightning, Fire, and the birth of children of the groups.

New dances related to these themes were performed on the inside ground. Then, instead of marching straight through the camp, the men first prepared stringybark brands. After lighting the brands, the Dhuwa men formed a line, each man carrying a brand, and walked through the camp to the beat of clapsticks. The men stopped at eleven family camps on the way to the Riyawarra Tree. At each place Wandhuma and Buwata called the *likan* names of Fire, Sea, and the *rangga*.

At the Riyawarra Tree a number of Dhuwa women lay covered by blankets. (Formerly, conical *nganmarra* mats of woven pandanus, used by women as pubic coverings and to protect infants from the elements, were used in this dance.) The men formed a U-shape around the women, with the open end of the U facing the sea, cast the firebrands over the top of the women towards the sea (they fell on the sand), and Wandhuma again invoked Fire as the men of both moieties performed the *marrawinydjun* chorus. The men then encircled the women, Buwata sang the Longicorn Beetle Larvae song, the men called the chorus, and Wandhuma invoked the *likan* names associated with Mangrove, Cloudy Water, and Thunder countries. The women were grouped accordingly, and during each invocation the group of women whose group was invoked wriggled about under the blankets to represent Longicorn Beetle Larvae in the rush-corms. After each invocation the respective group of women rose from under the blankets and danced as Longicorn Beetles flying, with movements precisely the same as those of the Charleston-like dance of 'Little Kingfisher' on all the previous evenings.

Later in the evening the Cloudy Water men and the leader of Thunder group sat in a group in Wurrpa_n's camp and sang Dhuwa-moiety *bi_lma* songs until dawn, following the journey of the Djang'kawu. Several of the young Cloudy Water women danced during the appropriate songs as the Djang'kawu walking along with their digging-sticks.

Day 33 was the last day of the ceremony, with dances related to those of the previous evening.

Early that morning men began painting up at the inside ground, and painting the women at the camp. Buwa_ta sang the Little Kingfisher song during this phase. All the men and all but two of the women were painted with the Little Kingfisher design. Two senior women wore the design that the senior men wore the previous day, although according to the men, the women did not know its meaning.

At about 9 a.m. at the inside ground the mature men lined up the older novices in front of the shade and revealed the completed *rangga* to them. After the other men had danced, two mature men explained the meaning of the *rangga* to the novices and enjoined them not to reveal what they had seen, not to steal from the *balanda* (white people), not to take other men's wives, and not to swear at other people (cf. Warner 1937: 356). When a mature man called for the young men to go to the shade to prepare it for the revelation of the *rangga* to the novices, he omitted calling the sons of Nga_tili, who some thought should have been leader rather than Wandhuma.

Wandhuma asked a Cloudy Water man from Galiwin'ku to explain the *rangga* to white visitors to the ceremony: myself, a Christian minister, and the Milingimbi craft-adviser. Reflecting both the greater Christian orientation of the Elcho Island Mission at that time, and the presence of the minister, he explained that the *rangga* were symbols given by God.

By mid-afternoon the body-painting was complete and the camps between the inside ground and the Riyawarra Tree had been emptied of people. The male Little Kingfisher dancers lined up at the end of the inside dance-ground, and Wandhuma invoked the Djang'kawu and the Ma:tjarra *ba:purru*—the set of Dhuwa groups created by the Djang'kawu. Then the Little Kingfisher dance began.

Holding a stick in each hand, the dancers walked from the inside ground through the camps, pausing on the way for twenty minutes for a shower to pass. They formed a circle, dug in their sticks, worked them back and forth, and called the *marrawinydjun* chorus, first at the shade and then at four family camps on the way along the dunes under the tamarind trees to the Riyawarra Tree.

At the Riyawarra Tree the Dhuwa women decorated with the Little Kingfisher design lay under blankets as they had on the previous night, but this time in six groups: Cloudy Water, Stringybark (associated with the Langarra country), Mangrove, Black-capped Tern, Stringybark (associated with the mainland country), and Dilly-bag. (The groups represented were Dhuwa groups with sufficient women present on the community.) Two mature women painted with the secret design and each equipped with a pair of digging-sticks, *wa<u>n</u>a* cords, and *ma<u>d</u>ayin* dilly-bags, acted as the Djang'kawu kneeling at the Tree, working their digging-sticks back and forth.

When they arrived at Wurrpa<u>n</u>'s camp the male Little Kingfisher dancers encircled first the huts and then the group of women under the blankets. They called the *marrawinydjun* chorus as the two women Djang'kawu dancers knelt, dug, and moved their digging-sticks.

Forming two concentric circles, the men moved closer to the women, called the *marrawinydjun* chorus, and danced, spearing the ground as before. Yirritja leaders joined in the chorus in this sequence. The two women danced as the Djang'kawu, accompanied by the clapsticks of the power-men and hand-clapping from the chorus, while the women under the blankets wriggled about as 'Longicorn Beetle Larvae in the rush-corms'. Buwa<u>t</u>a sang the Longicorn Beetle Larvae song and Wandhuma invoked the *<u>l</u>ikan* names associated with each group-country as the women of that group rose out of the blankets to dance 'Longicorn beetles flying' to the *marrawinydjun* chorus. The whole sequence of chorus and invocation was repeated for each group of women.

Dances to the water followed:

The ceremony had been timed to end at high tide. The Dhuwa men lined up facing the sea, with the Dhuwa women lined up behind them. Buwa<u>t</u>a stood in front of the men, then crouched down, also facing the sea. Accompanied by Wandhuma's clapsticks the group advanced to the water, with a crouching gait, as 'Pygmy Geese disturbed and flying'. Buwa<u>t</u>a led them dancing as Magpie Goose (Gurrumat<u>t</u>ji) of the Yirritja moiety, from time to time beating them back by hitting the ground with a paperbark roll representing Barramundi, also of the Yirritja moiety. The group proceeded in this way down the beach to the edge of the sea about 50 metres in all.

Each group of women entered the water and danced in turn Longicorn Beetle (identical to the Little Kingfisher dance at the Riyawarra Tree) to the invocation and chorus, then the men and women plunged into the water and washed off their body-paint, all the ochres and white clay mixing together in the sea. The bathing completed, the whole group of men and women dancers danced Butterfish up the beach to the Riyawarra Tree, as Wandhuma invoked the names of the Ma:tjarra *ba:purru* at three points on the way up the beach. A final '*gaaaa yay!*'

ended the ceremony, and everyone dispersed to their camps, or prepared to return to other communities.

There was a sequel to the ceremony of gift-giving and release from prohibitions.

Sequel Formerly Dhuwa and Yirritja leaders together would invoke the names of the *rangga* over cycad-palm nut bread (*ngathu*) which they ate at the inside ground (Warner 1937: 355). This labour-intensive food was at that period rarely prepared, and on this occasion Wandhuma told me that he and Wurrpan, his sister's husband, would eat flour-damper and smoke tobacco together. The money collected at night would go to the 'big men'—himself, Buwata, and Wurrpan, the senior male *waku* (wC/ZC) for the ceremony.

Thomson records that young men and others seeing the *rangga* for the first time had to give presents to the leaders who revealed them (Thomson 1975: 5). The young men on this occasion told me that Wandhuma was 'easy' and 'good' in not making such demands; in the old days young men had to make gifts of *madayin* and women. Wandhuma himself told me, 'I think we should set aside making gifts, because they [the young men] should learn our law and forget European law.' He did not seem to want to discourage them by making demands, a concern which emphasizes the increasing number of alternatives to participation in ceremonies. Wurrpan's eldest son, however, a man for whom the ceremony was 'mother', was pressed into giving a large bolt of cotton cloth to the Dhuwa leaders (used to wrap the corpse in a mortuary ceremony), probably because he had been absent for much of the ceremony. As a senior *waku* he should have been present.

Several days after the ceremony had finished, mature men of both moieties painted Cloudy Water and Two Women *likanbuy* designs on the chests of the novices to release them from the following prohibitions: talking to any but old people, eating shellfish, eating the meat of water goanna (*Varanus mertensi*), and eating the meat of large game-birds known as 'game of renown' (*ya:kumirri warrakan*, 'name-having game/meat'), including those revealed to them in dance.

After the ceremony Wandhuma told me: 'Everybody is very happy. They all weep for it [the ceremony] and want to learn more so that white people and black people can talk about it together and be equal, level. Everyone is overcome with happiness and they keep asking me when we will do it again.' (People often said that ceremonies induced 'happiness'.) Wandhuma added that he and Buwata were also *rrambangi* ('together', 'level', 'equal'); whereas a year before he had compared himself unfavourably with Buwata, he had now affirmed his position as leader of the group and his equality with Buwata.

Such expressions of co-operation and being 'level/together' (*rrambangi*) at the end of a ceremony were common. I have also heard a prominent man in an oration at the end of a Morning Star exchange-ceremony, for example, say that 'Dhuwa and Yirritja help one another', a sentiment that was greeted with enthusiasm by those present who shouted with one voice, '*Yuw!*' ('Yes!').

Interpretations of the Ceremony

As Chapter 4 pointed out, Yolngu rituals could bear many interpretations, including those of anthropologists, limited only by mimesis in some dances. (The Charleston-like step of Kingfisher could have many meanings, but it would be hard to interpret it plausibly as a fierce snake.) In order to compare several interpretations of the Ma_dayin I shall have to make my own, filling out implicit clues as well as drawing on explicit exegeses.

Dhuwa-moiety people explained episodes of the Ma_dayin ceremony, especially the daily dance of men from the secret ground to the Riyawarra Tree in the camp, with reference to stories about the Djang'kawu sisters of the kind we looked at in Chapter 1. The content of related songs, dances, and designs performed and made for other ceremonies and in the interstices of the main acts in the Ma_dayin ceremony, also referred to elements of the ceremony. I recorded a song-series at a *buku_lup* washing purification ceremony, which Wandhuma, the Cloudy Water leader, translated and helped me to transcribe, and upon which he commented. The series consisted of thirty songs, a number of them repeated several times, and each having about six verses or items (Table 8). I shall summarize Wandhuma's interpretations of public episodes in the ceremony and of related songs, and contrast these with other interpretations.

One general theme of the ceremony and the related songs seems to have been the transfer of powers of reproduction from the *wangarr* to men and women, in conjunction with the spirit world. The related songs describe the transfer of powers: in the songs the Djang'kawu are tired from their work, they walk heavy-laden and look at the sunset; Frilled-necked Lizard clings wearily to the tree. Yet the Longicorn Beetle Larvae gnaw vigorously in the rush-corms, and Pygmy Geese fly powerfully away. These tropes suggest not only the transfer of power from *wangarr* an-

TABLE 8. Cloudy Water Songs

1. Garfish	21. Pygmy Goose
2. Sea-water	22. Black-winged Stilt
3. Long Tom	23. The Djang'kawu travel
4. Black Butcher Bird	24. Longicorn Beetle Larvae in the Rush-corms
5. Little Red Flying Fox	25. Bustard
6. Shellfish	26. Black Cockatoo
7. Black Cockatoo	27. Red-collared Lorikeet
8. The Ma:tjarra group	28. Guwilirr Dilly-bag
9. Fresh and salt water mix	29. The Djang'kawu travel
10. Salmon Catfish	30. Poison Pea
11. Black Butcher Bird	31. The Djang'kawu travel
12. Spangled Gudgeon (?)	32. Honeybee (Yirritja)
13. Crabs and Shellfish	33. Frilled Lizard
14. Longicorn Beetle Larvae	34. Bush-bird (Yirritja)
15. The Djang'kawu travel	35. Spring Waters
16. Striped Butterfish	36. Mistletoe Tree, Lilies
17. White Ibis	37. Black Butcher Bird
18. Masked Plover	38. Little Red Flying Fox
19. Brolga	39. The Ma:tjarra group
20. Pied Goose	40. The Tide comes in

cestors to people, but the replacement of the old by vigorous young people.

Wandhuma explained the first part of the daily public dances with a story (recounted in Chapter 1) elaborating the central episode of the Djang'kawu story in which men came from the *nga:rra* ground and stole the dilly-bag from the sisters, a story he told me briefly on our visit to his country, Wa:langur:

The Djang'kawu hung up the dilly-bag in the tree and went to get nerite and telescope shellfish and crabs. The men came and stole the dilly-bag full of *rangga*.

'Hey! Where is the dilly-bag?' said the sisters. Then they heard '*binbir*', the men dancing at the Madayin ground. 'Maybe our elder brothers have it'. They put their fingers in their ears and went to see.

'They stole it, and now they are dancing,' they said, 'We will go back, work, and get food—the men can do the dancing.'

If men had not stolen the dilly-bag the women would still dance, and we men would have to cook! The men said, 'We'd better get it!'

Warner's instructors, who were perhaps of the Blow-fly group whose island country lay to the north-west of Sunrise country, interpreted this part of the ceremony, as performed at Milingimbi in the mid-1920s, with reference to the wider episode of the Djang'kawu myth in which the sisters hung up the dilly-bag and sacred objects (*madayin*) in the tree, went to catch shellfish, made spring waters, and gave birth. In their version of the story the men did not steal the dilly-bag, rather it burnt in a fire, represented by the Fire dance which took place near the end of the ceremony as a whole. Other aspects of the interpretation, such as the sisters' resigned acceptance of the role of 'worker', were similar to those of Cloudy Water men (Warner 1937: 350).

The ceremony re-enacted this or an equivalent event each day with reference to each country which it visited, for each participating group centred the loss of the dilly-bag at its own place. Although a single event, it appeared to have occurred at each group's country, but the ceremony was ambiguous enough to accommodate each group's point of view; the structure of the dances could accommodate the perspectives of all Dhuwa-moiety groups.

The country of the day was signified by the *likan* names which the power-man called from the tree as the other men called the *marrawinydjun* chorus. The conceptual journey of the ceremony was thus more 'objective' than the meaning of particular dances, since the *likan* names were common currency and their application widely known. The chorus connoting the incoming tide accompanied representations of the creative acts of the *wangarr* and images of the reproduction of the group. The *likan* names proclaimed, as it were, the results of those creative acts—the existence and identity of a group and country, and their connection with the *wangarr*.

The fish dances and songs were images of spiritual conception. In the related Cloudy Water songs, sung during the evenings in the Madayin ceremony as well as at circumcision and mortuary rites, the images of the Djang'kawu catching fish with their conical mats, collecting shellfish and putting them in their dilly-bags, were associated with the mixing of salt and fresh water, and connoted the processes of conception and reproduction, for the conical mat and dilly-bag were often interpreted as symbols of the vagina and uterus (see Keen 1991). But the symbols were multivalent; Warner's instructors saw fish as images of spirits of the dead (Warner 1937: 446). The ancestral action of catching fish could also connote the reincorporation of the dead into the ancestral domain and at least an implication of reincarnation.

In Wandhuma's interpretation of Little Kingfisher, the final dance performed each day in the camp, the bird/persons 'played' at the Tree as the name of each group's country was called, then men and women joined for the Little Kingfisher dance. In the dance:

> Little Kingfisher dives down and digs inside the termite mounds with his or her beak to get Longicorn Beetle Larvae [*ga:murung*]. We dance it because the Djang'kawu said, 'He/she plays at the tree.'

'Play' was often a euphemism for copulate, and given the equivalence of Kingfisher to 'the group', the implication was that the dance represented people copulating. But Kingfisher was also an analogue of the Djang'kawu, digging out larvae with their beaks/digging-sticks. Indeed, in the interpretation by men of Sunrise group of the very similar daily dance recorded by R. M. Berndt (1952: 18), the male dancers represented the Djang'kawu brother having intercourse with the sisters.

In Wandhuma's interpretation the Little Kingfisher dance through the camp (day 31) represented all the creative acts of the Djang'kawu sisters condensed into the one action of spearing the ground and working the sticks back and forth (see also Warner 1937: 350). The dance also embodied the male and female human reproductive roles. Wandhuma explained a similar dance at a Cloudy Water wake at Galiwin'ku, in which women and men danced through the camp together:

> The men are the group called Ma:tjarra and Bungunbungun [all the groups with the Djang'kawu as *wangarr* ancestor] and Cloudy Water group. The women are the Djang'kawu. The sisters went along, saw, and disturbed Bustard [*Ardeotis australis*], Water Goanna [*Varanus mertensi*], Black Butcher Bird [*Cracticus mentalis*], and Red-collared Lorikeet. When the dancers make the circle it means the Djang'kawu created groups—all the different ones right up to Cloudy Water country, the main place for outside and inside ceremonies. The two sisters walked and danced, and saw all the animals and fish, speared with the digging-sticks, and went on.

Other people commented that the dance represents the sisters making the springs at each country.

Dances on the final day of the ceremony appear to have signified the completion of the creative and reproductive processes on the part of the Djang'kawu sisters. In Wandhuma's interpretation, the line of men (and women in public ceremonies) at the public dance on the final day represented the Ma:tjarra group, that is, the string of groups created by the Djang'kawu, including Cloudy Water people. Two older women who danced by the Riyawarra Tree were the Djang'kawu. The circle of men dancers spearing the ground were the Djang'kawu creating Dhuwa groups and springs at each country. The striped body-painting meant Little Kingfisher but also had a secret meaning, the topic of a dance performed on the men's ground. The groups of women rising from under the *nganmarra* mats or blankets in the final days' dance represented Longicorn Beetle Larvae who changed their bodies and grew wings to be born like children. The Kingfisher design worn by the dancers who represented the Djang'kawu performing their creative acts (in the interpretation by Cloudy Water men) denoted 'the group'.

The related songs illuminate the significance of the dance. The song immediately before the Longicorn Beetle Larvae song in the series described how the Djang'kawu walked around digging in their sticks, creating the groups, and putting in the *madayin* in the dance. Between this and the Longicorn Beetle Larvae song, in the performance of the song-series, the men called the *marrawinydjun* chorus, which also accompanied the song describing the incoming tide (which mixes with freshwater springs in the mangroves) and another which proclaimed the names of the Ma:tjarra groups. The incoming and spring tides were associated with plentiful shellfish, especially at the beginning of the wet season (see Meehan 1982).[7]

Wandhuma commented about the Larvae songs:

Larvae lie in the mud inside the rush-corms, which they eat. The two women got them out with the digging-sticks and put them in the sacred Dilly-bag. The Larvae lay in the *Nganmarra* Mat as well. The Larvae grew wings [*wana*] and feet, and flew. The Dhuwa groups all the way

[7] The Yolngu term *ga:murung* denotes a large class of wood-boring beetle larvae, or 'witchetty grubs'. According to Rudder (1977), this taxon belongs to the more inclusive class of *maypal gongmirr* 'invertebrates with hands'. The Yolngu associate *ga:murung* with a wide variety of habitats, including freshwater billabongs, timber, and termite mounds, and consider it a very delicious food. The taxon is likely to include a variety of families of beetle (see Duffy 1960: 609). Swarms of termites also emerge at the beginning of the wet season, and the huge grub-like queen is, I believe, of the taxon *ga:murung*. Shellfish and ga:murung are included in the taxon *maypal*.

from the east up to here have the same song. Larvae are *gutharra* [wDC/ZDC] and are people—thus said the Djang'kawu. We follow the old men and the forerunners: son follows father, father and older brother die, and we copy them. But everything is the thought of the Djang'kawu; the Djang'kawu gave the meaning to Larvae.

The classification of children of the groups as *gutharra* (wDC/ZDC) seems to follow relations of reproduction among groups of the same moiety, in relations of mother's mother to woman's daughter's child (*ma:ri–gutharra*). Not only is this a matrilineal relationship but women of the *mari* group are potential mothers-in-law of men of the *gutharra* group (see Chapter 2). It is also striking that in the commentary images of the reproduction of the group are juxtaposed with images of the reproduction of knowledge from one generation to the next.

A repetition of the song later in the series described the Longicorn Beetles which emerged as a 'group of brothers flying west' towards the sunset. Wandhuma commented, 'The Larvae change their bodies and grow wings; they "turn", that is to say, they are born like children.' In this version of the song people with the *l̠ikan* name Wal̠awal̠a, of the Cloudy Water group, collected the Larvae.

A comparison of the final dances with the daily dances at the Riyawarra Tree suggests an equivalence. The Little Kingfisher dance (meaning 'the group') was identical in form to the dance of 'Longicorn Beetles flying' at the end of the ceremony as a whole, which explicitly signified the birth of new members to the groups. Furthermore, in the Little Kingfisher song the sisters collected Longicorn Beetle Larvae: the Larvae emerge from within the substance of the trees which form the habitat of the little kingfisher, and from within the rush-corms which the Djang'kawu gather in the song, as do Pygmy Geese, which represented 'the groups'. Wandhuma explained that the peregrinatory dance of Little Kingfisher denoted the group of people created by the Djang'kawu (i.e. the Ma:tjarra *ba:purru*). The Little Kingfisher design worn by the dancers who represented the Djang'kawu performing their creative acts signified 'the group', strongly suggesting an identity of people with ancestors.

According to Cloudy Water interpretations, the daily dance of Little Kingfisher, therefore, connoted the Djang'kawu themselves, the groups of people created by the Djang'kawu at each country, people (or the Djang'kawu) copulating, and children being born to the groups. It followed, logically enough, images connoting spirits of the dead, creation, and spirit conception. Analogous dances at the inside ground also connoted enthusiastic sexuality and fecundity. The ceremony thus moved through the end of the dry season, with Fire, Lightning, and Thunder Clouds characterizing the transition from the dry season to the early wet season, to the image of the emergence of the group's children (the Longicorn Beetles) associated with the coming of rain, at a time of general fecundity.

The dance of Pygmy Geese to the water showed the completion of the creative and reproductive process, for the bird signified the whole group of people, including all the Dhuwa groups, with the Yirritja moiety represented by a single Dhuwa dancer performing Magpie Goose.[8]

Washing together explicitly signified the unity of the group which the Djang'kawu created. As Wandhuma commented:

> All the groups right up to Wa:l̲angur join together with their *gutharra* [wDC] groups. Any group can do so because the Djang'kawu are the bosses. Any group can come and paint up, *ma:ri* and *gutharra* together. The paint is all mixed [in the water], and all are joined together. No one can break the old law—all are one in Little Kingfisher. Washing in the sea means we finish at Cloudy Water country.

In this instance the lake which the men crossed to steal the dilly-bag had as its analogue the sea near the Riyawarra Tree. The paint that mixed in the water was of the Little Kingfisher design, which united all the groups and their countries. (The secret meanings of the design had interlinking connotations also.) The design that showed links of common ancestry between the groups mixed with another interlinking medium, the sea, to signify the unity of groups of the same moiety, related through women. The 'mixing' of substances also connoted the creative processes which united the groups.

Exegeses of the final dances made by people of other groups reveal some similar themes and some differences. Explanations given to Warner (at Milingimbi in the mid-1920s) of the men's

[8] See also R. M. Berndt (1952: 133). In the version of the ceremony recorded by Warner (1937: 43) the people danced as 'Goose' and Black Duck (Yirritja).

dance from the secret men's ground to the camp was that it represented Kingfisher but also men who unsuccessfully attempted to have intercourse with the younger Djang'kawu sister. The two lines of men signified the moieties instituted by the sisters. The group of women (and boys) lying under the mats represented the camp of the Djang'kawu where the younger sister went to sleep, not realizing that an old man was going to try to copulate with her. The final washing in the sea depicted the young woman washing off the semen which went all over her body after the men's unsuccessful attempts at intercourse (Warner 1937: 346, 353).

Some 40 years ago at Yirrkala, Sunrise men interpreted the placing of women under the *nganmarra* mats as the Djang'kawu putting female babies under the mats for protection; males were left in the grass. The male dancers poking the mats with spears and sticks represented the Djang'kawu brother having sexual intercourse with the sisters, and the women wriggling under the mats was the movement of the foetuses. Alternatively, poking the mats with spears and sticks represented replacing the *rangga* in the *nganmarra* mats (equated with the dilly-bag). Casting aside the mats as the women rose was the birth of children or the removal of *rangga* from the mats (R. M. Berndt 1952: 18–19). It is evident not only that Sunrise men's interpretations differed from that of Cloudy Water men, but that Sunrise men gave more than one interpretation of the same movements.

The ceremony was shaped to accommodate the diverse perspectives as well as the commonalities between participating groups of the same moiety. People of Cloudy Water, Mangrove, Sunrise, and other groups were quite likely to attend occasions organized by people of each of these groups, as they did at the ceremony performed at Yirrkala in 1972, and were likely to interpret the episodes in which they danced together in different ways. Men introduced dances, especially at the men's secret ground, which represented *wangarr* specific to groups participating in that particular performance.[9] The content of a particular performance reflected its organizers' perspective on the meaning of the ceremony, but this is not to say that there were separate and stable local versions of the ceremony. Certain actions and effects in the

[9] See the film *Memorial to Mawalan* by Ian Dunlop (1984) for Film Australia.

ceremony, however, transcended local differences; these had to do with age and gender relations.

The form and content of a ceremony such as this accommodate a variety of anthropological interpretations as well as Yolngu ones (e.g. Munn 1969; H. Morphy 1991: 91–2). There seems to be little difficulty in interpreting the theft of the dilly-bag as an ideology justifying men's control of *rangga* sacred objects and secret aspects of the ceremony, and hence the exclusion of women. The Cloudy Water story implies, perhaps, that the sisters were at fault in carelessly leaving the dilly-bag where it could be stolen. The consequences as far as men and women were concerned was that men had control of the inside ground and sacred objects, and women 'worked', remaining at the hearth. The currency of the variants of this belief could be seen as hegemonic, for Yolngu women widely accepted both the truth of the particular version to which they subscribed, and its implications. Some women of eastern Arnhem Land complained that 'if those stupid Djang'kawu had not lost the dilly-bag, women would not have lost control of the ceremonies' (Frances Morphy pers. com.). As the dilly-bag is a symbol of the uterus, the theft can also be interpreted as representing male appropriation and control of women's reproductive power as an aspect of domination (cf. Lattas 1989: 459). However, I have suggested (Keen 1991) that an equally plausible interpretation of some versions of ancestral events, one that does not necessarily contradict the former, is that following primal creative acts by women, men's possession of the ancestral dilly-bag established heterosexual relations between people following creation by the Djang'kawu who took the male and female reproductive roles in their first creative acts. My point is this: if the ceremonies bore multiple interpretations from a Yolngu point of view, then no anthropological reading of the ideological significance of myth and ritual can be given 'objective' validity; they are among the manifold possible interpretations.

The Organization of the Ceremony

Echoing the theme of younger people 'taking the place' of older people, men worked their way up a hierarchy of age-related roles, and some gained in authority in the Madayin ceremony, just as in the *garma* and regional ceremonies. Men conceived of this movement as a process in which the young took the place of the old, as I said. Women were excluded from secret aspects of the ceremony and from control of the *rangga*, but they had a similar age-based hierarchy of roles in their own sphere. The network of core participants generally varied with the performance, but the core on this particular occasion focused around Wandhuma, his sister, and her husband.

Most participants in this performance were members of a few closely related groups of the Dhuwa moiety, together with their *waku* of the Yirritja moiety (Tables 9, 10). The Yirritja men and women who took part were all *waku* (wC/ZC) of the main participating Dhuwa groups, taking part on the basis of kinship links rather than group identity.

Men of the Cloudy Water and Thunder groups, together with their *waku* (wC/ZC) and *dhuway* (FZS/H/ZH), took all the leading male roles. The one exception was a renowned singer and dancer (but not yet *djirrikaymirr*, 'power-man') of a Black-capped Tern group, who led the dances at the Riyawarra Tree on most days. Most dancers were Cloudy Water people, but several other Dhuwa groups of the area were represented.

TABLE 9. Male Participants in the Ma<u>d</u>ayin Ceremony

Group	Novices	Dancers	Total attending
Dhuwa moiety:			
Cloudy Water	4	22	28
Thunder	0	3	4
Black-capped Tern	4	13	29
Mangrove	5	7	9
Sun	2	4	5
Stringybark	0	1	2
Lorikeet	0	1	2
Mosquito	2	2	2
Blossom	0	2	1
Digging-stick	0	1	1
Morning Star	1	1	1
TOTALS	18	57	84
Yirritja moiety:			
Honeybee	3		17
Lightning Snake	3		14
Black Duck	1		4
Spider	0		2
TOTALS	7		31

TABLE 10. Women Participants in the Madayin Ceremony

Dhuwa-moiety groups		Yirritja-moiety groups	
Cloudy Water	15	Honeybee	5
Thunder	3	Lightning Snake	1
Black-capped Tern	11	Spider	3
Mangrove	0	Black Duck	1
Sun	1		
Stringybark	6		
Lorikeet	0		
Mosquito	0		
Blossom	0		
Digging-stick	1		
Morning Star	0		
TOTALS	37		10

The organizers were of Cloudy Water and Thunder groups, and most numerous among other Dhuwa groups were the Black-capped Tern group, many of whom lived at Milingimbi. This group was *gutharra* (wDC/ZDC) to the Cloudy Water group. Most people of the seven other Dhuwa-moiety groups who attended the ceremony painted their bodies and danced.

The Cloudy Water men clearly established that it was their performance, while acknowledging the rights of others. When Mangrove men, who also owned the ceremony, arrived at the inside ground from their outstation, Wandhuma gave some recognition to their joint ownership by discussing with one of them his rights to be 'boss' for this particular performance. The Mangrove leader agreed that everything was 'good, all clear'. Men of other groups also demonstrated their recognition of Cloudy Water prerogatives over the ceremony: for example, when two mature Black-capped Tern men came to the inside ground, each asked those present if it was all right for them to paint up by saying, 'I'll paint up, all right?'

Cloudy Water people organized the performance, but in spite of expressions of group cohesion they were divided. Most people of the branch who lived at Galiwin'ku simply did not come, and there was dissent over the leadership, as we have seen. Neither the dead leader's sons nor members of the group's B lineage seemed to be involved in the decision to hold the ceremony, and none was present at the beginning of the ceremony. A rival contender for the leadership, Ngaṯili, was not living at the mission at the time, and indeed did not attend the ceremony at all, although two of his sons and two brothers were there. Only one or two of the deceased leader's sons went to the ceremony. However, apart from Wandhuma's individual rivals the group was united in the performance to the extent that some members of each lineage were present at the climax.

Only about a third of Cloudy Water adults participated, a remarkably small proportion of the group. The unit of greatest co-operation was the leader's lineage and the lineage-based network, not the patrifilial group as such. However, the ceremony extended the range of social ties beyond the everyday, with visitors from quite distant communities, and entailed the co-operation of a leader of a closely related group with the same *wangarr* ancestors, who jointly owned the ceremony.

In this Dhuwa Maḏayin ceremony, only Dhuwa-moiety men and women could wear body-paint and dance, but Yirritja people had other parts to play, especially as *waku* (wC/ZC). Age seniority applied to the Yirritja *waku* as it did to the Dhuwa-moiety people.

Three Yirritja-moiety women *waku* took leading roles in organizing the women for the last day's dances at the Riyawarra Tree. The senior male *waku* (wS/ZS) for the ceremony was Wurrpan̲, the husband of Wandhuma's sister Ma:tjarra, for whom the ceremony was performed 'to say goodbye'. Due to his age and infirmity he took little active part in the ceremony until the end, although he did offer advice from time to time, as did his oldest son when he was there. This 'mature' man was the son of Ma:tjarra and also a *waku* for the ceremony. He declared to other men at the inside ground on one occasion that the young people should not dance but only watch.

One of Wurrpan̲'s younger sons, whose mother and wives were Cloudy Water group, took a major part in refurbishing the *rangga* after Wandhuma's son left the settlement following the dispute described later. I think he was chosen because of his involvement with Wandhuma and some of the latter's 'sons' in the establishment of a Cloudy Water outstation at Langarra. Other *waku* made bird and other calls during the dances, looked after and instructed the novices, which included putting under-arm sweat on their eyes before showing them the dances, and telling them not to reveal the dances outside. The *waku* played the drone-pipe at the Riyawarra Tree, and collected money and food to 'feed the *wangarr*'.

Wurrpan̲, his brother D̲an.gi (a leader of the junior Honeybee sublineage), and other *waku*, helped to organize the painting of the women for the final day's dances, and led the *marrawinydjun* chorus in the final Longicorn Beetle Larvae dances at the Riyawarra Tree. The rivalry between the two brothers, and Wurrpan̲'s fading powers, were evident in D̲an.gi's constant efforts to upstage him, cracking jokes and stepping in to call the *l̲ikan* names.

There was evident inequality between men and women in the exclusion of women from the inside ground. Moreover, mature Cloudy Water men gave the women dancers instructions on the order of groups in the Longicorn Beetle Larvae dance. Wurrpan̲, the senior *waku*, upbraided Wandhuma's wife for not dancing at the Riyawarra Tree, commenting, 'You women have no *rangga*.' These kinds of exchange were not all one way, however, for Wurrpan̲'s elderly sister, a woman whom men acknowledged as being very knowledgeable because her *ngapipi* (MB) taught her 'everything', called out instructions to the male dancers at the Riyawarra Tree, on one occasion shouting to them to 'jab' with their spears. Similarly, women dancers commented to the men during evening singing, 'Get on with the singing and we'll dance.'

As with the public ceremonies, the growth of competence was linked with age. Girls and boys first witnessed the Mad̲ayin as they stood at the Riyawarra Tree with their mothers, fathers' sisters, and other women relatives. Girls and young women did not formally enter a secret sphere, but gained in knowledge of the public phases.

By contrast, men introduced male novices to the inside ground at some time after they were circumcised. These young bachelors saw the dances at the inside ground, but did not yet see the *rangga*. On each of eight days through the ceremony, beginning with day 12, the men introduced from one to four male novices (*yut̲a*, 'young') into the inside ground. Most were adolescent boys, although a few were middle-aged men including a Lightning Snake man and two men of Sun group.[10] Altogether I counted

[10] According to R. M. Berndt (1952: 14), males at Yirrkala saw their first Nga:rra ceremony well past adolescence.

fifteen Dhuwa and six Yirritja boys introduced to the ceremony for the first time.

The novices who took part in the ceremony for the second time began to participate in the dancing a few days after they entered. Others took part only in the final Little Kingfisher dance. On the last day of the ceremony eight Dhuwa youths in their late teens saw the *rangga* for the first time. According to Rudder (pers. com.) a young man's father should gain the permission of his own and his son's *ma:ri* (MMB) for the young man to see the *rangga*, and that these men had the duty to instruct him.

The role of young Dhuwa bachelor (*yawirriny'*) was both deferential and more active. They did most of the dancing and obeyed the directions of the mature men at the inside ground. Young men had to participate in the daily dancing while they still had body-paint on: one young married man told me that if he missed the dancing he would have to pay a money fine. Six of the Cloudy Water bachelors were 'knowledgeable' of the ceremony and *rangga*, including Wandhuma's elder son, who joined the mature men as *ma̲dayin-wa̲tangu* or *gong-ranggamirr* ('*rangga*-hand'). He was particularly bossy, ordering younger men about with commands such as 'Bring the white paint over here!' He tried to transcend his junior status in less acceptable ways, leading to his exclusion from the ceremony, after which Wandhuma's young *waku*, one of Warrpa̲n's sons, took over refurbishing the *rangga* (see below).

The Cloudy Water mature men (*wurrwiliny*, married men with children, not yet very grey-haired), while deferring to the old men, took the leading 'administrative' roles, as in *garma* ceremonies: they instructed the novices—saying for example that the dance was sacred (*dhuyu*) and should not be performed 'outside'—led the dances, sang, and made the appropriate calls during the dancing. Often in the company of the *waku*, they also discussed what should be done when problems arose, such as at what age boys should see the dances, or what to do when there was a fight near the store.

The mature men could go at will into the shade where the *rangga* lay, and one or two took over the work of the old men from time to time. Those among the mature men who were active, together with Wandhuma, shared the leading roles in the inside dances. Sometimes one played clapsticks, sometimes another; one led the dancers on one occasion, another on a different occasion. Most would comment from time to time during a performance about how things ought to be done.

Mature men deferred to older men if there was a difference of opinion. For example, Wandhuma's brother's son Wulakitj said to other men present at the inside ground on one occasion that his father taught him the ceremony—Little Kingfisher should now give way to Longicorn Beetle Larvae dance. Wandhuma cut in on him, correcting him. Wulakitj later told me that he realized that he should go more slowly and follow Wandhuma his 'father'.

The status of 'power-man' (*djirrikaymirr*) was related both to age and knowledge, and it was the prerogative of power-men of the organizing groups to take the lead. Leaders of other Dhuwa moiety-groups present did not take organizing roles but simply observed, and were shown the *rangga*.

In the ceremony the 'power-men' had a co-ordinating role. As one man explained, 'Wandhuma looks after the *maḏayin*; if there is trouble, he stops the ceremony.' Wandhuma, the Cloudy Water leader, called the names, played the clapsticks to co-ordinate the dances, and had the final word on matters of organization until the arrival of the Thunder-group leader. Wandhuma occasionally explained to young men what dances and movements to do. However, after joining the ceremony, Buwaṯa, leader of the Thunder group, assumed the greater authority and led the dances, although he did not make decisions about everything that happened. For instance, he painted a sample of the Kingfisher stripes on a piece of card and sent Wandhuma to the camp with it to indicate to the women how he wanted them to do their body-painting. He sent two mature men to check on the body-painting further up the inside ground. Wandhuma continued as power-man.

Three older women of the Cloudy Water group in particular played a leading part in the domain of the camp, as did two Stringybark women on the last day. A prominent woman of the Thunder group and a Cloudy Water women danced as the senior Djang'kawu on the last day, and some younger Cloudy Water women danced as the Djang'kawu through the night.

The Control of Precociousness

I have conveyed a picture of roles defined rather strictly along lines of age, gender, and competence. Indeed men did guard these boundaries: mature and older men combined to control attempts by younger men to climb the ladder of knowledge, competence, and authority too quickly.

One reason for the performance of the Cloudy Water Ma<u>d</u>ayin ceremony was for the leader's son Ga:murung to refurbish the *rangga* as *rangga*-hand. He had the task of winding the wooden core with banyan-fibre string and red-collared lorikeet feathers, a little each day. During the first two weeks of the ceremony he sat for an hour or two each morning at the inside ground, before the male dancers came to prepare. He performed this task adequately, but the older men found his behaviour in other respects to be unacceptable. Ga:murung's offences were that he took the prerogatives of a mature man in leading the dances and violated the norms of body-painting in order to draw attention to himself, 'making himself large'. Worst of all, he abandoned the ceremony half-way through by going to Darwin.

Ga:murung took it on himself one day to lead the inside dances in spite of the presence of older men; on several days he led the run forward at the end of each Fish dance at the Riyawarra Tree, the prerogative of a widely acclaimed song-man; he persuaded a mature man to play the clapsticks for the inside dances while the older men were at a meeting, and to lead the dancers to the Riyawarra Tree, and one day took the clapsticks himself; he began to paint stripes on his body half as wide again as anyone else's; he harangued the older men for acting improperly at the inside ground and for losing some feather-cords. Finally, after an open quarrel with his older brothers at the inside ground, Ga:murung took the plane to Darwin. His father would not permit anyone to send him money so that he could return, occasioning a dispute among Ga:murung's brothers. Ga:murung remained in Darwin, and did not return to see the finale of the ceremony. Wandhuma complained about being abandoned. He said that he was sick, he would just sit alone with the *rangga*, he was the last one, and all his brothers were dead. A few days later he announced that his *waku*, Wurrpa<u>n</u>'s son, was making the *rangga* for him and for everyone.

Wulakitj commented to me later that a leader and power-man must know all the *<u>l</u>ikan* names and the dances of all the Dhuwa-moiety groups. Ga:murung was knowledgeable about winding the *rangga*, but so were they all. He himself was still learning. Whoever was to take Wandhuma's place must have knowledge of all these things, and Wandhuma was a good man to teach them. Wandhuma was punishing Ga:murung because he made too much noise while he was working, and he should not have gone to Darwin leaving the *rangga* half wound. Ga:murung was not the only one to be excluded.

Another Cloudy Water man forbade his daughters from participating in the ceremony because he believed that they were having illicit affairs with Honeybee men. (When the Honeybee leader, Wurrpan̲, who was this man's wife's father, reported this fact to men on the inside ground, he was weeping.)

The Mad̲ayin ceremony described in this chapter incorporated a spatial division between secret ground, the domain of men, and the public camp referred to metonymically by men as the hearth, the domain of women. Men excluded women from the inside ground, and from seeing *rangga* and secret dances, revealing these incrementally to young men, who, like young women in the public aspects, gradually took on the prerogatives and responsibilities that went with age competence. In the Yolngu interpretations recorded above, this kind of ceremony was a re-enactment of ancestral creativity, and an enactment of continuing procreation. Interpretations of the ceremony equated humans and *wangarr*, implicitly superimposing ancestral creation and human sexuality and procreation, ordered by means of relations within the moiety between *ma:ri* and *gutharra* groups, and between the moieties. The ceremony re-enacted numerous parallel versions of ancestral events which legitimated male control of secret ritual, and incorporated male 'tricks' on women to 'make them believe'. Aspects of secret knowledge were encoded during the ceremony in the form of secret meanings of designs as well as songs performed in the camp during the evening and night. The next chapter examines this mode of secrecy in public.

7

Secrecy in Public

Yolngu men controlled access to religious knowledge not only through exclusion from, and admission to, esoteric aspects of performances in which secrecy involved the control of space, but also through secret interpretations of songs and designs. They performed and realized some of these in public, emphasizing perhaps the ignorance of those without access to knowledge of this secret significance and the privilege of those with inside knowledge. Just as distinct groups combined to enact common forms but projected different meanings on to them, in *garma* ceremonies men framed secret knowledge through the interpretation of ambiguous forms. Here the difference lay, not between one group's interpretation and that of another, but between levels

Putting the finishing touches to a sand-sculpture for a *d̲ad̲a* purification

of significance controlled by older people, and separating men's knowledge from women's. I cannot write in detail here about the actual content of esoteric knowledge, but the chapter will outline its structure.

Through the related attributes of ambiguity, condensation, metonymy, and analogy or metaphor in the broad sense that Lakoff and Johnson (1980) give to the term, the significance of a group's set of religious forms could be focused into a few simple geometric designs, simple song-phrases, and names which the knowledgeable were able to interpret (see also Wagner 1975; Weiner 1988: 123). Condensation, which refers to the possibility of interpreting one element such as a design or part of a design in many ways (V. Turner 1967: 29), rests on the ambiguity of the signs and analogies among their interpretations.[1] Though to a degree iconic, many designs and objects had a highly schematic form that made them ambiguous, and the same was true of dances and songs. These forms had clear referents when they were made the subject of exegetical statements or their significance was inferred discursively. Religious myths provided the co-ordinating propositions. Exegetical statements, which referred to or presupposed the events in an ancestral myth, assigned senses to the obscure or ambiguous forms, whether dance movements, elements of geometric design, obscure song-words, or phrases in song with a patent meaning.

Mimetic but highly stylized dances were accompanied by songs of the *garma* genre. The meaning of many of these dances, however, was not patent. Public as well as secret dances of the Ma<u>d</u>ayin ceremony were silent or accompanied only by sounds such as growls and hisses, and people did not routinely explain

[1] The use of the term 'condensation' to refer to encoding multiple meanings in one form derives from Freud's theory of dreams: 'latent' elements of a dream which have something in common are combined and fused into a single unity in the 'manifest' dream (Freud 1922: 144). Sapir distinguished between 'referential' symbols, which refer to known facts, and 'condensation' symbols, which are 'highly condensed forms of substitutive behaviour for direct expression, allowing for the ready release of emotional tension in conscious or unconscious form' and which are 'saturated with emotional quality' and which 'strike deeper and deeper roots into the unconscious' (cited in V. Turner 1967: 29). Turner distinguishes between the property of condensing many 'meanings' in a single form, so that they have 'economy of reference', and the emotional or orectic quality of condensed symbols (Turner 1967: 28). Munn refers to the 'multiple meanings' and 'semantic complexity' of Warlpiri design elements in conjunction with simplicity of form (Munn 1973: 166).

the meaning of the dances during a performance. As earlier chapters showed, interpretations of a dance or dance sequence could vary greatly. As for songs, whereas the language of songs in the *garma* genre bore a fairly close relationship to speech, men assigned secret significance to certain phrases with a patent sense, and once again interpretations varied through the metaphorical potential of images evinced in song.

I do not wish to suggest, however, that these forms had no significance apart from their mythical content, for they related in direct ways to everyday experience. At the most obvious level songs described the attributes and actions of species. At another level these were interpretable as spirit beings or things associated with them. People could also interpret the songs as analogues of experiences and processes such as birth, death, and bodily decay, and they were potential allegories for social relations and events. The song of longicorn beetle gnawing through a woody plant was a potent symbol of decay, just as the emergence of the beetle from the larvae exemplified new life. Natural symbols—what was known or believed about the properties and behaviour of species, objects, and substances—provided the raw material for analogical inference (see Keen 1989*b*). From the point of view of those subscribing to particular myths, the stories tied the whole body of religious practice together as instituted by, and referring to, the *wangarr* ancestors.

Yolngu men's exegetical statements were of various kinds. Some had the form, 'The song says that the *wangarr* did such and such, but the real meaning [*mayali'*] is that they did so and so.' Others simply juxtaposed a word, design element, or object and a being, thing, place, etc. or related the two sides with the term *mayali'*, 'means', for there was no verb 'to be' in Yolngu dialects. Such axioms established the conditions for an infinitely complex network of potential interpretative links among elements, and between these and experience. In this sense a group's iconography formed a coherent system, as Morphy remarks, so that a person could infer more information than they had been told (H. Morphy 1977: 251). In so doing a person could pre-empt his or her teacher through making interpretive inferences. The same or similar elements recurred within a group's body of sacra, and the elements cross-referred to others. Elements also referred to those of other, related groups.

Though coherent, a group's sacra should not be seen as a system of determinate (or discoverable) interpretations but as an open and strikingly creative system of potentially innovative interpretations. In one aspect of this process a person probably related what they saw and heard in one context to what was seen and heard in another. To take a simple example, hearing the secret name of a ceremony ground spoken while at the ground might enable a person to interpret a previously obscure song-phrase in which the name occurred. Moreover, while ancestral precedent fixed a body of knowledge, according to Yolngu dogma, people could incorporate new elements and reinterpret the set of designs, songs, and dances.

The productivity of ceremony elements, together with cognitive processes of evocation and implicature (Sperber 1975: 121–3; Strecker 1988: 48), led men to say that they 'found the meanings in thought', attributing such discoveries to revelation through the agency of the spirits of human ancestors and explaining the process of innovation in a similar way. Spirits revealed new songs, dances, and designs, perhaps in a dream. The potential for innovation was constrained, however.

Yolngu men had direct control over the means of communication, but there were more diffuse constraints on the questions people could raise about the implications of what was said or sung, a level of relationship between speech and knowledge which Parkin (1985: 54) sees as the source of innovation. A neophyte was not supposed to ask direct questions, but was enjoined to be silent, and to listen or look when directed. Furthermore, Yolngu men insisted, as we have seen in Wandhuma's comments on songs, that *wangarr* ancestors specified the meaning of elements, not people. Meaning 'found' in thought came from spirits of the dead.

The analogical productivity of Yolngu religion partly explains its power to captivate Yolngu people emotionally and intellectually, and it is important for this point that condensation and productivity were not only properties of the conjunction of secret elements with exoteric elements, but characterized the open dances and designs as well as the obvious significance of songs. Sapir, among others, has argued that processes of focusing, condensing, or distillation through analogical relations among elements increases the emotional load of an element (Sapir 1934; V. Turner 1967:

2). I would add that the elements required a high degree of interpretive participation on the part of the subject, maximizing a person's *engagement* in the system of knowledge (cf. McLuhan 1964: 22–3). The range of experience and knowledge of the interpreter, what Eco (1986: 68 ff.) terms the 'encyclopaedia', enables (indeed demands) a person to perform an indeterminate number of mental operations on its content, making links in thought among and beyond the related elements, primarily through analogy, bringing together the whole range of emotional and intellectual experience of, and beliefs about, the social and natural world. These open analogical possibilities not only enabled the individual to participate creatively and actively but implied a range of variation in interpretation as well as the possibility of transformations in interpretive trends over time. Yolngu ceremonies were 'machines for interpretation', each created as an artefact with an open content, capable of a multiplicity of interpretations, depending on the knowledge, experience, and dispositions of the interpreter, as well as the context in which the interpretations are made (Eco 1983).[2]

As Howard Morphy points out (1991: 76–7), a division of Yolngu religious knowledge into public and secret domains would be an oversimplification, for what was secret and what public changed over time. Nevertheless, esoteric knowledge was an aspect of all *garma* ceremonies, but I examine its structure in the context of mortuary ceremonies, each of which was specific to a different purpose (see Chapter 4). This chapter will first outline the wake and washing purification ceremonies of the Cloudy Water group, with descriptions of particular performances, then discuss the constitution of secret codes deployed in public and their control.

A Cloudy Water Wake

The earlier ethnographies of Warner and Thomson give the impression that the wake and burial was quite short in the 1920s and 1930s, with elaborations reserved for the hollow-log ceremony (see Warner 1937; Peterson 1976). With the prohibition of dis-

[2] Eco (1983) describes his novel *The Name of the Rose* in this way. Yolngu denied that the origin of creativity lay in the individual, however.

interment and reburial under mission-settlement conditions, Yolngu elaborated the wake, often incorporating songs and dances specific to the hollow-log second burial ceremony (see H. Morphy 1984 for an example). Wakes in the mid-1970s could last for up to a week. The Cloudy Water wake described here took place at Elcho Island Mission (Galiwin'ku).

The large Cloudy Water group of the Dhuwa moiety held a wake for its leader who died in mid-1975 at Galiwin'ku. Many people belonging to a large number of groups attended the wake in order, as one man said, 'to cry and say goodbye'. Many Cloudy Water people lived at Galiwin'ku, and several were living on Langarra at the time, the next island to the south-east of Galiwin'ku, as well as at Milingimbi. The death precipitated a crisis in the group leadership.

As the leader lay dying in his house at Galiwin'ku, his 'father' Wandhuma sang Cloudy Water songs while others watched at the sick-bed. When the sick man died, *ringgitjthirri* songs announced the death at communities from Yirrkala in the east to Maningrida in the west.

I travelled by sea from Milingimbi to Galiwin'ku with a group of relatives of the dead leader: a journey of several hours by dinghy and outboard motor. When we approached the south-west end of the island, we signalled our approach by raising a cotton cloth flag on a pole. On the beach a Cloudy Water man greeted us with a Pygmy Goose dance down the beach to the edge of the water, signifying 'the group' (see Chapter 6).

At the dead man's house, a timber and fibre-cement structure standing among trees on a headland in a group of similar houses, some two hundred people had gathered. The body lay wrapped in blankets lying next to a plywood coffin under a tarpaulin shade at one side of the clearing in front of the house. The wrapping of the body was usually performed by male relatives. Wandhuma sat at the head of the coffin, and the dead man's *ma:ri* (MMB), of the Sun group, sat with three of his brothers by the dead man's head. It was usual for a dead person's male *ma:ri* to look after the body in this way. A close male *waku* (wS, ZS) of the deceased had to be present for the wake to be concluded. Men of the deceased's own group and *ma:ri* (MM/MMB) group kept vigil nearby, and sang the group's *manikay* or *bil̲ma* songs. Men of the Sun group sang from time to time throughout the ceremony, night and day, sleeping for only a few hours each night. Four Cloudy Water men, including Wandhuma and his son, also sang, especially at dusk and dawn. Members of other Dhuwa and Yirritja groups sang separately, but often concurrently, and less often than the Sun men.

During the first three days of the wake various groups danced upon their arrival to 'visit' the dead man, and 'to say goodbye'. The visitors were from the settlement and from other communities, some as far away as Yirrkala and Maningrida. They sat and later slept in groups in and around the periphery of the open area of the camp. Those from other parts of the community gathered on their arrival at the perimeter of the camp and performed a peregrinatory dance to enter the camp, then danced up to the body. The men 'visited' the body in dance. The women threw themselves on the ground by the body, wailed, cut their scalps with knives and other sharp implements, and sang mourning-songs. According to one commentator people danced to cherish and thank the dead man because he was a man of renown (*ya:kumirr yolngu*, 'name-possessing man') and had looked after the *rangga* sacred objects.

As each group of visitors sat, its singers would begin to sing *garma* songs. Women stood to dance on the spot to the songs. Often during the ceremony two or more groups sang different songs concurrently. Men and women performed dances oriented towards the deceased to 'visit' (*guwatjman*) him or her, and to 'say goodbye' (*gupa-ngamathama*, 'make good'). Some singers continued through the night, resting for only a few hours.

The group of Cloudy Water people with whom I arrived, together with some of their relatives of their own and other groups, performed a dance of the Black-capped Tern group called Fighting (*mari*), dancing across the arena to the coffin. Women and girls followed to one side of the male dancers. Cloudy Water people were able to perform this dance because the two groups were 'conjoint' (*dha:manapanmirr*).

In the Fighting dance the men ran forward carrying spears to form a circle, speared the ground, and repeated the forward movement. The dancers performed this sequence several times until they reached the dead body, then made as if to spear it. According to some of those present, the dance meant that 'the trouble is finished', implying that any accusations of sorcery of a kind which usually follow a death, and any subsequent attempts to exact vengeance, were at an end. (On another occasion people explained to me that the men did all that spearing in the dances so that they should not spear each other, for feelings run high after some deaths. But note the similarity of this dance to the dance of spearing Rock in the Honeybee circumcision, which would have been differently interpreted in that setting.)

Let me break into the narrative at this point with a general remark. At some stage of a wake a person would usually raise the question whether or not someone had killed the person with sorcery. A person was likely to be accused of causing the death in that way unless the deceased was an infant or very old, or had died from an acknowledged chronic illness. Men sometimes held a post-mortem, in which they unwrapped the body for older relatives, including *marrngitj* healers and seers, to examine. They interpreted certain features, such as the position of the fingers on the corpse, as signs indicating the identity of the sorcerer. An adult male relative of the deceased then accused someone publicly by indicating his identity indirectly (people did not usually accuse women), for example by displaying the group design of the accused. If no post-mortem was held and no one made an accusation, then people were likely to assert specifically that no sorcery was involved but only sickness, or that even if a sorcerer did kill the person they 'forgave' or 'set aside' (*bay-lakarama*) the deed. Groups who espoused Christian beliefs were especially likely to follow this course in the mid-1970s. The sorcery accusation might be followed by redressive action, such as going after the accused with spears, but homicide had been inhibited, especially on the larger settlements, through the intervention by missionaries and police. Returning to the Galiwin'ku wake:

After the men had finally jabbed spears at the coffin, the women of the party ran forward and threw themselves down beside the body, some banging stones against their heads, others wounding their scalps with knives, and all 'weeping' (*nga:thi*) mourning-songs, a combination of weeping and singing. (See C. H. Berndt 1950: 307.) A younger brother of the dead man also ran up to the coffin and made as if to cut himself with a large bush knife in a rather theatrical manner, but others restrained him.

On the second day a Cloudy Water elder proposed that because the dead man was the leader and a man of renown, the women and children should leave the camp while the men performed a dance from the Dhuwa Ma<u>d</u>ayin ceremony (see Chapter 6). Cloudy Water men and several men of other Dhuwa groups, together with their *waku* (wC) and *dhuway* (FZC, WH/WHZ) from Yirritja groups (Whale, Honeybee, and Lightning Snake), assembled after sunset in front of the shade, formed an oblong arena, and performed the almost silent dances. The women and children returned when they were over.

On the third day the Cloudy Water members performed with their *waku* and *dhuway* a public dance from the Ma<u>d</u>ayin ceremony. The men and women prepared themselves at a house nearby, the men painting their own and the women's bodies with the Little Kingfisher design of horizontal stripes in white, yellow, and red ochre. Three mature men painted up with a design of white paint covering the chest, divided along the collar bones and down the middle with a red stripe bordered by yellow ochre. This design was also used in the Ma<u>d</u>ayin ceremony. The 'inside' meaning of the design was not supposed to be known by women and children; women knew the design as 'Little Kingfisher'. The striped Little Kingfisher design signified the Ma:tjarra group, and it too had another *wangarr* ancestor as its 'inside' meaning, which only the men were supposed to know (see Chapter 6). Some women wore *wa<u>n</u>a* cords attached to their arms just above the elbow, as in depictions of the Djang'kawu sisters, and each man and woman carried a pair of red-ochred sticks about a metre and a half long, one in each hand. Most of the dancers were Cloudy Water men and women, but some were their *waku* (wC/ZC), *dhuway* (FZC, WH/WHZ), and *galay* (MBC, WB/WZ).

The senior *waku* of the group was Wurrpa<u>n</u>, Wandhuma's sister's husband and the Honeybee leader (Yirritja moiety). This man, together with an old man of the Whale group (Yirritja moiety) and a mature Cloudy Water man, instructed and directed the dancers, while Wandhuma and a Sunrise elder (Dhuwa moiety) from Yirrkala played the clapsticks.

The dancers formed two lines by the house where they had prepared for the dance, then walked to the leader's songs and clapstick beat, via a circuitous route, towards the dead man's house, each dancer using the sticks as walking-sticks, as in the Ma<u>d</u>ayin ceremony described in the last chapter. Every so often they formed two concentric circles facing inwards, then 'speared' the ground with the sticks. Each dancer moved his or her two sticks alternately back and forth, each with the point as a fulcrum in the ground, as the leaders called the *marrawinydjun* chorus. The dancers repeated this movement of walking, circling, and spearing five times, until they reached the shade where the dead man lay. They then encircled the house and shade, finally speared the ground, and called the *marrawinydjun* chorus by the coffin, whereupon the women and some men threw themselves to the ground and wailed.

According to Wandhuma's interpretation, in this particular performance the men represented one of the groups created by the Djang'kawu sisters, the Cloudy Water group, whereas the equivalent dance in the Ma<u>d</u>ayin ceremony signified the Ma:tjarra or Bungunbungun string of groups which they created, again in Wandhuma's interpretation. The women represented the Djang'kawu sisters travelling from country to country, spearing the ground with their walking-digging-sticks to make springs bubble up, collecting shellfish and grubs, and creating the groups. The *wa<u>n</u>a* cords represented the *ma<u>d</u>ayin* which the sisters left at each country. The dancers 'followed' the Djang'kawu with their digging-sticks and dilly-bags, so that the dance 'visited' several places in its imaginary journey just as in the Madayin ceremony: the countries of Mangrove, Two Women, and Thunder groups, and finally Cloudy Water country, the country of the dead person.

The same afternoon members of the dead man's own *ma:ri* and 'mother' groups placed the body into the coffin, as a leader of the Thunder group (Dhuwa moiety) sang the Mistletoe Tree song, which signified the *rangga* sacred objects. Men of both moieties called the *marrawinydjun* chorus, during which Wandhuma invoked Red-collared Lorikeet, whose feathers adorned sacred objects of the Dhuwa moiety, the names of the *rangga*, and the *likan* names of the country. During the invocation several men lifted the body into the coffin while a number of women danced Longicorn Beetle Larvae. As the power-man invoked Red-collared Lorikeet, other men put some *wana* cords, decorated with red-collared lorikeet feathers, into the coffin. The chorus and invocation were performed whenever men moved the body or the painted coffin.

Flag and Tobacco dances of the Whale group (Yirritja moiety), depicting Macassan trepang-collectors, followed, danced in the cleared arena in front of the shade where the dead man lay. Whale was the dead man's mother's group. At the end of each dance the dancer placed a gift of a flag or tobacco inside the coffin. Then Cloudy Water and Up-river men (Up-river group comprised Honeybee and Lightning Snake groups, Yirritja moiety) performed dances of the Stringybark and Blow-fly groups, their 'mother's mother' and 'mother' dances respectively.

Meanwhile in the house, out of sight of women and children, the Thunder-group leader and a Honeybee man painted the coffin lid with Cloudy Water *likanbuy* designs depicting springs at Cloudy Water, Thunder, and Two Women countries. Whale men (Yirritja moiety) painted their Octopus design, 'mother' to the dead man, next to the Dhuwa paintings. In former times men painted designs on the dead person's chest. (The Yolngu adopted the plywood coffin when the authorities started to require a post-mortem examination, usually in Darwin, before burial.)

When the paintings were finished, the men in the house called Whale and Cloudy Water invocations over the designs, wrapped the lid in cotton cloth, and carried it to the men at the coffin. During this short journey they danced Octopus as a man of Crocodile group sang songs of the Whale group, the singer's 'mother's mother'. Two men then nailed the coffin lid down, with the paintings on the inside, as song-men sang from the Whale- and Crocodile-group song-series about the Macassans. These included songs of Boat-building, and Knife.

At dusk some of the women would 'cry' their mourning songs. Later that evening in the shade where the coffin lay, after people had finished eating and were sitting or lying by their fires, Cloudy Water and Honeybee men stood in a huddle to sing Gudjika songs from the Ma<u>nd</u>ayala ceremony of the regional genre. These songs belonged to the Stringybark group, 'mother's mother' to Cloudy Water, and 'mother' to Honeybee group.

The following morning the dead man's close women relatives wailed for the last time. Wandhuma invoked the *rangga* and *<u>l</u>ikan* names related to the Cloudy Water country as other men called the *marrawinydjun* chorus, signifying the incoming tide, and lifted the coffin into the back of a station-wagon. Others drove with it to the burial ground, where the Christian minister held a short service. They had 'given the body to the mission'. The minister told those present that in order to be real Christians they should give up the traditional ceremonies, which were the work of the Devil; nevertheless they should be Christians in a Yolngu way. A Yolngu man standing next to me commented wrily that the minister was *mirithirri <u>d</u>a:l*, 'very hard' or 'very strong'.

The house in which the dead person had lived was probably vacated for a while, as this was the usual practice. Customarily, residents abandoned bush camps for a season or so after a death and closed the country around to foraging when the wake was completed, to be opened later with purification ceremonies.

Generally, in the wake people 'said goodbye' to the dead, they 'sent them on their way'. Milingimbi people did not say to me, as they had explained to Warner, that the purpose of the ceremony was to separate the soul of the dead from the living and guide or transfer it to the 'clan' waterhole (Warner 1937: 444). However, many songs and dances represented journeys towards the group's country. In this case the Little Kingfisher dance was the most significant. Some dances represented the examination of the corpse by the *wangarr* in order to see if it was of the correct affiliation, and many songs and invocations were symbols of the incorporation of the individual with the *ma<u>d</u>ayin* (cf. Warner 1937: 414, 446).

On this occasion people did say that the ghost (*mokuy*) of the dead man damaged a boat during the ceremony, for when one family was preparing to leave, the gear-cable of the large fishing boat in which they had travelled from Milingimbi was found to have been burnt by the exhaust pipe. A few months later, during a visit to Cloudy Water country, Wandhuma explained to me that the dead man's spirit had made a small hole which he found in the scum on the surface of a waterhole (Chapter 3), which suggests that the return of the spirit to the person's country was the expected aftermath of the wake.

The form of wakes was similar to circumcision ceremonies in including songs and dances representing journeys to the country of the focal subject. Another similarity was that the 'making good' painting (*gupa-ngamathama*), the invocation, and (in this example) the inside Maḏayin dance, identified the focal subject of the ceremony with the *rangga* sacred objects, the *wangarr* ancestors, and the country. The body in its coffin occupied the same place in relation to the 'inside' dances as a *rangga* sacred object, and was decorated, like the *rangga*, with *ḻikanbuy* designs and *waṉa* cords, the public equivalents of *rangga* sacred objects. All these ceremonies projected the person as a part or phase of a larger process of transformation between ancestor, place, and incorporeal being. In the Cloudy Water wake the dead leader's relationship to his 'mother' group, as well as to his own group, was reiterated in ceremony, although he was most closely identified with the latter. Unlike circumcision initiation, men and women were treated in a similar way in the wake.

Purification

A death had effects which called for purification. Purifying with smoke got rid of the smell of decay and chased away dangerous *mokuy* spirits of the dead. Purifying with water and red ochre released close relatives of the dead from prohibitions arising from a polluted state called *wukuṉḏi* which they acquired at the time of death, making people 'clear' (*rebal'yun*, 'clean up', 'burn off undergrowth', 'tidy up').

Since it had no *manikay* songs and no Fire song the Cloudy Water group possessed no smoke purification ceremony (*ḏaḏa*). Cloudy Water people had to ask another Dhuwa moiety group to perform the ceremony when they needed it. A smoke ceremony usually followed immediately on the burial, but after this wake the Sun and Morning Star people (Dhuwa moiety) performed a washing ceremony a few days after the burial at the dead man's camp. Members of only seven groups took part, some people from Yirrkala and Lake Evella (Gapuwiyak) already having left.

Five days after the washing ceremony at Galiwin'ku the Cloudy Water people who usually lived at Milingimbi put on a washing ceremony there, at Wurrpaṉ's camp. (Wurrpaṉ was the senior *waku* at the wake, as well as being Wandhuma's sister's husband.) Through the course of the day the washing ceremony followed in song the journey of the Djang'kawu sisters, from the same series as the men sang at the wake and in the evenings of the Ma<u>d</u>ayin ceremony. It ended with groups of people standing in turn in a sand-sculpture which represented the springs at Wa:<u>l</u>angur created by the sisters, as water was poured over them and men performed the *marrawinydjun* chorus and invoked *<u>l</u>ikan* names.

Secret Domains and Codes in garma *Ceremonies*

The mortuary ceremonies outlined above illustrate several ways in which men's exclusive elements were woven into a public ceremony, and the contexts in which people deployed songs and designs with secret meanings. Exclusion through spatial segregation and barriers to vision and sound in public ceremonies were similar in structure to barriers segregating esoteric from exoteric domains in revelatory rites.

In the Cloudy Water wake the men decided to perform a secret part of the Ma<u>d</u>ayin ceremony because the dead man was important, and excluded the women from the camp while this was performed. In post-mortem examinations men unwrapped the body but protected it from view while the *marrngitj* healers or other men examined it for signs of the identity of the sorcerer. Men painted the coffin lid with *<u>l</u>ikanbuy* designs within the privacy of the house, from which women and children were excluded. And I think that men without close relationships to the groups whose paintings were being completed would not have entered without an invitation. Once painting was completed, men wrapped the coffin lid before carrying it out to the coffin and nailing it in place.

When men painted an initiand during a circumcision, he lay within the group of men in the public camp but some way away from where women and children sat. Before the initiand moved back into everyday interaction, men would smear the design to erase the cross-hatching which was the mark of esoteric designs. The same stricture applied to a dancer wearing a *<u>l</u>ikanbuy* design.

Secrecy was constituted in the public domain not only through the control of space and by restricting sensory contact with the medium, but through secret codes.

Secret Codes in *garma* Songs

Men maintained secrecy in Yolngu *garma* songs by encoding secret meanings in ambiguous forms and restricting access to information about those meanings. All the groups participating in the Cloudy Water wake described above sang and danced to long and complex song-series. I shall pick out elements from songs of the Cloudy Water group used in a washing purification ceremony, and similar to those they used in the wake (see Chapter 6).

Like the Ma<u>d</u>ayin ceremony, people construed the songs as a journey through various topographical categories—through the sea to the mangroves, on to the salt-flats, into the forest, back to the plains, swamps, and lakes. They also conceived of the songs as a westward journey equivalent to that of the Djang'kawu from Burralku, or as part of that journey.

The language of Yolngu *manikay* and *bi<u>l</u>ma* songs was related to speech, but consisted of cryptic phrases and clauses, and included lists of proper names and special song-words (including verbs) and archaisms. Unlike myths, songs were not strictly narratives but strings of descriptive expressions, as well as lists of names. But they indirectly implied a narrative through the succession of topics related to a myth. Singers or others often referred to the myths during a performance or at other times. The exegetical statements by the Cloudy Water leader (Tables 11 and 12) show how a cryptic song-text was linked with the co-ordinating story of the Djang'kawu sisters, although the song did not itself have a narrative form.

There were a number of means by which to frame secret interpretations of *manikay* songs:

1. Interpreting a song-word or archaism which had no everyday sense or referent with a word or phrase in everyday speech. Here, different male performers sometimes interpreted the same word differently.
2. Assigning an 'inside' meaning to an action described in a song which involved an object such as a plant. The 'meaning' (*mayali'*) of the plant was, say, a sacred object. Thus the song described the *wangarr* digging up, say, a root, but this meant that the *wangarr* put a sacred object into the ground.
3. Assigning a referent to a proper name whose significance was not supposed to be common knowledge. For example,

TABLE 11. Little Red Flying Fox Song of the Cloudy Water Group

gapu ngakul ngurruyunawuy	water heard mixed together
dharrayigatinyawuy,	tide came in powerfully,
malwawa Gunbulapula gurrungal,	grandfather [Flying Fox] above,
gapu nha:ngal bulul'yunawuy	water saw open (?)
dharrayigatinyawuy,	tide came in powerfully,
Malwawa Gunbulapula Man.girrikirr Gawulayulay/	grandfather [names of Flying Fox]/
warrakanmirr nhe rirrakaymirr	game-possessing you noisy
gulikulimirr nhe,	fluffy you,
warrakanmirr nhe rirrakaymirr	game-possessing you noisy
gulikulimirr nhe,	fluffy you,
Djarraman.gungu nhe,	created by Flying Fox you,
Malwawa nhe gurrungal	grandfather you up above
Man.girriki nhe,	[Flying Fox] you,
rirrakaywuy nhokungu	pertaining to noise from you
wulma djalyalyun,	thunder nears,
bay' nhungu Dja:pul	Oh your [place name]
Yawurryawurr Balmanga Djunumu Djarramangungur Yambungal/	[place names]/
Bay' nhe gurrungal rraygurru'yun,	Oh you above tired,
liyanydja nhe wurrulwurrul,	head (of) yours (of) old man
dhiyali nhe gurrungal	here you [are] above
Mulanggatjngur	[at place]
Gaymanangalangur nhe,	[at place] you,
Gupagathar warrakanmirr	[name of tree] game-possessing,
gulikulimirr nhe	fluffy you
Yirriyirrimirr Ganbalatjmirr//	Lorikeet [feather] possessing//

the proper name of the men's secret ceremony-ground might be in common use as a person's name, and as a place-name in songs, but the fact that the name did signify the men's ground was secret. People might apply one proper name or set of synonyms multivocally to a wider set of objects, connecting them through the same proper names.

4. The topic of the song, such as an animal/*wangarr*, was also the topic of a secret men's dance. The actions or attributes of the animal/*wangarr* described in the song connoted the dance.
5. The topic of the song, or some aspect of it, was the same as the subject of a secret design or object which only men saw, or the meaning of which only men (of a particular group-centred network) were supposed to know.
6. The topic of the song, which had a secret meaning in one of

TABLE 12. Salmon Catfish Song of the Cloudy Water Group

guya wali, guya wali,	salmon catfish, salmon catfish
guya wali, guya wali,	salmon catfish, salmon catfish
dimirri-da:lmirri/	sharp spines [da:l, 'hard']/
dunhunbu nhe guya ma:ri,	[name] you fish [my] MM/MMB
dimirri-da:lmirri guya,	sharp-spine[d] fish
wali wali dimirrinydja,	[name] [name] spines
guya nhe-e-e/	fish you . . . /
ba:ni gara dhambakungal,	lying-in-water spear turned [in dance]
rapurapu miliyarr gala-pungal,	[spear names] held-up [in dance]
dhukarr [?] dhunupa-e-e/	path straight/
gapan nhungu gurrungal	white clay [of] yours above
Nupurray,	[at] Nupurray
gurranygurrany mawuda-e-e/	[names of white paint]/
guya djimalinya,	fish [name]
wali wali weduwuy,	[song and common name of salmon catfish]
dhumdhurr wirringur	cloudy-water cloudy-water
gabubumirr//	scum-possessing//

Exegesis

The Djang'kawu sisters paddled along, scattering the salmon catfish, called its names and sang about it, calling it *ma:ri* [MM/MMB]. It has sharp spines near the tail. Salmon catfish dances with his spear, turning it, as the Djang'kawu see it and sing. He got his paint [white clay] from Muwalangal. As the fish swim through the sea they make the water cloudy.

the ways set out above, also elicited esoteric mythic statements, such as that the certain *wangarr* made and gave a ceremony ground to another *wangarr* or to a person.

The inclusion of many special song-words and archaisms no doubt lent the songs an aura of mystery. The Salmon Catfish song-item in Table 12 includes the verb forms *dhambakungal* ('turned') and *galapungal* ('held up in dance'), which were not in used everyday speech. Esoteric nouns and names in the song include *wali* and *djimalinya* which denote 'threadfin' (the common name was *djulurrpi*), *dhumdhurr* ('cloudy water', though this may be a place-name), *rapurapu*, and *miliyarr* (names of spears).

A key phrase in the Cloudy Water songs occurred in a song about Little Red Flying Fox, among others (Table 11). The song describes the animals as 'tired old men' hanging in the trees, and describes the trees with the phrase, 'game-possessing, noisy, feathery/fluffy'. This phrase connoted many complexes of meanings in the ceremonies: the feathery flowers of the tree; the image of Little Kingfisher 'playing' at the tree in the daily dance of the

Ma<u>d</u>ayin ceremony; the feathered dilly-bag hanging in the tree in the Djang'kawu story; noisy mobs of flying foxes in the trees; and the Djota digging-stick, named after the tree, adorned with the group's orange lorikeet-feathered dilly-bag and feathered *wa<u>n</u>a* cords standing in the sand-sculpture in the *garma* washing-ceremony. Such phrases had secret meanings relating, for example, to dances in the inside ground and to sacred objects, but other associations would be available to anyone who was familiar with the songs and related ceremonies, creating cross-cutting webs of significance, the more so as secret dances of the Ma<u>d</u>ayin ceremonies also had equivalent public songs. Thus the songs did not have an analogical significance only for men who were supposedly privy to the secret meanings. Indeed, women's knowledge of men's songs closely paralleled that of men (Franca Tamisari pers. com.). McGowan (n.d.) points out that girls were exposed from the earliest age to the public meanings of songs, concerned for the most part with the properties and activities of species.

Secret Codes in Geometric Designs

Yolngu art encoded meanings in the context of a revelatory system of knowledge in which knowledge of certain 'levels' of meaning was restricted to a few individuals, and in which knowledge of the meaning of paintings varied according to an individual's status (H. Morphy 1991).

As with song, men maintained the secrecy of designs by restricting sensory contact with the medium, and by encoding meanings in ambiguous forms. Morphy shows that in a given context people could easily interpret figurative representations, such as designs painted on to hollow-log coffins, intended to be interpreted to refer to particular objects. The limited number of figurative representations in the geometric *<u>l</u>ikanbuy* designs, however, reduced the number of easily interpretable signs within them. Men painted designs of this kind on to the coffin lid in the Cloudy Water wake, and on initiands and dancers in circumcision ceremonies (Chapter 5). They appeared in simplified form as sand-sculptures in purification ceremonies, where they were on public display.

The Yolngu acknowledged that the properties of geometric art helped to maintain secrecy. One man informed Morphy that if people looked at a geometric painting they would think that it meant nothing. Another man offered the opinion that such a design looked as if it did not have many meanings whereas in fact it had more meanings than the figurative paintings. One of the functions of the geometric component of *likanbuy* paintings, then, was to restrict the number of people who could interpret them (H. Morphy 1977: 228).[3]

The levels of 'inside' and 'outside' interpretation were relative rather than absolute, according to Morphy. A particular interpretation was 'inside' relative to another only until further information had been revealed, when it was 'outside' in relation to that new information. However, it remained 'inside' in relation to the information revealed first (H. Morphy 1977: 253; 1991: 78). Morphy shows that Yolngu learned the significance of a set of paintings at one 'level' before progressing to another, although some designs were more restricted than others. The most publicly available interpretation of one Manggalili painting referred to the activities and journey of Koël together with certain other *wangarr* ancestors. A second level of interpretation revealed that the painting was a map of the group's country. This level was known to the group leader's sons and brother's sons over the age of 15 (H. Morphy 1977: 253, 261, 266).

As an individual learned new meanings, he gradually developed knowledge of an abstract 'template' common to all paintings of a set, and which encoded the interrelations between the meanings of the paintings, represented in concrete form by the *rangga* sacred object. The template consisted of several 'feature blocks', on which could be located multivalent geometric designs, or figurative equivalents of aspects of the significance of those designs. This template could also be superimposed conceptually on the form of the *rangga* (H. Morphy 1977: 275–7). I might add that the *wana* cords, public equivalents of the *rangga*, could also be interpreted in the same way, as could certain versions of a group's hollow-log coffin. This most condensed form was, I think, equivalent to a particular song-phrase. In the Cloudy Water

[3] However, Morphy does not think that the objective of maintaining secrecy was the major explanation for the predominance of geometric motifs, rather, 'The geometric art, because of its multivalency and potential ambiguity, encodes meanings in such a way that it can be used to make complex statements about the relationships between things that cannot be made in other ways' (H. Morphy 1977: 228).

songs, the equivalent phrase to the design 'template' was probably the one mentioned earlier, 'Mistletoe Tree, game possessing, noisy, feathery/fluffy'.

This outlines the structure of secret codes. I turn now to the ways in which men controlled admittance to encoded knowledge.

The Control of Secret Religious Knowledge

Several dimensions of the control of knowledge can be distinguished: who was allowed to produce or perform an item, who could hear and see the item, who told whom about it, and what information they imparted. As Morphy shows, the right of a person to produce a particular ritual form in a given context did not always coincide with the right to divulge information about its significance (H. Morphy 1991: 58). In considering these dimensions we will see that relations among kin belonging to different groups were intermingled with age and gender relations.

Age, gender, group identity, and kin relation to a group were important determinants of who could impart information about elements of ceremonies, especially secret meanings, and to whom. As I suggested earlier, while the control of esoteric knowledge did not work as a formal structure of grades based on age and gender, legitimate levels of knowledge divided into two broad categories of people, on the one hand, females and uninitiated males, and on the other, initiated males, although there was some overlap and much variation in knowledge within these broad categories (see H. Morphy 1991: 84). The circumcision ceremony marked the passage from the first of these categories to the second.

We saw in earlier chapters that men and women had distinct roles in *garma* and other kinds of ceremony. In the former, men sang with clapsticks; such singing could be called 'tapping' (*d̲ar'ṯaryun*). Women 'wept' songs with similar texts. Men's and women's dances differed, and only men could paint *l̲ikanbuy* designs, and only senior men produced designs which represented those placed on the *rangga* sacred objects. While some women were involved in the production of bark-paintings and other forms of art for sale, especially at Yirrkala, such women claimed only limited knowledge of their significance. Morphy reports that the daughter of one Yirrkala group's leader would only impart the references of figurative paintings and the general location it referred to. She did not have the right to discuss its detailed topographical associations (H. Morphy 1977: 262).

Everyone could actually hear *manikay* songs, for they filled the auditory space of a ceremony, but knowledge of esoteric interpretations of the songs was (by definition) restricted, and even the sight of some types of design was limited to men. As in the case of the painted coffin-lid in the Cloudy Water wake, men often covered a *likanbuy* design when transporting it through a public area. When painted on the chests of circumcision initiands or Madayin ceremony novices, *likanbuy* designs were desensitized, as it were, by smearing the cross-hatching, leaving the broad outlines intact. These designs gained their special status from the fine cross-hatching. Such smeared designs were similar in form to the designs of sand-sculptures used for purification ceremonies, open to public view, although men did not tell everyone their detailed interpretations. Thus people who were familiar with *the form* of a design did not necessarily know its inner interpretation. Indeed, like *manikay* songs, knowledge of the form of sand-sculpture designs was available to all.

Turning now to groups and kin relations to groups, in general terms I showed in Chapter 4 that people could use elements belonging to their own group, their mother's group, mother's mother's group, and, in the case of dances and women's mourning-songs, MMM's group. These categories extended to the elements of groups with the same *madayin* as each of these categories. A male mother-holder or *waku* (ZS) had the obligation to 'work' for his mother's group by producing designs (but not singing) in a ceremony. (For example, during the mid-1970s senior *waku* of the Honeybee group, who were themselves of a Black-capped Tern group, made its sand-sculptures for purification ceremonies.) But male *waku* could not claim the full range of rights as a member of the group, for they could not sing their mother group's songs (the women could) and usually refrained from imparting secret knowledge without the permission of men of the group.

Some men such as a group's senior *waku* (ZS) could claim the right to be consulted before designs were used in a ceremony, even though such men never made them themselves. In spite of such assertions, during Morphy's research male *waku* were sometimes unwilling to impart information about a design in the absence of an older man of the group. Even the senior male *waku* of one group would not discuss the group's paintings with Morphy in any detail, and advised a son of the leader to impart only 'outside' meanings of a painting to Morphy in the absence of the leader. Indeed, the second oldest man of one group at Yirrkala deferred to the leader, even with respect to paintings he made himself (H. Morphy 1977: 66, 255; 1991: 63).

Some attempts to constrain the dissemination of information were made among group members. According to Morphy the eldest son of an eldest son was acknowledged to have authority over a group's painting in relation to other members of his generation. As with group leadership, ideals of primogeniture and generation might clash, leading to conflict between a leader's son and men of his father's generation (H. Morphy 1991: 61–3).

The oldest active man of a group might claim or be accorded the right to be consulted, or at least informed, when information about religious elements, secret or public, was imparted. For example, a Honeybee man had been helping me for some days to transcribe and translate *garma* songs belonging to the Spear-grass group but used by Honeybee people, some recorded by people living at Nangalala, others performed by him. When we passed the Honeybee leader's camp one day my instructor told him that he was telling me about the songs, perhaps to ensure his approval (see also H. Morphy 1977: 66–7).

However, people did not always inform or consult a leader in this way. I have already described how a Cloudy Water man showed me and told me about a *l̠ikanbuy* design over which he had particular rights. As far as I know he did not inform his 'father', the group leader. But in the case of secret information, failure to inform a leader could lead to conflict, as the following events about *rangga*, the topic of secret designs, show:

In 1976 the Milingimbi school headmaster presented Department of Education plans for a new school-building to the Milingimbi Council. One proposal was new buildings on the existing site. An alternative was a new school on a site to the south of the existing school. The Council was prevaricating because the proposed new site was uncomfortably close to a sacred place in the forest behind Bottom camp. A man aged about 50 of the Honeybee group, the son of the sixth wife of the leader's father's younger brother, and for whom the country was *ma:ri* (MMB), took several young married Yolngu men to see the sacred site and the sacred objects secreted there, so that they would understand why the school should not be placed near the site—there would be a danger of children straying too near the place. One of the young men was a teacher in the school, another was a teaching assistant, and one was an active intermediary between the Yolngu community and the missionaries (and later became a community adviser). The leader of the Honeybee group was reportedly very angry at what his brother had done without consulting him.

We have to bear in mind here that leadership was not an office, and that people did not always agree about who the proper leader was (Chapter 2). Chapter 5 showed that men of one group had at least temporary effective control over the religious life of other groups of the same moiety who were said to lack competent singers. It seems that Honeybee men maintained this control by refraining from performing the other group's songs, except in a token way. The decline of the dominant Honeybee group's fortunes after the death of a prominent male leader in the early 1980s was followed by a resurgence in the competence of at least one of these other groups. In this case the group seems to have taken on and adapted the songs of closely related groups (Franca Tamisari pers. com.). Honeybee people may have done something similar in the past, for they sang mainly songs which they called their *ma:ri* (MM) songs and attributed to the Spear-grass group (cf. H. Morphy 1977: 86, 1991: 63).[4]

Turning now to the imparting of knowledge and competence, men were most likely to teach close kin, especially sons and sisters' sons, both how to produce forms, and the meanings of those forms. Men passed through many stages of knowledge and competence: seeing the 'inside' dances and later performing in them; sight of at least one of the sacred objects; participation in making or refurbishing an object; gradual acquisition of competence to sing by participating over many years; learning how to paint designs, and learning the meanings of those designs; coming to know the order of ceremony and all the *likan* names, and so on. But people did not admit others to competence strictly on an age-

[4] This kind of right was evident in the co-operation of Dhuwa moiety groups with the Djang'kawu as *wangarr* in the performance of the Madayin ceremony (Ch. 6). Another nice example can be seen in the film *Waiting for Harry* by Kim McKenzie (Australian Institute of Aboriginal Studies) in which Yan-nhangu men from Milingimbi visit a Burarra hollow-log ceremony at Cape Stewart and assess and criticize the paintings on the coffin.

categorical basis, for not all young men had the opportunity, say, to wind the *rangga* in the Maḏayin ceremony. The leader's son and one of the sons of the senior *waku* were permitted to do this in the performance described in the last chapter.[5]

Young people learned the public dances and songs first; adults encouraged children to participate in dances as soon as they could walk, and young boys would sit with their fathers as they sang. Learning to paint and make sacred objects, and learning esoteric elements, came later. Yolngu men emphasized the role of a man's 'father' (*ba:pa*, father and father's brother) in the transmission of knowledge and skill. For example, Wulakitj, a mature man of the Cloudy Water group, told another man of the group in my presence that he had told his teenage son 'everything' about the group's *maḏayin*.

Howard Morphy generalizes that a young man gained the right to produce his own paintings first by watching his father paint, ideally after he had been circumcised. (This might be actual father or father's brother.) A son had to gain his father's permission to produce a painting for sale that he had been taught, and was under an obligation to share the proceeds with him, although this was not necessarily done at Yirrkala. An eldest son was supposed to be taught paintings and their meanings first. Ideally, a man learned the most restricted designs just before, or soon after, his father's death; a father might set out formally to hand over the designs to his sons before his death, or an older brother would teach a younger brother after the father's death. If the eldest son showed no aptitude, then a man taught a younger son in preference (H. Morphy 1991: 61–3). Thus the leader of one large group complained to me that his sons were not learning the group designs. I asked him whether or not he was teaching his oldest son, a man of about 40, about the paintings. This man was regarded as socially rather incompetent, but he produced many small bark-paintings of figurative representations of plants for sale, but not the more important paintings. The leader replied

[5] Morphy reveals that at Yirrkala during the mid-1970s young men, and some women, learned how to paint a design before being told about its meaning. They first made 'outside' paintings, such as designs of the group *mokuy* ghost. One young man produced his first bark-painting 3 years after circumcision, at the age of about 13. Men in their early twenties produced their first 'inside' designs. They learned the 'outside' story of the painting first and only later the 'inside' meanings (H. Morphy 1991: 58 ff.).

that he had tried, but to no avail. I suspect that ability to produce a design was prerequisite to full knowledge of its significance.

People attributed the absence of knowledge and skill (the verb *marnggi* denotes the possession of both) in religious matters to the failure, for whatever reason, of fathers to teach their sons. And they testified that certain versions of ancestral events as well as interpretations of sacra were true, on the grounds that certain senior relatives (fathers, mothers' brothers, and so on) had been their instructors. If a knowledgeable man died without imparting knowledge to sons, or if the group had no other competent men, then a man of another group, such as a *waku* (ZS, mother-holder) or *gutharra* (ZDS, mother's mother-holder) of the group, 'looked after' the knowledge until the group again had men of an appropriate age to be taught. Such custodianship sometimes led to conflict among claimants to the position, just as there was conflict over country. Moreover, as we saw earlier, custodians of the same moiety were sometimes reluctant to hand on knowledge. Men would also refuse to hand on knowledge as a sanction against sons' offences, such as drinking too much alcohol, or refusing to share proceeds of the sale of paintings in which the father had rights. For reasons such as this, a man would teach his group designs and their significance to his sisters' sons (*waku*) as well as his own. If a group had no competent male *waku*, the son of a group's *waku* could be taught his father's mother's designs, or the *waku* of a group with the same *maḏayin* might take on the knowledge (H. Morphy 1991: 63). We have also seen that men learned their *ma:ri* (MM) group's designs, and that men and women sang their *ma:ri* songs. Presumably the men also had access to knowledge of the secret interpretations of these designs and songs.[6]

This chapter has outlined means of encoding secret religious knowledge in publicly disseminated forms, and patterns in the control of secret knowledge. The conclusions of this chapter will also provide the opportunity to draw together the three chapters of Part II.

[6] One man told Howard Morphy that his *ma:ri* (MMB) told him everything when they held a ceremony. Because making paintings implied rights in the associated country, men rarely produced their *ma:ri* paintings for sale unless the group had no male members (H. Morphy 1991: 60, 65, 67 ff.).

Conclusions

To control access to religious knowledge and maintain secrecy is, by definition, to exercise power over others. But to what end or with what effects did Yolngu men withhold or reveal secret religious practices and information? Many discussions of 'secrecy' as a mode of activity link it to power relations more generally, especially those between men and women, and between older and younger people.

Arguments linking Aboriginal religion to gender relations, and indeed debate about Aboriginal gender-relations more generally, have shown a tendency to make a tally of women's as against men's power or status (see Merlan 1988). A tick in the women's column for autonomy in foraging, a tick in the men's column for a high level of polygyny; add up all the ticks for an overall assessment. Thus we might balance the exclusion of women in the Ma<u>d</u>ayin ceremony with Gunapipi, in which women had their own particular interpretations of songs that were similar in form to men's (see Chapter 8); or compare men's ability to appropriate goods from women, both in the form of cycad-palm nut bread (*ngathu*) in the past and in the 'feeding the *malagatj*' sequence, with the phase of the circumcision ceremony in which women were licensed to snatch food and other resources from men.

But notions of 'power' and 'domination' are crude instruments, implying simple asymmetries and a promise of quantification, impossible to gratify except in terms of 'more' or 'less'. For the record, it is patent in my opinion that Yolngu men had more control of the conditions of women's lives overall than women had of men's. The very high levels of polygyny, as well as the greater use of violence by men against women than the reverse, are indices of this asymmetry (Gerard n.d.). That being said, the reality was both more complex and more interesting. For instance, many older women increased in knowledge, power, influence, and independence with age (see H. Morphy 1991: 76); some wives dominated their husbands and daughters' husbands; some men had more power than others; some relatively powerless younger siblings became more powerful having survived their older siblings.

This book represents Yolngu men's practices more than women's. Consequently, even if such a thing were possible in principle, a balanced assessment of the relative power of Yolngu men and women in the mid- to late 1970s cannot be made here, and this book does not pretend to be a complete or balanced account of Yolngu social life of that period. What can be done is to consider the organization of males through religious means, some of the ways in which men benefited from their hold on religious secrecy, and how this hold may have fitted into a wider nexus of processes, practices, and relations. These matters will have implications for gender relations.

Part II began with the ritual separation of boys from women. Circumcision initiation marked the detachment of boys from women and girls, symbolized in Ma<u>d</u>ayin by the division between secret ground and hearth, and more generally by the distinction of 'inside' from 'outside'. At circumcision initiation boys became 'bachelors'. Though not precluded from the hearth, they had to sleep in the young men's camp, and they avoided their sisters. And they could, other things being equal, be admitted to men's secret aspects of rituals and encoded knowledge. Participation in circumcision and secret rites was necessary for a young man to enjoy all the prerogatives of adult life, such as the right to marry and to use hunting-techniques and technology.

Chapters 6 and 7 have outlined Yolngu technologies of religious secrecy. The two main modalities were the control of space and of information encoded in ambiguous forms, characteristic both of the control of access to religious knowledge between group-centred networks and between age and gender categories. In both *garma* and revelatory rites spatial barriers were set up and guarded, but the kinds of sounds used, especially the relative silence of inside dances in Ma<u>d</u>ayin, were also a mode of minimizing the range of communication. Men restricted sight of some designs or types of design, and revealed 'inside' interpretations of multivalent designs and expressions to young men. Hierarchies of 'inside' to 'outside' interpretation were not absolute, however, but relative and changeable.

The constitution of religious secrecy as a mode of control requires that people value the promise of knowledge. Simmel and others have proposed that secrecy itself enhances the value of what is hidden (Simmel 1950: 330, 333; Barth 1975: 217–18; Luhrmann 1989). However, I suggest that secrecy alone does not produce value; an additional requirement for information to be a source of power is that it has some intrinsic worth, increased by scarcity. Notions of intrinsic worth were involved in Yolngu religious secrecy, for people believed that the most intimate representations of powerful *wangarr* ancestors were imbued with their power, which could be tapped. The most important secret objects, designs, songs, and so forth were connected with the most powerful creators such as Crocodile, Djang'kawu, Olive Python, and Rock. People believed the most secret manifestations of ancestors—*rangga* sacred objects, 'inside' ground and dances, and *likanbuy* designs—to be the most powerful. Howard Morphy (1991: 83) makes a similar point; that the 'inside' is closest to ancestral creation, the core of Yolngu metaphysics.

Because people believed them to be manifestations of the *wangarr* ancestors and their power, the elements of the Madayin ceremony described in the last chapter had an intrinsic value to people. Songs were words that the ancestors sang. At least some believed the inside ceremony ground and the sacred object to be imbued with ancestral power; for example, a young man was warned that if he wrongly left the ceremony the body-paint would 'bite' him. Such beliefs were not determinate, but relative to an overall context. This is shown by the Cloudy Water leader's concern that younger men may have set less value on the Madayin ceremony because of their interest in white people's things, apparently leading him to waive the usual gifts. The possible devaluation was due to the existence of alternatives which were not formerly available.

Given this general value, however, it was the fact of being secret that differentiated 'inside' religious knowledge from 'outside', for as a rule Yolngu celebrated the same *wangarr* ancestors in public and secret forms. One name of an ancestor or object was publicly known, another secret. A certain dance was secret, but the same topic as a song was public. During the public wake, men might perform a dance that was secret in the context of a Madayin ceremony (cf. H. Morphy 1991: 76). Moreover, the content of secret knowledge could change. As Simmel remarks, 'what originally was manifest becomes secret, and what once was hidden later sheds its concealment' (Simmel 1950: 332, 335). Yolngu men sometimes 'brought out' (or claimed to have brought out) former secret proceedings into the public arena, while others remained secret. This led to sometimes bitter disagreements with people of more distant areas who still held a similar form to be secret.

Herdt (1990: 362) raises the important question of how much secret information or procedure the excluded actually 'know'. What is important with respect to power according to Lattas is not so much actual secrecy or knowledge but the 'ideology' of knowledge and secrecy; male deceptions are responded to *as if* they were real (Lattas 1989: 453). Morphy discusses this issue in some depth. Knowledge of inside and restricted matters always extended well beyond the boundaries of the men's inside ground. Inside things were always linked to outside ones; thus Yolngu men produced messenger forms of sacred objects carried in public which were very similar to the restricted objects (Morphy 1991: 80). In many respects men's and women's access to knowledge was equivalent and in private contexts men acknowledged this to be the case, although women were less willing to admit to knowledge. Women and children often saw revealed in public what men produced in secret. (A Morning Star pole for the exchange ceremony would be an example.) Women sometimes uttered secret names during mourning, and men at Yirrkala sometimes included outside names in invocations so that the women would know what the invocation was about. But knowledge of what it was that men controlled enhanced the power of secrecy; the mask of secrecy was 'an illusion and a tease', not the restriction of knowledge but an assertion of control (98–6).

The Yolngu concept of knowledge provides a clue here; *marnggi*, 'know' or 'knowledgeable' implies ability and the right to do something. It was what a person could do that was important rather that what they could think or remember (*guyanga*). Yolngu varied less in what they 'knew' in an intellectual sense than in the degree to which a person was 'knowledgeable' (*marnggi*) in the Yolngu sense—whether someone could legitimately divulge certain information or act in certain ways. After all, simply knowing would not have been a threat, but to demonstrate illicit knowledge would have been subversive, undermining the value of secret knowledge. Women could not generally admit to being knowledgeable (*marnggi*) of inside things or demonstrate their knowledge. Indeed, women concurred that certain words, names, forms, meanings, and performances were for males only. Conversely, men disclaimed knowledge of, and refrained from talking about, matters to do with childbirth which were 'women's business'. Women's and men's interpretations of ceremony elements differed, and men saw this difference as a matter of women's ignorance (excepting a few older women) and lack of possession of *rangga*. Men's use of physical force and threats of sorcery undoubtedly inhibited those who had illegitimate religious knowledge from demonstrating the fact.

Like the rituals of the bush Kaliai of New Britain (Lattas 1989: 452), deceit of women and the young had a central place in most major ceremonies, although my impression is that the Kaliai engaged in more outright deceit, as opposed to revelation of layers of meaning, than Yolngu men. Like older Kaliai men (Lattas 1989: 452), Yolngu elders expressed their dominance of younger males as well as women by telling them false secrets, specifically in the circumcision ceremony. One of the most dramatic aspects of the asymmetry of gender relations acted out in the Ma<u>d</u>ayin ceremony was the men's tricking of women 'to make them believe'. Many women no doubt played along with men's 'tricks' while some sincerely believed in the visits of spirits. However, Lattas suggests that a focus on the question whether women are 'taken in' by male trickery ignores the fact that women do not know all the secret operations of male ritual, and that the cult of secrecy 'articulates a cultural positioning of men as superior to women by articulating an ideology of women as foolishly naïve' (1989: 453). This is not quite true of the Yolngu, for although men goaded women with having no *rangga*, they also admitted that some older women were very knowledgeable.

Together with revelatory rites and the encoding of secret knowledge in public forms, circumcision initiation has been interpreted as a mode of control of the young on the part of older men (Chapter 5). There was no standard graded hierarchy of rites through which every man had to pass, for rites overlapped in functions with much redundancy. It was clear, none the less, that older, more knowledgeable men guarded the prerogatives of age and experience against the onslaught of precociousness. Yolngu religious secrecy was as much about the authority of older men in relation to younger men, as men's exclusion and domination of women.

PART III

Our Great Father

*In Christianity the symbol of a tree denotes the cross on which Jesus was crucified. The iconography of one Yolngu group identified the 'tree/wood' (*dharpa*), which was the group's sacred object, with the Christian Cross.*

The theme of this final section of the book is recent transformations in beliefs and the scope of co-operation in religious practices, in response to changes in the broader social context. As the Introduction argued, Yolngu have recently tended to favour universalistic religious forms which united groups and moieties (whose differentiation was the topic of Part I), attempted to overcome the separation of men and women through religious secrecy (the topic of Part II), and which linked Yolngu to Balanda (white people). The Gunapipi ceremony, which appears to have been increasing in popularity (Taylor 1990), incorporates symbols of a wider scope than Ma_dayin, and seems to be organized on more inclusive principles, while remaining a rite of revelation and exclusion. Christianity challenged the very separation between 'inside' and 'outside'. However, the challenge was met in part by the incorporation of Christianity into religion as a whole.

Following a discussion of Gunapipi, Chapter 8 describes the 'adjustment movement' of the late 1950s and the Christian Revival which began in the late 1970s, and goes on to draw implications of transformations in the control of religious knowledge for power relations and the wider nexus of practices and relations. The title of Part III alludes to a commonality between a Stringybark interpretation of one of the men's sacred objects in Gunapipi as 'our great father', and the Christian God as father.

8

To Make One People

The first part of this chapter describes the universalism of Gunapipi, the ceremony of the regional genre with the widest participation in the mid-1970s. These ceremonies, said to be 'for everyone', were built around symbols which had a very wide distribution among Aboriginal people. Their structure enabled

A cabbage palm at Langarra; a transformation of Olive Python ancestor

people of both moieties, from a wide area and with a variety of *madayin*, to participate on an even footing. Elements of the ceremony such as dances and body-decoration were assigned to subsections and moieties (see Chapter 4), allowing people with a wide variety of ancestral affiliation and somewhat different modes of social organization to come together in the ceremonies, for subsections were widely distributed among people whose organization differed in other ways, and subsections and patrimoieties could be translated into sections, matrimoieties and semi-moiety systems.

Many Dhuwa and Yirritja groups found an important *madayin* among the dances and designs, moreover those who did not directly own the ceremonies could take part through group kin-relationship to owning groups. These commonalities enabled many people of a broad region to attend the same performance. In spite of this extended co-operation, local groups asserted their identity through performing dances related to their own *wangarr* ancestors and mapped their own myths on to common ritual forms, as they did in the Madayin ceremony. As a ceremony 'for everyone' the universalistic symbols of Gunapipi united moieties, groups, and languages. The main categories relevant to the structure of the ceremony were subsections and patrimoieties.

Gunapipi

In the Madayin ceremony recorded in Chapter 6 the participants were all speakers of Yolngu languages, and subscribed to distinct versions of the same mythology. Attendance at the Madayin did sometimes cross language boundaries (Elkin 1972), but while it shared with the Madayin ceremony features of revelation, exclusion, and separation, Gunapipi tended to attract people from a wider region, who spoke languages of a variety of language-families and who had distinct mythological traditions as well as social organization. In order to keep faith with the Yolngu, I am unable to describe secret aspects of ceremonies in detail in a work of this kind, the more so since people now express outrage about descriptions that have already appeared in print, such as the one by Warner.[1]

[1] Gunapipi has been more fully described by Warner (1937: 290–311) and the songs of one version by R. M. Berndt (1951).

During the 1950s and 1960s people at and near Milingimbi performed Gunapipi every few years, but after 1976 something of a resurgence of the ceremony occurred, coinciding with the beginning of the evangelical Christian Revival, during which the ceremony was performed nearly every year in the region near Milingimbi. In the company of the Stringybark leader, I witnessed a Gunapipi in full at Balpanarra, a lake and swamp in the Gunadba-language area, the location of the Gunapipi described by Hiatt (1965: 63–7), some 50 kilometres west of the Yolngu region. The ceremony differed in its details from the one which Warner saw, although it was similar in outline (Warner 1937: 290–311). I did see the first stages of a Gunapipi at Milingimbi at the beginning of 1977, which was very like the one described by Warner.

Most if not all Yolngu people related the Gunapipi and other ceremonies of this genre primarily to long epics about two women, the Wagilak sisters, similar in some ways to the Djang'kawu stories. In the Stringybark group's version of the story, two women of the Dhuwa moiety travelled to a place on Stringybark country, where they camped. To summarize the story:

> A mature woman and a young woman made pregnant by her incestuous relationship with men of her own moiety, travelled from the south down the river to Stringybark country. There they camped, and built a bark hut. They tried to cook the plants and animals which they had gathered on the way, but each one jumped up and went into the waterhole, to become *maḏayin*.
>
> The younger sister bore a son, and allowed birth-blood to enter the creek in which lay the giant Olive Python (*wititj*). The Python rose up out of the water, made a cloud, and caused there to be thunder, lightning, and rain.
>
> The sisters tried to stop the rain by singing all the songs for the regional ceremonies, but in vain. The Python coiled itself around the hut, swallowed the sisters, their children, and all their belongings, regurgitated them, and swallowed them again.
>
> Other giant Snakes and Pythons at other Dhuwa-moiety countries in the region stood up high in the sky, and asked Olive Python what he had eaten. He told them about the two women.
>
> Men who had been listening to the two women as they sang saw what happened. People now follow those events, and sing those songs in the regional ceremonies. (Warner 1937: 250; R. M. Berndt 1951: 19ff.; C. H. Berndt 1970*b*: 1307)

Like the Djang'kawu myth, the Stringybark version of the Wagilak story ends with a catastrophe which caused the women to lose the ceremonies and the men to take them. In the Djang'kawu story, the men stole the sacred objects, and the women were unable to retrieve them because they could not enter the men's secret ceremony-ground; in the Wagilak story, the Python swallowed the women (who were not ancestors of the group, whereas the Python was), men took the songs and re-enacted the events. Both myths set out reasons for male control of the ceremonies. In both, the dances were taken to be re-enactments of the events which gave rise to them, and people understood the songs to be those which the *wangarr* ancestors sang.[2]

In public stories about what the Gunapipi ceremony accomplished, a giant Serpent swallowed and regurgitated the novices, as in the myth—a potential analogue of ritual death and the incorporation of novices (and initiands in the Mandayala and Djungguwan circumcision ceremonies) into the category of knowledgeable males in a *rite de passage*, and the exclusion of females from this category.

Organizers timed the ceremony so that the last night of the performance coincided with the full moon, which rises at sunset and sets in the west at dawn, opposite the rising sun.

In 1976 the Stringybark leader was invited to a Gunapipi at Balpanarra, where people performed the ceremony more or less yearly. My family and I travelled by four-wheel-drive vehicle from Nangalala with him, his wife, and his 'son' (his deceased brother's son).

Two Burarra owners (Dhuwa moiety) of the ceremony had arranged a performance, with the agreement of members of their own and of other owning groups, and their sisters' sons. Previous organizers of the ceremony had handed on the paraphernalia to the sisters' sons ready for this performance. People of outstations and parts of Maningrida settlement had initiated their own preparatory songs and dances, in which bachelors went into seclusion and were made 'sacred' (*dhuyu*). All had arranged to assemble at the main ceremony-place about a week before

[2] Munn's (1969) commentary on the Gunapipi points to similarities between the Gunapipi ceremony as interpreted by Stringybark people and their neighbours, and Madayin, especially that of ancestral loss and the transfer of ritual control and ancestral power to men.

the full moon. They computed the time as a given elevation of the risen moon at sunset.

The ceremony ground consisted of two camps situated near Balpanarra swamp, on the border between open scrubby country and forest, some distance from an outstation. The public camp stood in open country. In the centre lay a ground-sculpture called Mol̲k, consisting of two trenches about 3 metres long and half a metre deep, with a long mound between them. This was to be the focus of the women's songs and dances. Groups of people from various nearby communities camped in clusters, generally centred on the ground-sculpture. The men's ground was sited in the forest, about 200 metres from the public camp. The focal point here was a specially made pit about 3 metres square and one metre deep, with mounds on two sides. In contrast to what happened in the Mad̲ayin ceremony, novices and other bachelors remained at the inside ground day and night. Mature and old men slept at the public camp with their families until the last night of the ceremony, when they remained at the secret inside ground all night.

Male messengers called Gakawarr, who were young married men, took the news to nearby communities that the ceremony was to take place, and sought novices for induction. The Gakawarr messengers ritually 'captured' novices at the settlements and outstations for the Gunapipi, then took them to the main inside ground or to an inside ground at one of several other settlements and outstations where a parallel preparatory performance was under way. Many prospective Gunapipi novices were unwilling: in early 1977 there was a rumour at Milingimbi Mission that several boys had escaped to another settlement disguised as girls rather than be inducted into a Gunapipi. At the inside ground the men painted the novices with red ochre, making them 'sacred' (*dhuyu*). This 'sacred group' (*dhuyu mala*) could not leave the inside ground until the ceremony was over, except to move from a satellite ground at an outstation or settlement to the main ceremony-ground.

Once inside the men's ground the novices were subjected to peremptory and curt instructions from the mature men: to refrain from laughing, to hide under blankets, to get up and watch. One middle-aged man told me that Gunapipi was a very 'bad' (*ya:tjkurru*) ceremony for this reason. The older bachelors and the mature men had control over the novices, who were onlookers for most of the ceremony but took part in the final dances. At the public camp mature and old women of the owning groups (Dhuwa moiety), with their daughters (Yirritja moiety), danced and sang women's *gudjika* songs at night at the ground-sculpture.

Every day for about two months men decorated themselves at the inside ground with cotton-wool during the day, and from time to time called falsetto to women at the public camp, who ululated in reply. Each day at dusk and dawn a noise emanated from the ground, interpreted by some as the voice of the Snake. At sunset the decorated men danced on the inside ground, and after sunset mature and old men of both moieties sang over the prostrate novices, then performed dances. Meanwhile the Dhuwa and Yirritja women performed their Gudjika songs and dances at the Mol̲k ground-sculpture, re-enacting according to Stringybark traditions the attempt by the Wagilak sisters to stop the Python making rain. The women frequently ululated to calls from the inside ground. Each night after the singing and dancing the mature and old men returned to the public camp, while the bachelors remained at the inside ground.

The same daily sequence was enacted at several other communities by people who intended to assemble for the final phase at Balpanarra. During the final week, as the full moon approached, people arrived from communities within a radius of about 80 kilometres of the camp. Mature men had captured novices at each of these communities, making them 'sacred' by taking them into seclusion and painting them with red ochre. The visitors established camps around the core public camp, except for the bachelors, who went into seclusion on the inside ground. (Not all the 'bachelors' were in fact unmarried men; some mature married men took the 'bachelor' role in the ceremony.)

On the penultimate day, Yirritja-moiety workers made another ground-sculpture on the inside ground, this time a long trench, and made sacred poles (see Chapter 4). During the day the Yirritja men revealed these formally to the Dhuwa men and youths. Then the dancers prepared their body decorations. The men performed various dances representing *wangarr* ancestors, and sang through the night till dawn, when the full moon rose in the west, opposite the rising sun. The women danced and sang all night in the same way as before.

On the morning of the final day the male Yirritja workers made a large wooden Djapanbani (or Djapanmani) trestle of two forked poles and a cross-piece, erecting it at the edge of the forest between the inside ground and the public camp. At the public camp one or two workers ushered the group of women and young boys, who were led to the trestle by two women beating the ground with digging-sticks (women used to do this to make the ground flat and soft for a camping-place), where the 'sacred group' of bachelors were lined up, painted red all over their bodies. Two men sat on top of the trestle. The group danced around the women and young boys, who were to be novices in a later Gunapipi, taking the boys from the women. The sacred group and the older men then marched in file out from the inside ground.

The final act was to return the ceremony's paraphernalia to the sisters' sons, who were to hand them on to another group for another ceremony.

Exclusion and Separation

As in the Madayin ceremony, men were believed to have greater access to manifestations of the *wangarr* ancestors in Gunapipi than women, and did have more control of the course of the ceremony. The men set their ceremonial ground at some distance from the main camp, and the boundary was imbued with danger for the women. This separation allowed the women and children to hear and observe only what the men intended. The men's call and the women's ululated reply maintained contact but allowed little information to pass. The women saw the men's body-decoration as they emerged from the inside ground, but they were not supposed to know the meanings.

In contrast to what happened in Madayin, Yolngu women excluded men from their songs and dances at night in the camp, but even in the ceremony described above the men directed the movement of women in the public parts of the ceremony rather than the reverse. A degree of asymmetry was to be found in the secrecy also, for women and children did not experience men's restricted aspects at any time, whereas uncircumcised boys, as well as girls, were present in the women's camp in the Gunapipi. Furthermore, as Merlan points out, men were able to apply more severe sanctions for breaches of secrecy than women (Merlan 1988). Men also had greater acknowledged access to elements of women's performance and their interpretation. In 1976 a Stringybark woman allowed a mature man of Cloudy Water group to record some *gudjika*-style songs which the women had performed, and which he played at his house in the presence of a male Black-capped Tern leader. He asked an adult-education worker, a white woman, to make a copy of it for him to take on a trip to Bali as a present.

As in other revelatory ceremonies, men admitted younger men to the inside ground, and gradually introduced them to roles of greater responsibility. Older men protected the exclusiveness of the inside domain by applying three kinds of sanctions: the threat that if novices revealed the secrets the older men would die, the threats to the young men themselves of death or illness by sorcery; and violence. Leaders earnestly told the novices in all the regional ceremonies that what they saw was not a game, and that they must not reproduce in play what they saw and heard, nor tell women and younger people about these things. If they did, they would die by sorcery or the spear.

Revelation and Belief

On the face of it the difference between public accounts of what went on in the men's secret ground—such as that the sounds from the inside ground were the Snake's voice—and what was revealed to young men, would create a problem for belief. However, there is evidence that adult men subscribed to beliefs that spirit-beings of various kinds were present at ceremonies and that their powers resided in objects and enactments. In the mid-1920s men attributed the men's ground of the Gunapipi, especially the old men's camp, with power (*d̲a:l*, 'strength') (Warner 1937: 264). Men explained to me that the spirits of the Giant Snakes inhabited objects in the Gunapipi ceremony, and returned to the water when the Poles were broken up, and that men could gain power from them (cf. Meggitt 1966: 210). Women said to a white woman observer during one 1976 Gunapipi that all but older women had to sit with their knees together during the nightly singing and dancing at the Gunapipi in order to prevent impregnation by supernatural agencies which they believed to be present at the inside ground.

Belief in the potentially destructive power of the Giant Snake was dramatically demonstrated at the Gunapipi described above:

> Two men arrived drunk at Balpanarra during the Gunapipi ceremony described above and had a fight. Gul̲wirriwuy, the Stringybark leader, chased one of the malefactors, who ran into the nearby lake right up to his knees. After this, Gul̲wirriwuy was weeping, and insisted that I drive him several kilometres away from the ground because he was afraid that the Giant Snake would cause a torrential downpour. It is likely that Gul̲wirriwuy believed that the Giant Snake would detect the man's presence because he was in a polluted state from fighting, just as the Olive Python had detected the Wagilak who polluted the water with blood and because he was 'sacred' as a participant at the inside ground. Gul̲wirriwuy returned only when the other leaders told him that the place was 'good' again. The malefactors had 'paid back the earth' (which had been polluted) to the tune of twenty-five tins of tobacco, which were probably shared among the leaders.
>
> Leaders told the two men that they would ensorcell them. I heard about a month later that people were saying that one of the fighters had

become very ill with pneumonia, and that some attributed this illness to the leaders' sorcery.

As I said, in public versions of the events at the inside ground a giant Snake was present (cf. Warner 1937: 261, 293). Appropriately enough, the women at Balpanarra wailed when men took their sons into the ceremony. I do not know what proportion of women believed the public version literally, but some older women claimed to a white woman observing a Gunapipi in early 1977 that they knew what really happened, for their husbands had told them. As was suggested in the last chapter, what people could publicly admit to knowing was more significant for the control of religious knowledge than what they could recall privately.

Enactments in the ceremonies tallied with Yolngu conceptions about the nature of the *wangarr* ancestors as persons with the identity of animals and other species. Taylor (1990: 333) reports that younger Gunwinjgu men believed literally in the existence and powers of the serpent Ngalyod, moving to a more abstract level of understanding when older. I suggest that, especially for the young, the ceremonies, in enacting ancestral events and depicting *wangarr* ancestors in a context of excitement and great beauty, enhanced the credibility of propositions about the beings. The aesthetic and emotional effects of the ceremonies were perhaps taken as evidence of their spiritual origin and associations. As Geertz expresses this kind of idea, religious symbols express and shape the world, by inducing in the worshipper a certain distinctive set of dispositions.

> In a ritual, the world as lived and the world as imagined, fused under the agency of a single set of symbolic forms, turn out to be the same world, producing thus [an] idiosyncratic transformation in one's sense of reality . . . (Geertz 1966: 28)

In Gunapipi the symbolic forms united participants, but they interpreted them in their own ways.

Universal Categories and Symbols

The control of knowledge through exclusion and revelation were aspects of broadly similar modes of activity transcending regional differences. Not only did the organization in terms of subsection and patrimoieties enable people from different areas to join together, but versions of the ceremony current in different areas had several similar *ma<u>d</u>ayin*. Designs of men's body-decoration used in both performances of Gunapipi which I attended in 1976 included Giant Reptiles, Cabbage Palm, Stone Knife, and Brolga. People of most if not all groups believed that Giant Reptile lay somewhere in their country, so that the content of the ceremony had relevance to all, just as the bases of organization were inclusive rather than exclusive. Other designs expressed local affiliations, for participants could choose designs made with wild cotton (or a substitute such as cotton wool) for the 'inside' dances relevant to their own group *ma<u>d</u>ayin*. Most (perhaps all) participating patrifilial groups in the regional ceremonies which I observed had at least one important *ma<u>d</u>ayin* represented. Individuals could perform dances of their own, their mother's, or their mother's mother's groups and wear the related design.[3] Styles of the ceremony in different regions were sufficiently similar for people to co-operate, but tension arose out of local variations. People occasionally grumbled, fought, protested, and argued about the proper form of a ceremony, generally regarding their particular variant to be the proper one.

People from an extensive area attended regional ceremonies, but numbers probably varied with the prestige of the organizer and the length of the performance, and the type of ceremony was also a crucial factor. Table 13, which compares the estimated number of people at four performances with the approximate adult populations of the areas upon which each performance drew, gives the impression that performances of Ma<u>nd</u>ayala and Djungguwan were of interest mainly to the relatives of the initiands, as well as the main owners of the ceremonies in that community. The largest funerals drew a somewhat higher proportion of the local population than these ceremonies. But a performance of Gunapipi clearly involved nearly the whole of the adult population of the region on which it drew, and attracted more people than the Ma<u>d</u>ayin ceremony. The catchment area for the performances overlapped; many of the people at the Balpanarra Gunapipi went to the performance near Milingimbi, and many of those who went to the Ma<u>nd</u>ayala went also to the Djungguwan.

[3] See Keen (1978) for details of the designs.

TABLE 13. Estimated Proportion of the Local Adult Population Attending Regional Ceremonies

	Ceremony			
	Djungguwan	Mandayala	Gunapipi 1	Gunapipi 2
Approx. adult pop. within 50 km of ceremony location	500	450	375	600
Approx. adult attendance	60	120	300	500
% of adult pop. attending	12	27	80	83

Performances of Gunapipi then, more than any other kind of ceremony until a Christian Revival began in 1979, united the population of a very wide area in common ritual, extending the range of sociality beyond the everyday. A woman wrote to me after a Gunapipi near the mission in 1977 that everyone was very happy that they had got together to make the old law. Expressing a similar mood, a person told Warner in the 1920s that the 'making of one mob in a dance is all the same as making one people out of all those people from the Gunabibi' (Warner 1937: 296–7), a mood reflected in the universalism of the themes and symbols in the ceremony.

As some Yolngu people interpreted it, Gunapipi does not seem to have been directly concerned with species increase or with ensuring human fertility in any direct way, themes which R. M. Berndt considered to be common to the Gunapipi irrespective of region (R. M. Berndt 1951: 14, 33). The Stringybark leader insisted to me that it was not necessary to perform ceremonies in order for food species to reproduce, for the *wangarr* provided game and vegetable foods as 'a gift'. However, at least one 'increase' site has been recorded in the country of Djinang-speakers. Moreover, Warner was told, after a performance, that after the next rain a man could come out and see all kinds of animals. He would say, 'My word, we sang him good in that last Gunabibi' (Warner 1937: 311). The eastern Gunwinjgu certainly made this kind of interpretation in their performances of the ceremony. Not only did participation make young men feel 'fat' and healthy, but performing the ceremony was believed to make the sap ('fat') run in the trees and make the Serpent disgorge food-animals such as flying foxes. Men associated the ceremony with the potentially destructive power of the Serpent, who could create wind, thunder and lightning, and floods, and with energy such as that felt when a person runs fast (what Yolngu call *ganydjarr*). Gunwinjgu believed that 'clever men' could tap this power (Taylor 1990: 333–4). In Munn's analysis of Yolngu interpretations of Gunapipi (as recorded by Warner), belief in the transfer of ancestral power is central (Munn 1969).

It is clear from Taylor's analysis that recent Gunwinjgu interpretations of Gunapipi and the related Lorrkon mortuary-rite have very similar themes to Cloudy Water interpretations of the Ma<u>d</u>ayin ceremony. Dominant symbols include: salmon catfish, the mouth-brooding species celebrated also in Ma<u>d</u>ayin; female crocodile, which carries its infants in the mouth; barramundi, which changes from male to female with age; and kangaroo, in which the strong attachment of the young animal to the mother's nipple epitomizes maternal attachment (Taylor 1990: 335). Taylor detects themes of 'cyclic rejuvenation' through the re-enactment of ancestral swallowing and regurgitation, and ancestral creation of groups and species (Taylor: 336–7), very similar to themes of the Yolngu Ma<u>d</u>ayin described in Chapter 6.

As people explained the ceremony to Warner during the 1920s, related themes in Gunapipi (at least in the men's domain) were proper marriage and legitimate reproduction in a context of erotic expression. In the 1920s a man told Warner, 'Black men walk about single, and by and by catch a mate belong his and by and by *yoto*' (*yothu*, 'child'); if the Wagilak had not done wrong, everyone and all the animals and plants would have stayed single: the Wagilak 'made this new law for everyone' (Warner 1937: 297). The oldest leader of the Gunapipi in the area explained the origin of ritual copulation to Warner. Two Wagilak men 'played' with the Wagilak women, who made the Gunapipi 'higher' than Djungguwan, and laid down that people should 'play with each other' in Gunapipi. The sisters said, 'It is better that everybody comes with their women and all meet together at a Gunabibi and play with each other, and then nobody will start having sweethearts the rest of the time.' Sanctions were brought to bear against men and women who would not accede (Warner 1937: 308). (It is not clear from Warner's description whether ritual copulation actually took place during the event he witnessed or whether it was something that ideally took place. Men told me that it was not done in the Glyde River area but that it was done further south.)

During the mid-1970s sexual roles in the dances were mapped on to the moieties, just like preformances recorded by Warner (see Warner 1937: 296; Keen 1978). Men divided by moiety into 'owners' and 'workers', and the women divided into the two patrimoieties in the dance at the Mol̠k trench and mounds. One man's exegesis of part of the Wagilak myth also stressed the sexual roles of the moieties: the reason why the Snake became sick and regurgitated the women and their children was because both eater and eaten were 'one body', Dhuwa. 'If they had been Yirritja it would have been alright,' he explained to me (cf. Warner 1937: 257). This theme recurs in the episode of the story in which a Wagilak sister tried and failed to cook the birds and animals, and in the Mangrove group's myth in which one Djang'kawu sister cooked shellfish of her own moiety and found it full of mud (Chapter 1).

Some interpretations of the Gunapipi ceremony thus projected categories and roles entailed in the regulation of sexuality and marriage on to subsections and patrimoieties, and represented them as cosmogonically necessary. As one of Warner's informants, commenting on the Djungguwan songs, said, 'No man made these songs. Those two old women made them to keep the two lines (Yirritja and Dhuwa moieties) sacred to keep marriage straight' (Warner 1937: 280). Thus in Yolngu interpretations in the 1920s and perhaps 50 years later, the central symbolism was sexual, both overtly in the content of the secret dances, and implicitly. The ceremony celebrated and commemorated the institution of sexuality and the proper control of sexual relations in terms of moieties and subsections, and for some participants was a means of inducing the fecundity of species, whereas the Mad̠ayin ceremony was concerned with relations of reproduction between groups of the same moiety, and perhaps the institution of a male role in reproduction, and referred to ancestors of more limited scope. They shared the re-enactment of initial creation and continuing reproduction. In spite of differences in beliefs and interpretations, most if not all groups were believed to have a giant reptile in the waters of their country, so that the content of the ceremony was relevant to all, just as the bases of organization were inclusive rather than exclusive.

In contrast with these inclusive themes and modes of organization, people from different areas interpreted the ceremony in even more diverse ways than Ma_dayin. But these were two sides of the same coin; common themes and aims united the groups, who were able to join together while retaining doctrinal differences.

Polydoxy in Gunapipi

At the performance described above, participants elucidated the ceremony according to several ancestral traditions, although there were thematic commonalities among interpretations even of different genres of ceremony. Some men even construed the same song-text, which they were singing together, in different ways. Most Dhuwa groups of the Yolngu region seemed to agree that the Wagilak sisters were eaten at Stringybark mainland country, but each appeared to have a somewhat different version of the myth. A man of Blossom group agreed that a giant reptile swallowed the Wagilak at Stringybark country but substituted his group's giant King Brown Snake for the Olive Python of the Stringybark story (Mirritji 1976: 62). People of groups whose countries lay on the track of the Wagilak sisters probably located the main events at their own country, as in the case of the Djang'kawu myth. In the 1960s speakers of the Djinba dialect, whose country near Mirrngadja lay to the south of Stringybark country, related the Gunapipi ceremony to a Snake that came out of a swamp and travelled south (Nic Peterson pers. com.).

Speakers of the Djinang dialect of the Yolngu language-family who were members of groups whose countries lay to the west of the Glyde River, associated the regional ceremonies with the Djang'kawu sisters. At the Balpanarra Gunapipi I asked an older male Djinang participant, a man whose country lay between the western limits of Yolngu speakers' land and Balpanarra, what the ceremony meant:

> This is the number-one ceremony. The Djang'kawu made the swamp with bamboo spears, created the groups and gave them all the ceremonies. Aunt and her daughter came, put the wells in the swamp, made the 'tribes' and skins [subsections].

The Djang'kawu came from Mosquito-group country. The Djapanmani Trestle [erected at the threshold of the men's ground, where at the end of the ceremony the men surround the women and the next season's male novices] is their bark hut. Then the warriors struck. The two sisters said, 'Very well, they have the power, we'll go to the bush.'

Everything is from Djang'kawu. They came from the east, speared the water-holes where fresh water now lies on their path. They made Balpanarra and said this is Dhuwa. They made the kapok tree [for wild-cotton body-designs] and pandanus.

The sisters crouched down. Leeches [Yirritja moiety] played with them and opened them, and there was water, wild cotton, and pandanus. 'Where shall we sit? This is Yirritja,' they said [implying that they could not remain there because it was of the opposite moiety]. They made [two places to the west]. They came to a big water, put their footprints there, and planted rushes.

They cast [the sacred objects]. Which side [i.e. secret or public]? Secret perhaps. They said, 'This is ours.'

Another man at the ceremony, who was speaking about the country and ceremony in the presence of the Stringybark leader as well as myself, said, 'This is from Wagilak, but it is all of ours, like Ḏupun [hollow-log ceremony].' He switched the sisters' identity, out of deference to the Stringybark leader perhaps, for during the ceremony the same man interpreted one song as meaning the fish-trap which the Djang'kawu made, identified with the Trestle built for the conclusion of the ceremony. Or perhaps he sought names known to Gul̠wirriwuy and myself which referred to female ancestors similar to his own and which he knew by different names.

The Rembarrnga-speakers, whose country lies to the south toward the Roper River, and who participated in the same performance, singing in a separate group on the inside ground, probably associated the Gunapipi ceremony with the Nagorrko men (Peterson n.d.; R. M. and C. H. Berndt 1970: 139; Maddock 1976). Maddock recorded this myth about Nagorrko current among Dalabon and Rembarrnga people of the Roper River in the 1960s:

A man and his son, known jointly as Nagorgo, used to perform the Gunabibi in company with Lumaluma or Gunabibi. She was a serpent who stood as mother to all Aborigines and lived in the salt water. The father and son would call out and sing as now is done in Gunabibi performances, but when they went to the *djabanmani*—a bough structure situated halfway between the secret ceremonial grounds and the women's camping ground—to meet Lumaluma she would vomit the bones of men she had swallowed. She should have brought them up whole and alive. The father and son decided not to give her more men to swallow, but to put bullroarers in her place. When next she spoke a left-handed nocturnal bird caught her voice by swinging a bullroarer . . . These events happened at Balbonara [Balpanarra].

The father and son, accompanied by their wives, the bird and two snakes, then set off south to spread the Gunabibi. Lumaluma was left behind in the sea, where she remains to this day. As the party travelled along, the father and son called out and sang, the bird swung the bullroarer and the two snakes stood up. The two wives are said by some to have sung as men now do in the Gunabibi, but others say they just followed and listened to their husbands sing. A third opinion is that the wives sang in the way women do now while their menfolk are at the secret ceremonial grounds.

The party passed through the Bulman area and skirted Mainoru Station before changing direction westwards to cross Mangarai country and Elsey Station. In the vicinity of Warlock Ponds on the Stuart Highway, direction was changed again to south and the journey ended at last in the general area of Newcastle Waters in Mudbura or Djingili country. People of the Beswick reserve are hazy in their minds about the exact whereabouts of the southernmost point reached on the journey, but usually it is said to have been a waterhole named Anggulala or Wanggarala. Some say that after reaching this point the father and son and their remaining companions turned back to Balbonara, but no interest is evinced in the return journey. The party had diminished during the second southwards stage of the journey, because the father and son or the bird killed the two women—hanging them or cutting their throats—in order to keep the secret parts of the Gunabibi exclusively for men. (Maddock 1976: 166)[4]

Gunwinjgu people have interpreted the ceremony with reference to stories about Yingarna Serpent, who travelled with the Nagorrko father and son, arousing the anger of the Serpent Ngalyod, whom they met and who swallowed them, transforming them into features of the country. Ngalyod was thus the more local and transforma-

[4] See Maddock (1976: 167–70) for an analysis of oppositions set up between the Yolngu myth of the Wagilak and the Nagorrko story: oppositions in direction of travel, gender and moiety of the travellers, reasons for exclusion, sexual relations, and so on.

tive; Yingarna more generalized and incorporative (Taylor 1990: 331).

People of course knew that many co-participants interpreted the ceremony in ways different from their own, and a certain tact is necessary to avoid ill feeling. Such tact was not always forthcoming:

At the men's ground on one evening during the Gunapipi ceremony, Gul̲wirriwuy announced from the mound by the trench, 'I'm teaching you all. Don't trust any other meaning [*mayali*]. These are all my people, but Ye____holds it [the ceremony] here.'

Gul̲wirriwuy got rather drunk on the last evening of the ceremony and, again standing on the mound by the trench, he announced that Gunapipi, Ngul̲marrk, Man̲d̲ayala, Djungguwan, and all the washing and smoke purification ceremonies were his, but those present were all joined together. It was all very good. He then sat on the mound and said that all this was his own 'business'. This remark provoked some unrest and a few sceptical comments.

Same Song, Different Meaning

The exegeses outlined above applied to episodes of dance complexes and the ceremony as a whole. Multiple interpretation occurred at a more fine-grained level as well. Men and women from different regions could sing together in the Gunapipi and other ceremonies of the genre while according different significance to them (see Keen 1977).

Men's *gudjika* and *nyira* songs of the Gunapipi occurred only in the men's secret ground in the presence of adepts and neophytes. Women normally performed *gudjika* songs only in restricted sessions in the women's camp in the Gunapipi, although I heard that women performed them in the presence of Yolngu men to farewell a white missionary at Milingimbi airstrip.

Just as there were regional styles in the ceremonies, so people recognized distinctive styles of songs. At the Gunapipi which I attended, three groups of male singers, each from a different area, sang concurrently on the men's secret ground somewhat similar series of songs but with rather different tunes. At the same time, the 'bachelors' were singing *nyira*, a repetitive unaccompanied chant, with an enthralling musical effect. Comparing their own songs with those of another group, men said that the words of the song were 'turned' by another language. The songs with boomerang clapsticks (*gaḻigaḻi*) employed in the regional ceremonies consisted of word-like sounds with conventional meanings. Each short verse was formed from a repeated formula. Some verses were simply one or two words or names repeated which, with the exception of proper names for elements of the ceremony (such as the pit, trench, and poles), did not belong to any specific language. They could only be interpreted with special knowledge either of the referents of proper names in the songs or of the conventional interpretations of otherwise meaningless sounds. Access to knowledge of these interpretations was restricted, although there were probably similarities between the texts and interpretations of men's and women's songs of this genre. Yolngu interpreted men's and women's songs alike as following the journey of the Wagilak sisters.

Here are examples of Man̲d̲ayala songs, which are in a similar style but public, followed by Guḻwirriwuy's explanation:

> *warragagabaga warragagabaga*
> *warrabiginydjura warrabiginydjura*
> Just what the sisters sang to try to stop the python making rain
>
> *Linbidjirrirri linbidjirrirri*
> *Ngalimagamarrka ngalimagamarrka*
> The Wagilak sisters

Some verses had no meaning, according to the men who discussed them with me. Other verses incorporated the proper names of elements of the ceremony; one for example was explained to me in this way:

> The Djapanbani rail and two forked poles, but [it means] a bark house.

Others were assigned conventional meanings such as:

> It is lightning. Those two looked up and saw the thunder clouds following behind them.

Even within a single style, identical words had distinct significance for different people. I was standing with Guḻwirriwuy, the Stringybark leader, who was singing with Blyth River men of the Gidjingali language. He told me that his father had given his songs to the father of one of these men, and that is why they sang together. The men uttered the same words, sometimes one of the older men initiating a verse, sometimes another. In spite of this concord, the Burarra men gave me very different accounts from those of Gu<u>l</u>wirriwuy. They said that some of the songs were about the sea, whereas Gu<u>l</u>wirriwuy contradicted this. One man told me that a verse meant 'lighting from the Snake', but Gu<u>l</u>wirriwuy asserted that it was 'cloud'. A verse said to be 'cloud' by a Burarra man was without any specific meaning according to Gu<u>l</u>wirriwuy. Such divergent senses occur even among closely related men: when we were listening to a recording of the songs back at Nangalala a man of Gu<u>l</u>wirriwuy's sister's husband's group interpreted a song as 'Barramundi', contrasting with Gu<u>l</u>wirriwuy's gloss which was 'Lightning'. (Of course, verses which included the widely used names of objects in the ceremony would receive a similar interpretation.) Thus, even where co-operation was most intimate, as in singing, men attributed, usually in private, different meanings to the same sounds. Such radical variability in interpretation was possible because the songs were not anchored in ordinary language. Conversely, the fact that the songs were conventional formulas allowed the speakers of different languages to sing together and to project their own distinct beliefs on to them, just as the weakly iconic dances were consistent with a plethora of significations.[5]

Interpretations of their songs by women of Stringybark group and their *waku* were similar to those of men in that they traced the journey of the Wagilak sisters through country of Stringybark group, but they also took them across the Yirritja-moiety country of Ganalbingu as well as that of Djinang-speaking groups. Other interpretations were specific to women: the menstrual cycle, pregnancy, customs to do with childbirth and ritual purification.[6]

The ability of Gunapipi and other regional ceremonies to accommodate a variety of interpretations as well as incorporate a variety of elements made it possible to extend ties of amity through co-operating in a performance well beyond the social network of everyday affairs. It was far more inclusive in this way even than

[5] Meggitt (1966: 26–8) notes that the meaning of Warlpiri Gadjari sons, which are similar to Gunabibi sons, are not anchored in ordinary language, and are liable to reinterpretation.

[6] I am grateful to Sue Harris for this information.

the Madayin ceremony.[7] The Christian Revival which began at Galiwin'ku in the late 1970s was an attempt to extend this amity still further, to even more distant Aboriginal communities, and, like the earlier 'adjustment movement' at the same settlement, to extend ties to the white community while again retaining local identity and control. The Christian movements were attempts to legitimize new social relations and modes of organization and to incorporate the white world.

God wangarr *Father: The Christian Movements*

Two religious 'movements' arose at Galiwin'ku (formerly Elcho Island Mission): the 'adjustment movement' of 1957, and the so-called Revival of 1979–81 (R. M. Berndt 1962; Bos 1988*a*, 1988*b*). Both appear to have been attempts to achieve a syncretism of Yolngu and Christian beliefs, to establish greater unity within Yolngu social relations, to legitimize new forms of community structure and authority, and to extend relationships of amity to whites through shared sacra. This motive of extending amity through universalistic beliefs links the movements to the regional ceremonies, especially Gunapipi. Unlike the indigenous cults, Christianity did not include an esoteric domain.

The Elcho Island Mission had a history of a greater commitment to Christianity on the part of its Yolngu residents than the other mission stations. The Elcho Island Mission was founded with the support of a few Yolngu who identified as Christians early in the history of the Methodist Mission in Arnhem Land (Introduction). Bos points out that unlike the leaders of other missions, the long-term superintendent of Elcho was not trained in theology, even though he was eventually ordained minister, and it is perhaps for this reason that an evangelistic style of Christianity was encouraged (Bos 1988*a*: 167).

R. M. Berndt and Robert Bos explain each movement as a response to rather similar changes in the social fabric, all the greater by the time of the second movement. Firstly, there were external threats to people's security and well-being, including the discovery of bauxite and then mining at Gove, and in the 1970s

[7] The universal appeal of the serpent symbol is tellingly exemplified by a performance by urban Aboriginal people at the 1988 World Expo in Brisbane using a Gunwinjgu design of a Rainbow Serpent as a backdrop (Taylor 1990: 329; Michaels 1988: 43).

the Proposal by the Northern Territory Government to extend the Arnhem Highway across Arnhem Land to the Gove Peninsula, as well as its opposition to land rights legislation. There were also increasing numbers of incursions into coastal waters and estuaries on the part of commercial fishermen (R. M. Berndt 1962: 29; Bos 1988*a*: 84).[8]

A second set of changes were the existence of contradictions and discontinuities in settlement authority, and a desire for a new social order. Berndt writes of the loosening of internal controls without the integration of introduced ones. Bos points to the increasing secularization of life in the township, as well as conflict between the male leaders who formed the older Village Council, and the younger, western-educated people of the Town Council. The Aboriginal Community Council and its Departments were very inefficient, so that community services were breaking down in the late 1970s after the withdrawal of missionary supervision (R. M. Berndt 1962: 17–18; Bos 1988*a*: 77, 210, 217–18).[9]

Bos also remarks on the great differences in wealth in the Galiwin'ku community by the late 1970s. Members of some families had positions in the Church and community affairs, giving them access to the control of equipment and money, and the possibility of travel overseas, while others lived in relative poverty, with only family allowances and pensions as income.[10] These changes, together with the background of Christian evangelism, seemingly provided some of the motives for the movements.

The Adjustment Movement

What R. M. Berndt called an 'adjustment movement' took place at Elcho Island Mission in 1957. Several male leaders of groups publicly displayed a number of *rangga* sacred objects, one incorpor-

[8] Galiwin'ku Aboriginal people organized a conference on sea rights through the Adult Education Centre. In the 1980 Northern Territory election for the Legislative Assembly 80% of the Galiwin'ku voters voted for the Labor Party, even though the candidate for the party in power, the Country Liberal Party, was a Yolngu man from Galiwin'ku.

[9] Bos notes that the mission structure had echoed Yolngu practices in linking activities generally to religious beliefs and related ethics. With the separation of roles of superintendent and minister, and the taking-on of responsibilities by elected Community Councils, there came a division between township affairs and the Church.

[10] The Council controlled the equivalent of unemployment benefit in the form of Community Development Employment Program grants (Bos 1988*a*: 216).

ating a Christian cross, in the form of a 'Memorial' set centrally in the mission village within a (normally secret) Nga:rra ceremony-ground. Two of three main leaders were of Yirritja-moiety groups, and in a relation of distant older brother–younger brother, and one was Dhuwa. One of the former was the mother's brother of the latter. They all had a history of close co-operation with the missionaries, and one was quite well educated in western terms, and was literate. The older Yirritja leader and the Dhuwa man were regarded as *yindi bunggawa*, 'big bosses'. As Bos points out, none of these men belonged to the group on whose country the mission was situated, but one was of a group with *ma<u>d</u>ayin* connections with the land on which the mission stood. His son was said to have affinal connections with the extinct land-owning group, and the father's junior wife of another of the three leaders was of that group (Bos 1988*a*: 240; R. M. Berndt 1962: Appendix B). The three men invited a number of leaders from Yirrkala to go to Galiwin'ku and display some of their objects in public, but only four agreed. Two other Galiwin'ku leaders also took part (R. M. Berndt 1962: 23–4, 36, 66).

The men had several related motives for this action, which they outlined to R. M. Berndt. One was to achieve a *rapprochement* between Christianity and Yolngu religion. One leader explained that the Bible spoke of graven images: 'We thought of our *rangga*. The word of God made us ashamed. Could we keep our laws? Could we hide our *rangga*? The Bible was here. Instead we must show these *rangga* to all: to men, women, and children.' One leader, together with the son of another, gave similar explanations to Robert Bos some 20 years later (R. M. Berndt 1962: 40). However, according to Berndt's interpretation, this was also a response to anthropologists taking photographs and showing films of sacred objects, taking away Yolngu power to hide the *rangga*, their only wealth. If they were to be shown publicly, then men should receive something in return.

A second motive was to achieve unity between men and women, as well as the separate groups, by overcoming exclusion. One man commented to Berndt:

> Now women as well as men can use the feathered string of the *rangga*. Let all the men and women assemble as one. We shall bring everything into the clear [i.e. open, public] place: there will be no jealousies. (R. M. Berndt 1962: 40)

The two Yirritja leaders asked Ronald Berndt rhetorically, if the Djang'kawu were not women? Lanytjung (a Yirritja *wangarr*) and the Djang'kawu made the *ma_dayin* for all: 'They are the true leaders. The people after them distorted what they meant.' It would be better for people to look to their ideas and be 'brothers and sisters in their ideas'. In this way the leaders were ostensibly attempting to reverse the male appropriation of control of ceremony depicted in the Djang'kawu myth, and the secrecy and divisions inherent in that control. The leaders reported to Ronald Berndt that women were frightened and that some had left the community for Milingimbi and the mainland. The women eventually became accustomed to the presence of the Memorial, but no woman entered it unaccompanied by a man (R. M. Berndt 1962: 66–7).

Thirdly, the men intended the Memorial as a gift to the superintendent, Harold Shepherdson, for bringing new knowledge and skills, 'so that white and dark colours intermingle—especially those who teach us', a sentiment similar to that expressed by Wandhuma about the Ma_dayin ceremony described in Chapter 6. Several inscriptions on the Memorial commemorated the founding of the mission (R. M. Berndt 1962: 41).

The unification of the mission community under the leadership of one of the three men was a fourth motive, underwritten by the *rangga* whose revelation he had helped to instigate. The three leaders of the movement presided over the Village Council. The leaders were trying to make one of them, whom the missionaries regarded as the leader of the Aboriginal community, a big leader over everyone. Bringing out the *rangga* gave him 'honour', and now they would have to listen to him (R. M. Berndt 1962: 41). To achieve this aim it was necessary to overcome the differences and rivalries between the groups. Two of the leaders commented to Berndt that people with the same *rangga* and songs were none the less jealous of one another and becoming divided as enemies, stealing an elder brother's wives and joining another group, whereas Lanytjung and the Djang'kawu 'joined everybody and made all friendly' (R. M. Berndt 1962: 72). These themes of unity are just those propagated in the Ma_dayin ceremony and Gunapipi.

A fifth motive was to achieve independence in new enterprises and the running of the community. One leader stated that he wanted to 'buy' a better school by displaying the *rangga*, so that the children could learn new economic skills such as farming and engineering, and govern themselves. He typified the *rangga* as the only wealth possessed by the Yolngu. Indeed, the leaders had various development-plans, including pearling and forming the mission into a regional centre. Learning from, and co-operating with, Europeans was seen as essential to this end. But the men also thought it necessary to keep uninvited whites out of the Arnhem Land Aboriginal reserve. The desired transaction, then, was a transfer of benefits in return for displaying the most valued religious properties, in a similar way to payment on the part of novices to a leader for revealing secret religious knowledge (R. M. Berndt 1962: 40, 79–81). The leaders, then, gave rather diverse accounts of their motives.

Not all of the other leaders in the community co-operated with them. Moreover, the leaders of the movement compromised by not bringing out all their *rangga*, and keeping perhaps the most important ones hidden (R. M. Berndt 1962: 68). Nevertheless, in breaching religious secrecy the leaders were (temporarily) removing some elements of the media which strongly differentiated group from group and women from men and which formed the basis for controlling young men. The display of all the different *rangga* together in public, one incorporating a Christian cross, implied unity between Yolngu groups, men and women, and Yolngu and Balanda, through the shared symbol of the cross.

The Christian Revival

Like the earlier movement, the Christian Revival, which began at Galiwin'ku in 1979, was an attempt to extend amity to whites and to legitimize leadership, according to Bos's interpretation. It was a blend of the various styles of religious action introduced into Galiwin'ku, and it was a response to external threats in emphasizing the perceived common spiritual links between black and white communities. It emphasized values of settlement co-operation; and it legitimated the roles and power of settlement leaders in a way that the traditional religion did not. There were some immediate problems. The community was afflicted by a spate of minor illness in the year preceding the onset of the movement, and was having great difficulties with alcohol consumption and petrol sniffing, accompanied by fighting and a number of deaths (Bos 1988*a*: 77, 219–20).

The style of the movement derived in part from the recent religious history of the settlement. The lay missionaries had promoted evangelical crusades, with the intention of encouraging 'revivals', including a visit by a black American member of the Billy Graham organization, and visits by Yolngu Christians to conferences.[11] Many of the white Christians working at the community practised a particularly fervent variety of the faith, including speaking in tongues, exorcizing demons, and the laying on of hands to 'slay the spirit'. Some Yolngu Christians copied aspects of these rituals, and elements were incorporated into the Revival meetings (Bos 1988*a*: 79, 82).

The two main instigators of the Revival were sons of one of the Yirritja-moiety leaders and the Dhuwa leader of the earlier movement.[12] Like their fathers, they were in a relationship of *ngapipi–waku* (MB–ZS). One was an ordained minister of the Uniting Church with 5 years experience in Papua New Guinea, while Bos characterizes the other as the 'evangelist' of the movement. Both were Uniting Church elders, and they were also in positions of power in settlement affairs, despite being members of two of the smaller groups. One leader was Deputy Principal of the school and the younger brother of the Council President. Another brother had been Town Clerk and member of the Legislative Assembly of the Northern Territory, and another was a lay pastor (Bos 1988*a*: 232–5). About half of the Uniting Church elders at Galiwin'ku were closely related to these leaders. Others represented groups with closer ties to the country. The Church leaders were monogamous (Bos 1988*a*: 247).

Thus leadership of the movement and the Town Council overlapped at Galiwin'ku. At Milingimbi, however, the Revival leadership was separate from the official Uniting Church leadership

[11] A charismatic Uniting Church leader, the Revd. Dan Armstrong, visited the community to encourage evangelism on the part of Yolngu Christians. Some Yolngu Christians, including the Council Chairman, visited a National Charismatic Review Conference (Bos 1988*a*: 51).

[12] Six more Yolngu elders were elected the next year, then informal 'leaders' were elected. The Yolngu elders were of the Gupapuyngu, Wan.gurri, Dhurili Djambarrpuyngu, and Warramiri groups in 1979; and Golumala, Wan.gurri and Galpu in 1981 (Bos 1988*a*: 106).

(Bos 1988*a*: 106, 231, 242).[13] Indeed, leaders of Revival meetings in other communities were sometimes younger men who were not leaders of patrifilial groups, and hence did not have great authority in traditional religious matters. In this way the Revival overturned the age-related authority of other ceremonies.

The Course of the Movement

Nightly Christian-fellowship meetings, begun by the two leaders after the return of one of the men to Galiwin'ku in March 1979, proliferated until two to three hundred people were involved nightly, and half the population attended regularly (R. M. and C. H. Berndt 1980). The movement began to spread after visits by people of other Aboriginal communities to Galiwin'ku and by Yolngu leaders of the movement to other Aboriginal settlements. But by 1981 its former scale and intensity had diminished. However, some elders continued Bible-study classes, smaller 'fellowship' meetings involving up to twenty or so people became a regular feature of settlement life, and an annual Thanksgiving meeting revived something of its former spirit (Bos 1988*a*: 124).

Fellowship meetings included speaking in tongues, a practice engaged in by some lay mission-staff, teaching and preaching, and kneeling in prayer.

The central ritual was the evening meeting, at which people sat in a circle on sheets or blankets in a sandy place outside a house. Song-groups with guitars led songs about God involving hand actions and bodily movements (such as swaying) while women danced on the spot, Yolngu style. After the meeting was formally opened the leader called for 'testimonials' from those attending.

The testimonies concerned commitment to the Lord, changes in the person's life due to the power of the Holy Spirit, and being healed through the power of God. These expressions were greeted with cries of 'Praise the Lord' by others present. The songs during this phase, in English and Yolngu dialects, addressed God directly. Then the group formed a tighter circle, and the leader would give an extempore prayer. In the third phase a Church leader would read a passage from the Bible,

[13] Positions in the Galiwin'ku Town Council and its Departments tended to be monopolized by descendants of the father's brother of one of the leaders, an influential man during the twenties and thirties, people with uterine connections with the country, people of a large clan which claimed sites on the country, and relatives of the former headman (Bos 1988*a*: 257).

> paraphrased in Yolngu dialects, and comment on its relevance to present concerns—land rights, mining, the greed of white people for money and power, Aboriginal identity and self-determination.
>
> The fourth phase was the emotional climax of the evening. The people formed a tight circle again, and the song-groups performed again. A few people stood or knelt in the middle, praying and some weeping while the leaders spoke and prayed over them, sometimes speaking in tongues and laying on hands. The subject would sometimes keel over in a 'trance'. Adults were seeking 'healing' in this phase, while white commentators interpreted this practice as 'praise' or 'worship' (Buchanan, quoted in Boyd 1986). The meeting closed with a prayer (Bos 1988*a*: 88–99; R. M. and C. H. Berndt 1980: 3).

Another form of ritual was the baptism, for which leaders invited people from other Anglican and Uniting Church communities in Arnhem Land. Several hundred visited for this purpose each year. The number of people at baptisms at the nearby communities of Gapuwiyak and Yirrkala reached about a hundred in 1981. Its format was similar to that of the fellowship meeting, or was modelled on the *bukuḻup* washing-ceremony (Bos 1988*a*: 99–100).

Some people claimed to experience 'visions' during the movement, for example that a fire engulfed the town, and that a rock in the form of a map of Australia would be found. During one fellowship meeting people claimed to see the face of a little girl who was lost without trace at Langarra. A *marrngitj* from outside the Yolngu region was brought in, who reported that the child had been taken up to heaven. Some people foresaw a great flood, while others thought that Jesus would soon to return to Jerusalem, or that a nuclear war was imminent. Video tapes distributed by fundamentalist groups persuaded others that the Bankcard logo was the 'mark of the beast' (Bos 1988*a*: 107–8, 113–14).

The movement spread to other Yolngu communities. Leaders flew with the aid of the Mission Aviation Fellowship to Aboriginal settlements in Arnhem Land, the State of Queensland, Central Australia, and the State of Western Australia, financed in part by collections taken during meetings. Uniting Church agencies organized six 'teams' of Arnhem Landers to visit six other Uniting Church Synods to create an awareness of Aboriginal issues, although the Yolngu interpreted these trips as extending the movement to whites (Bos 1988*a*: 118–19, 121).

While the intensity of the meetings died away by 1981, the formation of the Uniting Aboriginal and Islander Christian Congress, proposed by the Reverend Charles Harris and supported by Galiwin'ku elders, was seen as the outcome of the movement. One leader of the movement became Vice-President as well as a full-time officer of the Bethel Presbytery (formerly the Northern Synod) of the Uniting Church (Bos 1988*a*: 130).

Beliefs and Motives

The discourse of the movement drew heavily on Christian symbols, but the symbolism was syncretic at several points. People referred to God as *God wangarr ba:pa* or *God gunhu wangarr* (God *wangarr* father). Leaders likened *wangarr* such as the Djang'kawu to Adam and Eve or to Moses. One of the instigators of the revival drew an analogy between the *rangga* in its shade on the *nga:rra* ground and the Ark of the Covenant, and the group to which the other belonged represented its *rangga*, which was formerly somewhat cross-shaped, as having the shape of the Christian cross. Indeed, some Christian symbols were decorated with lorikeet feathers like other sacred objects. Bos shows that many of the songs were similar in form to *manikay*, and the way in which baptism was modelled on the *bukul̲up* washing-ceremony (Bos 1988*a*: 285, 288, 294, 301).

Nevertheless, the Yolngu version of Christianity was morally somewhat distinct from the indigenous cults; people did not regard God as aggressive and avenging like the giant Snakes (*motj*), but as peaceful like the Djang'kawu sisters. Furthermore, leaders of the movement found Gunapipi to be antithetical to Christianity, and opposed its performance at Galiwin'ku. A Cloudy Water leader claimed to have acquired rights to Gunapipi in recompense for a killing (Bos 1988*a*: 126–9, 178). A leader of another group thought that the Revival was defeated by Gunapipi, and someone claimed that Jesus visited the Gunapipi ceremony to expose its secrets. One reason for this opposition may have been the threat to the leaders' status, since they did not have rights in the ceremony and feared losing adherents to it. It is perhaps significant that the ceremony was first performed at Galiwin'ku (or performed again after many years) in the year of the movement's decline.

The unity of Yolngu with whites was a dominant motive in the movement. One adherent wrote:

> Lot of people is saying—why is it only for Aboriginal people—what's wrong with us. I will tell you—it is not only for Aboriginal people. Holy Spirit and God is for both black and white no matter who you are, no matter what colour. What nationalities, what race. In Christ we are all the same and we believe in one Christ-Jesus, one God. (Yunupingu 1982: 2, cited in Bos 1988*a*: 102)

One of the Revival leaders saw the movement as bringing blacks and whites together 'as brothers and sisters in Christ'. The songs tended to stress unity, with phrases such as 'Father make us one', and 'We are one big happy family, God's family'. Recurrent themes were God's love, loving one another, and being brother and sister as the children of God (Bos 1988*a*: 102, 317, 320).

Adherents to the movement deployed arguments validating Aboriginal claims to land. God gave the various people of the world their lands, and God entrusted the ancestors with the countries of Yolngu groups. Some likened the struggles of the Yolngu over land rights to those of the Israelites. Yolngu people were thus hoping that religious arguments would lead the white community to understand their point of view and bind them in a way that arguments solely related to the *wangarr* ancestors would not (Bos 1988*a*: 329–30, 333–4).

Another motive was independence: some Yolngu claimed that they were dealing directly with God in the movement, and did not need white missionaries or other Church authorities to mediate between them and God (Bos 1988*a*: 82). One man saw the movement as a reversal: the Gospel came from the *balanda*, but now he was ministering to the white community as well as black (Bos 1988*a*: 102, 320).

But as in all ceremonies, values of co-operation with others were stressed, together with autonomy, expressed by adherents of the movement by the terms 'love', 'friendship', 'brotherhood', and 'sisterhood'. Adherents stressed co-operation within the community and the ending of conflict, as well as values related to settlement order, especially keeping the environment clean (R. M. and C. H. Berndt 1980: 5). It was claimed that the cult broke down group and kin-network boundaries: people attempted to lessen the constant accusations of sorcery against the one particular group, and there was a feeling of freedom to talk to outsiders (Bos 1988*a*: 102). I heard from Milingimbi that one old leader wept out of relief that hostility and sorcery accusations among former enemy-groups would cease.

The sexes were not separated in the Revival, or indeed in Church affairs more generally, to the degree that they were in the indigenous ceremonies. About a third of the elders of the Church at Galiwin'ku were women. Women sometimes led the singing during meetings, gave testimonies, and laid on hands. There were no divisions between esoteric and exoteric religious knowledge, apparently, although women did not preach or lead the longer songs, and seldom played musical instruments. No Arnhem Land women were ordained ministers at the time of the movement (Bos 1988*a*: 265–6).

Fellowship members were not supposed to drink alcohol. Indeed, according to some adherents, some drinkers gave up alcohol, petrol sniffing lessened, marital problems were resolved, and sick people were healed. People interpreted these supposed effects as the results of the power of the Holy Spirit. Health workers noticed a decrease in the number of visits to the hospital for minor complaints, although some observers accused some mothers of neglecting their children. Visitors thought the township to have a more open, friendly, and co-operative atmosphere, while some individuals became more confident and forthright, willing to confront and talk to government officials (Bos 1988*a*: 76, 83–4).

Bos believes that, as with the earlier adjustment-movement, one motive behind the Revival was to legitimize the authority of Town Council leadership positions. The Council was limited in its effectiveness because its authority was not generally accepted, and Councillors were caught between conflicting values. In the event the most prominent leaders of the movement moved to Darwin to be employed by the Uniting Church and Nungalinya theological college (Bos 1988*a*: 324).

Gunapipi and Christianity

In the period of the mid-1970s to the early 1980s Gunapipi and other revelatory rites continued to be practised even while the Christian Revival burgeoned. As I suggested in the introduction to Part III, Gunapipi was more universal in its symbols, themes, and organization than the other rites, but Christianity took this movement further. Like the regional ceremonies, the Christian Revival employed symbols that were universalistic but chosen in an attempt to integrate whites and to extend a degree of control over their actions, by finding a common moral order. This moral order differed from the one associated with traditional ceremonies, especially in the stress on peace and forgiveness, in contrast with the harnessing of ancestral power in order, among other things, to harm others.

The causes and motives behind the Christian movements were complex. In the growing adoption and transformation of Christian symbols, beliefs, and practices by Yolngu, we should not underestimate the effects of mission activities, especially schooling of the young, including Sunday School, and of imported evangelists. But the adjustment movement and the Revival seem to have been responses to particular circumstances such as changes in settlement organization and the Aboriginal Land Rights Act.

In creating religious links with the white community through the Revival, Yolngu were also asserting autonomous control (soon after moves to indigenize local control of the Church). This echoed the structure of indigenous cults, in which wider links through co-operation in shared forms of practice went hand in hand with local autonomy in belief and control. The incorporation of Christian forms into indigenous structures of belief and practice, fostered by the activities of Nungalinya College in Darwin, where Aboriginal pastors trained, was also an important aspect of affirming autonomy.

How far was the adoption and transformation of Christian beliefs and rituals likely to inculcate changes in social practices and relations more generally? Gunapipi included men's and women's separate and secret aspects, but the move towards inclusiveness through Christianity threatened to erode the power of older men, which rested on the constitution and control of secret religious knowledge and practices, for Christianity challenged the very separation of 'inside' from 'outside'. It is significant, therefore, that Yolngu people adopted Christianity alongside the older ritual forms rather than replacing them with new beliefs and practices.

The Christian movements cut across the grain of Yolngu social organization. However, the adjustment movement did not promise fully to undermine male power, for the leaders compromised in bringing out only some of the sacred objects, and reserving the more important ones. Moreover, we can infer from Chapter 7 that while *garma* ceremonies continued relatively unchanged, so did the differentiation of religious knowledge encoded in their elements.

The continuing popularity of Gunapipi, which itself seems to have undergone a renewal about the same time as the Christian Revival, as well as other regional ceremonies such as Djungguwan, implies the continuity of that regional co-operation that underpinned power relations along age and gender lines, and of the beliefs that underlay these relations. But the preference for Gunapipi over, say, the Ma<u>d</u>ayin ceremony, was perhaps a compromise between the strong divisions implied in the Ma<u>d</u>ayin, as well as secret knowledge embedded in the *garma* ceremonies, and the universalism of the Christian Revival. In Gunapipi, women had a more equal part to play in having their own secret aspects, while the bases for age-related authority and the division of initiated men from children and women remained, and local identities could be expressed. Nevertheless, its popularity also implied a greater degree of unity over the region than could be expressed in *garma* ceremonies or the Ma<u>d</u>ayin ceremony. Thus, processes of change in Arnhem Land may have been similar to those encountered by Kolig in the Fitzroy region of Western Australia, where religion has recently placed greater emphasis on interconnections than on local differences (Kolig 1981: 2).

The concluding chapter relates these attempts to overcome differences between groups, male and female, old and young, Yolngu and Balanda, to other transformations in the control of religious knowledge and to effects of these changes. First I review what earlier chapters have had to say about the themes with which the book began: ambiguity, variation, and change.

9

Conclusions

The book began with ideas about language as an open project in the creation, negotiation, and imposition of meaning, and with the attitude that no measure of agreement should be taken for granted. The consequences of this starting-point are rather different from the results of structural models. Language has a constitutive role in social practices; people try to put their constructs into practice out of a variety of motives and subject to a multitude of conditions. Systems of practice do impose constraints in that people try to hold others to them and have to use what is to hand, but they are built on indeterminacies and are fuzzy at the margins. With the heterogeneous cultural tools provided, people move towards or away from each other, accommodating to one another or striving to create and maintain difference. Systems do not exist in an 'objective' domain but are modes that people make, share, disagree about, interpret through, and try to impose on others.

Disagreement is possible, however, only where there exist common terms in which it can be expressed, and shared tacit assumptions underlying those terms. After all, meaning is forged by the need to live together and to interact with things in the world. Concord about proper practice, at least in its outlines, was greatest among close neighbours, although neighbours in one direction may have subscribed to different modes of practice than those in another, making it necessary to accommodate to more than one perspective. In just such a way Djinang people had to adjust to the kin categories and definitions of marriage among Yolngu of the east, and to the contrasting Burarra categories and definitions to the west and north. The density or 'thickness' of the body of shared assumptions, as well as more explicit social forms, decreased with social distance, although similarities among speakers of Yolngu languages were probably greater than those between, say, Dhuwala-speakers and Gidjingali people of the Blyth River. Indeterminacies of meaning are greatest where the actions, processes, and entities constructed in discourse are imaginary, and least when constrained by interaction with beings and things which exist independently of their constitution in discourse.

Ambiguity, Heterogeneity, and Control

Elaborating a pervasive obliquity in social interaction, Yolngu religious practice was founded on systematic and elaborated ambiguity in the constitution and control of public and secret religious knowledge alike. Ambiguities were built into the discourse structure of myths, such as the gender ambiguity of pronouns (*ngayi*, him/her), and indeterminacy in the number of people referred to by proper names. Although people of a wide area agreed about many attributes of 'the same' ancestors—Djang'kawu travelled along the coast following the sun, created groups, and made springs with their digging-sticks—in constructing ancestral events people of different groups with 'the same' *wangarr* had their own rather ethnocentric and relative perspectives on them. People of the various groups placed focal events at their own country, disagreed about ancestral routes, and differed as to quite what sort of people the ancestral names referred to (two sisters, two sisters and a brother, one man or several men, one woman or several women, and so on). Yolngu recognized that others lay on a 'different path', but framed their differences in terms of shared, if ambiguous, categories. Difference multiplied with social distance, while tolerance of difference decreased (Chapter 1).

In discourse about *wangarr* ancestors and ceremony, agreement about the reference of proper names was often illusory, for speakers might use names with an agreed sense but divergent reference (to put it in the language of Frege). One party to a discussion about the journey of a being *x* would interpret the name to refer to an ancestor of one kind, the other as an ancestor of quite distinct identity. They might also agree that the name referred to a sacred object of a particular pattern used in a shared type of ceremony (Chapter 1). The ancestral character given to countries varied greatly.

Yolngu groups did not fit a stereotype model of 'segmentary' society whose constituents are 'corporate' elements of a like order arranged in a hierarchy of increasingly inclusive groups. Groups were heterogeneous in structure, and people disagreed about their constitution. If a group's identity was anchored in anything, it was connection with country and ancestor identified as species. But the multiple attributes and many names of groups, together with the sharing of attributes seen to be (or negotiated as) 'the same', gave groups a somewhat ambiguous and shifting identity.

Commonalities of language reflected the intensity of interaction. At the same time, people picked out and reproduced differences of speech in order to mark group and moiety identity and difference. This differentiation has been elided more recently in that younger people in larger settlements came to speak a common language regardless of moiety difference and group identity.

People sometimes disagreed about the group or even moiety identity of country, although the identity of ecologically rich areas was clearer than that of forest. Several groups had rights in the songs connected to forest places but people sometimes differed about who were the true owners. In at least one case these indeterminacies conflicted with attempts to control new resources in formerly resource-poor country, and gave rise to a new need to fix identity and boundaries (Chapter 3). However, it is not likely that conflicts over the interpretation of country, intimately connected with ancestral identity and the performance of ceremony, have suddenly erupted with the appearances of new resources; the parts of country richest in food resources were associated with the most important (the most 'heavy') *wangarr* ancestors, sacred objects, and ceremonies, suggesting that people have long attempted to anchor control of resources in ancestral identity and religious practice. Furthermore, religious practice itself can be seen as a 'resource', so that individuals and groups competed for control of religious practice as a resource. Then, the possession of one implied the possession of another; a person or group who identified with country through making the *l̲ikanbuy* design or performing the song and ceremony *ipso facto* identified with the country, as Chapter 3 showed. Control of country and economic resources did not imply exclusive use, however, but rather 'the right to be asked' (Williams 1986: 85–6; Myers 1982).

I have written of common 'forms' receiving variant interpretation. It is evident, however, that even though the shape of a given performance retained distinctive features of a genre or type, it emerged out of a process of negotiation and accommodation. Ceremonies of all genres provided 'slots' in which people of different ancestral affiliation could place their particular songs, dances, or designs, and display their ritual objects. Ritual forms were similar (or accommodating) enough for people of a wide region to co-operate in a performance, sufficiently shaped to be interpretable as re-enactments of ancestral events, but of such ambiguity that participants could interpret them in their own ways, and co-operate in spite of differences in their modes of practice (Chapters 4 to 8).

The chapters of Part II reveal an interplay between the control of creation of religious forms and dissemination of religious knowledge between group-centred networks and between age and gender categories. People restricted the information they made available to those who were not closely related to them by group identity or kinship about publicly displayed forms, at most disseminating elided accounts of their significance. Being granted more detailed information was a privilege marking a social relationship with someone of the group. This 'vertical' control of religious knowledge separating gender and age categories was modified by inter-group politics, in which the politics of ceremony was tied to the politics of resources. An example in Chapter 5 suggested that men of one group seem to have striven to prevent people of another group from learning their own songs.

It is perhaps not putting it too strongly to say that the ambiguity of religious and other forms of discourse entailed both variation and continual transformation: variation because the common shapes co-operatively produced were variously 'read'; transformation because people could alter their readings of similar forms through time as well differ, as between individuals and groups, and because those 'forms' were the products of contexts.

We found some diversity in modes of kin classification and marriage practices across the region. Although people agreed about the core of classification and the basic expression of norms governing legitimate claims (Chapter 2), they were somewhat manipulable and formed the basis of negotiated agreements. Translation procedures came into play on the margins of very different languages. The patterns of marriage networks tended to make strings of connections between groups in the west but cycled among Arnhem and Buckingham Bay groups. Accordingly, people's meta-theories about marriage varied, reflected in anthropological models of 'marriage systems'.

As well as productivity and innovation in ritual forms and interpretation, we found great diversity in the types of ceremony current across the region. The complement of genres and forms possessed by groups varied greatly as well. In parallel with language similarity, people living close to one another and who interacted most frequently tended, as one would expect, to share a common style of a given type of ceremony, such as Gunapipi, distinct from that of other regions.

Similarities among song-series deemed to be 'the same' were hard to pin down, but the similarities and differences seem to have been those of polythetic groups. Songs in the regional ceremonies were more standardized, but even here singers could arrange topics in different sequences, and make their own interpretation of songs. Moreover, people of different regions had somewhat distinct singing-styles and very different forms of words of 'the same' kinds of songs, such as Gudjika.

Variability across north-east Arnhem land, together with the extension of social networks and practices across differences in language family, creates difficulties for notions of relatively clearly bounded and homogeneous 'communities', 'groups', or 'classes' bearing language, culture, or *habitus*. Moreover, judgements about similarity and difference of 'culture' or *habitus* depend on the context in which contrasts are drawn. It is possible to pick out a cultural 'commonality' belonging to Aboriginal people of cities and country alike, by implicit comparison with the values, assumptions, and practices of white Australians (Sansom 1982). By the same token it is appropriate for some purposes to compare, say, the way Yolngu people do things with the way Pintubi do them (Myers 1986: 295). The reasons behind particular patterns of variability are of greater moment than the fact of variation, however.

Heterogeneity in assumptions, beliefs, and practices reflects patterns of political control, for orthodoxy has to be imposed. But a simple model of Yolngu political relations is an impossibility, whether a reconstruction of pre-colonial structure or of Yolngu in articulation with missions and the Australian State. If I were to deploy a single metaphor for an abstracted Yolngu domain of practices and relations it would be an image of intersecting, focused networks.

Within the network of qualified agreement as to the constitution of 'law' or 'right practice' (*rom*), it was up to people to attempt to enact their own interpretations of law. With the support of kin or others with common interests, the aggrieved themselves had to take remedial action against those who they believed had committed an offence. These supporters might include others with shared rights in sacra, if the offence were a religious one, such as improperly seeing a sacred object or fighting on the ceremony ground. Thus it appears that it was those with the greatest ability to muster support who had the best chance of enacting their particular interpretation of 'law' or right practice in justification of their action.

Even setting aside differences in power among individuals, the social network did not have a continuous web of intersecting and unbounded kindreds like an unwoven fabric, but was made lumpy or grainy, as it were, through patrifiliation. Groups, *mala*, or *ba:purru*, were the locus of greatest orthodoxy in belief and religious practice, although I have commented on the degree to which group identity and structure was ambiguous. Group-centred local networks strove to maintain their autonomy within an order of 'law' (*rom*) more or less agreed with their neighbours, while sending out tentacles of co-operation or attempted domination, or coming under the temporary sway of others. The most direct control of religious affairs lay within the network centred on a patrifilial group or part of one, and including close uterine relatives of people of the group. This control was modified by the ability of individuals of the same group to make their own interpretations of ancestral events, the meaning of ceremony, the structure of the group, and so on; and by factions cross-cutting people of the same group-identity, as well as by power exerted by men of other groups of the same moiety over a group lacking effective leaders. It was modified, too, by the fact that one group came under the scrutiny of people of other groups with common sacra; this power was exercised within the context of that co-operation.

Widely shared practices such as regional ceremonies, and long distance leaps by means of exchange ceremonies, gave people a common if limited language and matrix for sociality beyond the everyday range, extended further perhaps during the last few hundred years by a lingua franca derived from transactions with the Macassans on their annual visits (Urry and Walsh 1982), and extended further still in the 50 years, following the establishment of missions, with the learning of English (with varying degrees of skill), travel to cities, and involvement in institutions such as schools and the Church.

It seems, then, that the network metaphor reflects Yolngu patterns of interaction more closely than 'society' or 'community'. Individuals worked outward from nodes in the network, having most in common with their close kin and fellow group-members. Their view of the whole was relative to their position in the dynamic web. People could accommodate to differences as they moved away from familiar forms, interpreting what they encountered in their own terms, and aided by the long-distance projection of some of these forms through exchange. The 'objective' social reality, therefore, was one of a kaleidoscope of relative viewpoints within a matrix of co-operation and conflict.

The book's focus has been on techniques in the control of religious knowledge. The encoding of information in ambiguous forms of dance, design, and song formed one basis of control of religious knowledge between men and women, old and young (Chapter 7). The other was the guarding of space, which, like obliquity and ambiguity, was also a more general aspect of social interaction. Weakly iconic and ambiguous geometric designs related to more figurative patterns, together with related song-phrases and names, were the subject of complex layers of 'outside' and 'inside' exegetical interpretations, revealed incrementally by and to males and, ostensibly at least, hidden from females. Another technique of control incorporated within religious practice was initiation (Chapter 5), in which boys were formally detached from the category of females and converted into 'young men'. I suggested that circumcision initiation, in conjunction with revelatory rites, produced a change in *habitus* through 'radical resocialization', reinforcing propensities to defer to the authority and interests of older men.

In addition, aspects of Yolngu myth and interpreted ritual can be read quite unambiguously as ideological. Chapters 1 and 6 illustrated the pervasive theme of the loss of control of the 'inside' of religious practice on the part of women, and all that it implied. The interpretation of Djang'kawu myths (and similar ones) and of related songs and ceremonies as implying men's control of reproduction, however, is contestable. The hegemonic effect of beliefs depended on the 'level' at which assumptions worked. It may be that most women accepted without question the truth of those ancestral events; Frances Morphy recounted the comments of some Yirrkala women that if the 'stupid' Djang'kawu sisters had not lost the dilly-bag, women would have control of the ceremony and sacred objects. There is nothing inevitable about this, of course. We saw earlier, in Chapter 8, that (male) leaders of the 'adjustment movement' reinterpreted the same ancestral events, accusing those who followed the ancestors as having distorted what they meant. At a deeper level, in the recorded versions of the Djang'kawu events the stories take for granted the exclusion of the ancestral women from the men's secret ceremony-ground—which was the reason why they could not retrieve their dilly-bag and sacred objects. This underpinning of men's power was indeed hegemonic, unchallenged by men bringing out some of the *rangga* sacred objects during the 'adjustment movement'.

This brief discussion brings us back to the Comaroffs' concern over the limits of indeterminacy in the light of tenacious power-differences. What Marxian scholars refer to as a 'social formation' (as a mode of social relations rather than a social unit or 'society'), in its diverse manifestations, transcended local differences and relative perspectives. As Bern (1979) has pointed out, co-operation of a regional scope among older men underpinned the particular form taken by local age- and gender-related power-differences. People of different groups and local networks may have interpreted ceremonies in the light of their particular ancestral traditions, but they were able to co-operate in performances which had similar pragmatic effects in different places. Moreover, ceremonies of the distinct and unevenly distributed genres were not dissimilar in their import. The power relations entailed in this social formation, however, were undergoing transformations.

Transformations in Power and Religious Practices

Like those of the Murinbata people (Stanner 1963), Yolngu institutional forms had evidently been, if not in continual ferment, then at least simmering, notwithstanding a religious ideology of ancestral creation and subsequent stability. Subsections were new to the older people of the 1940s, and not completely adopted at Mirrngadja in the 1960s (Nicolas Peterson pers. com.). Yolngu traditions recorded the introduction of Gunapipi from the south; Ngulmarrk was in all probability borrowed from the west; and traditions of Gunwinggu people held that they received Madayin from the eastern end of Arnhem Land. Mandayala had affinities with similar ceremonies of the Gulf country to the south-east. The annual migration of the Macassan trepangers, which began probably not more that a few hundred years ago, inspired a whole genre of *garma* songs and ceremonies, more recently incorporating Christian interpretations. Yolngu assimilated introduced ritual forms to their own mythology, and interpreted old forms in terms of mythologized newcomers. The enterprise should be, therefore, not to record changes to a 'traditional' order (Cowlishaw 1987), but to trace trajectories of transformation in relations, powers, trends, events, and the forms into which people try to shape their worlds. Some of the directions in which modes of action and relations were moving in the 1970s have been followed throughout the book. I have treated mission and State as an external environment; a fuller analysis of Yolngu–Balanda relations would examine Yolngu social life, the missions, and the State equally.[1]

The Introduction showed that Yolngu opposition contributed to the failure of the pastoral industry in Arnhem Land, and indirectly resulted in the establishment of missions. By comparison with other government and mission policies in Australia, the Methodist Mission was relatively benign, electing to gain its ends through persuasion rather than force, and without radical cultural surgery such as the dormitory system, banning of ceremonies, prohibition of arranged and polygynous marriages, and enforcement of the use of English found elsewhere. Arnhem Land was encapsulated within the nation; inside its formal boundaries the State and its agents intervened in Aboriginal social life, but they did not dissect and reassemble it in the way that occurred in other parts of Australia such as Queensland.

The missionaries and the later state-run schools imparted elementary trade-skills to some individuals, knowledge of English to a majority of the younger generation, and literacy and basic numeracy to some. The missionaries succeeded too in establishing an institutionalized settlement mode of life. Individual Yolngu could move in and out of the mission regime more or less at will, but they were subject to coercion, especially in relation to medical

[1] The current research of Ingrid Slotte at the Australian National University has this focus.

treatment and homicide. Their movement outside the region was restricted until the Welfare Era, and they were relatively powerless until the 1970s to control incursions by mining prospectors and others. Restrictions were greatest during World War II, although Yolngu gained new abilities, such as earning wages with the armed forces.

In the Welfare Era we saw that Aboriginal people had increasing involvement in the money economy. At the same time, the monitoring and control of movement, births, health, and mortality was magnified, as was the presence of government agencies. With the succession of generations a new *habitus* manifested itself as people were increasingly inclined to live on settlements, adopting all that went with that life. By the 1970s Yolngu were increasingly involved in the administration of settlement affairs, while still relying on the advice and participation of whites, as well as their own skills in literacy and numeracy. The Yolngu also took over the local church. New powers in settlement governance gave rise to uncertainties and to conflicts within communities, especially between those whose authority rested on the control of religious practice and those with new skills.

One of the most significant of changes in the overall social context was the provision of alternatives to which the individual could turn. Some people travelled in past centuries to the Celebes with Macassan traders, but white colonization opened up many more possibilities. The means by which an individual could partially transform his or her life-pattern also become available within Aboriginal townships, including the opportunity on the part of women to earn salaries and wages. The degree to which the social order persisted, however, is testimony to the high social and personal costs of leaving the community for long periods. Thus during the mid-1970s the practices that sustained age-related authority remained current if somewhat transformed. I shall focus the discussion here primarily on men.

The Erosion of Age-related Authority

In the long term the importance to communities of new skills and European education to meet those needs may tend to undermine the value of secret religious knowledge by rendering it less relevant. Full-time boarding at secondary colleges on the part of a number of young people (more recently discontinued) removed them from the normal participation in the religious life and the processes of learning. A proportion of young married men in the mid-1970s at Milingimbi were more involved in, and more skilled at, occupations such as Town Clerk or community adviser than in ceremony. Not only were there local alternatives but young men could move away from their home settlements to live on the fringes of towns, and a few worked in those places for sustained periods. Women gained the possibility of taking up alternative modes of life on the fringes of white communities, offering the opportunity of escape from Yolngu power-structures. Employment as nurses and teachers or teaching assistants, and potential relationships with white men, increased the possible alternatives for some. Men's use of force declined, and women gained in their ability to evade marriage contracts. A minority of people committed themselves to monogamous Christian marriage, although men often reneged under pressure from affines or out of personal desire.

Many older people at Milingimbi in the mid-1970s were worried about introduced attractions, including school, seducing the young. In 1974 anxiety of this kind led one man to try to compete with the attractions of rock music by singing *manikay* songs on the stage of the community hall for one or two evenings, and the reader will recall the Cloudy Water leader waiving the obligations of neophytes to make gifts to the older men in order, it seems, to encourage them to take part (Chapter 6).

The religious authority of older men, with its concomitant restraint on the part of younger men, was a condition for age-related polygyny. Genealogies collected in the mid-1970s revealed no great shift in the age of marriage. However, widow inheritance was declining and some young women were refusing to accede to bestowal arrangements, resulting in the setting-up of several single women's camps; nor were the inflated levels of polygyny of an earlier period being matched. We saw that since the 1970s growing numbers of young men have sought wives outside the region to evade the constraints of bestowal. Such shifts must have affected the way in which dominant men could build networks of influence, as well as the dynamics of group growth, competition, and decline resulting from high levels of polygyny and the politics of marriage.

Polygyny had been focal in male power-games. A man with

many wives and children was at the node of a network of obligation, exchange, and influence, and at the head (or 'at the front' in Yolngu metaphor) of growing numbers of descendants. His own obligations to his wives' kin lessened at he got older, but the obligations of his actual and potential daughters' husbands towards him increased with his age. Men also saw sons (and brothers) as a source of support in conflicts. This nexus between religious authority, polygyny, and the dynamics of group competition continued in a modified form into the mid-1970s in the Milingimbi area. However, with the establishment of missions and increasing intervention of the State in Yolngu social life, the roles of male leaders had shifted from connections with success in fighting and raiding to mediation between Yolngu and Balanda, and peace-making. With the move to a welfare economy centred on settlements, and new resources such as positions in settlement administration (and control of budgets), mining royalties and leases, other arenas of power replaced to an extent the connection between marriage, political support, and control of economic and religious resources.

Control of religious knowledge had been a key element in the political economy of marriage, country, and ceremony. There was a direct link between religious prerogatives and power, namely through the sacralization of forces of production, producers, and the product itself. This aspect of men's power was undergoing profound change in the mid-1970s. The ability of older men to control production, distribution, and consumption through religious means within a foraging economy (Chapter 2) became largely irrelevant in the context of settlement economies through the decreasing use of the technology and food categories involved. New contexts for the exercise of power based in religion included the production of bark-painting for the market and the control of monies, leases, and other matters connected with mining. Processes of settlement affected the ability of some groups to control access to their country and its resources, some becoming powerless, others gaining in power.

Turning to transformations in religious belief and practice, some evidence suggested a trend to a more dualistic cosmology as people speculated about the relationship between Christian spiritual beings and the *wangarr* (Chapter 1). The ancestral constitution of country was under some threat through settlement of many groups on missions far from their countries. However, the outstation movement, which began to burgeon in the mid-1970s, brought a larger proportion of younger people into contact with their own places (Chapter 4). Yolngu had continually incorporated new ritual forms into existing dogma and given new interpretations to old forms. Some rituals had been dropped from the repertoire, others were coming in, others were in the process of being adapted to new conditions (Chapter 5).

Part III has traced trends towards more universalistic symbols. Christianity moved further along this trajectory than Gunapipi and other regional rites, not only in trying to transcend differences between patrifilial groups, men and women, old and young, but in attempting to forge relations of greater equality with the white community. The stress on peace and forgiveness in Christianity went against the grain of endemic conflict among people of competing patrifilial groups. Although the apparently growing popularity of Gunapipi may, like Christianity, have engendered stronger regional ties, I argued above that it reproduced the structure of gender separation and age-related authority which Christian rituals tended to oppose. However, Gunapipi seems to have been the only major ceremony in which women had their own secret domain, from which young bachelors and adult men were excluded, so it may be that the increasing popularity of this ceremony reflected a shift in gender relations.

It could be argued that the acceptance of their interpretations of Christian beliefs was hegemonic in effect (Trigger 1992). Some Yolngu saw the Bible as the key to the understanding of Balanda law, just as *ma<u>d</u>ayin* sacra and stories were the foundation of Yolngu *rom* ('law'). This 'accommodation' to the power of the State was brought about especially by the intervention on the part of the missions in socialization, especially through schooling and Sunday school, which effectively brought about a belief in the power of Christian deities and symbols, along with visits by charismatic evangelists and so on.

It is not clear if, and in what manner, Yolngu accorded 'legitimacy' to the authority of the mission and the State in accepting aspects of Christianity, a problem discussed by Trigger (1992) in relation to people of Doomadgee in Queensland, for I do not know in detail how people conceived of the State. Some individuals expressed very strong hostility to the missionaries, especially when drunk, for example arming themselves with spears and surrounding the mission office at Milingimbi. On the other hand, some individuals expressed gratitude to the missionaries for bringing peace to Arnhem Land. Typically, people discussed government in terms of its obligation to 'help' Aboriginal people, and of the activities of some other sectors of the Balanda community in terms of 'pressure', for example by potential miners, and those proposing a highway through Arnhem Land. Yolngu had experienced the presence and power of Australian authorities, and their readiness to intervene in people's lives, over many decades. It may be that Yolngu accepted Balanda power as a fact of life, rather than consenting to it (as the concept of 'legitimacy' implies), and sought to live as effectively as possible within its shadow.

Certainly Yolngu did not cede total control of Christian matters to Uniting Church authorities. During the mid- to late 1970s active Yolngu participation in church services was small in scale. It was not until Yolngu gained control of the local churches at the end of the decade, and the related Christian Revival, that the expression of Christian belief became more widespread. The strong assertion of a desire for local control of the church in the late 1970s and early 1980s within the wider religious practice centred around the organization of the Uniting Church, and in the context of discussions with Nungalinya College in Darwin about the development of a Yolngu theology (Bos 1988*a*). Christian religious practice shared with the indigenous religious complexes a structure of assertions of local autonomy within the bounds of co-operative practice.

Given the constraints of State power, Yolngu had maintained a vigorous, if transforming, domain. Their ability to do so was contingent upon the failure of the pastoral industry to establish itself in the north in the face of Yolngu resistance as well as other factors, on a long association with a relatively non-interventionist mission, and more recently the self-determination policies of the Whitlam Government and subsequent administrations, which affected the Northern Territory very directly, especially through the role of the Department of Aboriginal Affairs, and the effect of the Land Rights Act. The latter was enacted in part as a result of Yolngu opposition to mining at Gove Peninsula.

A senior man of Milingimbi, whom I have called Wurrpa<u>n</u> in this book, expressed confidence to me about his power. He explained that a child's vitality was not just 'wild' (*wakinngu*); the child had power (*ma:rr*) such as the ability to run because the *rangga* lay in the ground and waters. The heart-blood was the source or condition of vitality and the power of blood was also from the *madayin*. Control of *rangga* gave men particular power. Explaining the *likan* names which he called at the circumcision ceremony described in Chapter 5, Wurrpan said to me:

> The big paperbark fell at Honeybee country. I called that. It is my knee or elbow for the young people lying down with blood on them. They lay down, and I called the *likan* name—the [elbow or knee] joint for everyone—of paperbark falling. *Dalkarra* is power from the *madayin*. This, the knee or elbow, is power. I have a lot of power; no one gets the better of me, white or black. I have the law for everyone. I am the elbow. I have the *dalkarra*, the power.

GLOSSARY OF YOLNGU WORDS

Note that as the order of the following list of Yolngu words follows Yolngu phonemes, it does not appear to follow strict alphabetical order. The long vowels are grouped with the equivalent short vowels: *a:* with *a*, *e* with *i*, and *o* with *u*. The suffix forms are of the Dhuwala'mirri (e.g. *-irri*) and Dhuwal'mirr (e.g. *-irr*) dialects.

balanda	light-skinned person
balang	subsection or 'skin' name, male
balanya bili	the same, similar
baman'	long ago
bangaḏi	subsection or 'skin' name, male
bangaḏitjan	subsection or 'skin' name, female
ba:pa	father, father's brother (including father's parallel cousins), man's own (or person's brother's) son's son's son
ba:pa'mirringu	closely related 'father'
ba:purru	relation between *wangarr* ancestors, places, sacra, the living, and the spirits of the dead
barrkimirr/i	in the relation of *galay* (MBD, MMBDD) to *dhuway* (FZD, MFZDD), wife to husband, married
barrku	far off
bathi	dilly-bag (oval basket woven from pandanus fibre), container
batjimurrungu	angular ear shellfish, *Cassidula angulifera*; *ba:purru* name, Yirritja moiety
bay-lakaram/a	forgive, set aside
bilinydjan	subsection or 'skin' name, female
biḻma	clapsticks or tapsticks with which singers accompany themselves; category of songs in which clapsticks only are used and not the didgeridoo
birr	particle signifying great distance in time or space, hence *barrku birr!*, 'exceedingly far'
birrimbirr	person's spirit which survives them after death
birrkuḏa	honeybees and their nest, *Trigona* sp., identity of a major Yirritja-moiety *wangarr* ancestor
bukubuṯ	'head-fly' disinterrment ceremony
bukuḻup	'head-wash' purification ceremony

bul'manydji	adult shark of *Carcharhinus* sp., identity of a major Dhuwa-moiety *wangarr* ancestor
bulany	subsection or 'skin' name, male
bulanydjan	subsection or 'skin' name, female
buma	make, create, hit, kill
bundurr	generic term for 'knee' or 'elbow' names, linking place, group, and an attribute of a *wangarr* ancestor; synonymous with *l̲ikan ya:ku*
bunggawa	'boss', leader (from Macassan)
bunggul	generic term for ceremony
burralang	subsection or 'skin' name, male
buthalak	yellow-ochre pigment
buthuru-bitjun	listen; *mali'-buthuru-bitjun* — 'listen-to-the-spirit' ceremony (see *ringgitjthirri*)
d̲ad̲a	ceremony, purification with smoke
d̲a:l	hard, strong
d̲alkarramirr/i	ritual leader who calls *l̲ikan* names, Yirritja moiety
d̲ambuy-wal̲uy	mid-day
d̲an.gi	sulphur-crested cockatoo
d̲ar't̲aryun	'tap', sing with clapsticks
d̲ilak	elder, old person
d̲iltji	back, 'back country', hinterland or 'bush'
d̲et̲i	native almond, *Canarium australianum*
d̲upun	hollow-log coffin for final disposal of a person's bones; ceremony incorporating the coffin
dha:dhud̲ipuy	'bottom country', mouth of the river
dha:gupapuy	'top country', up-river
dha:manapanmirr/i	joined together, conjoint (of groups)
dhangang	full
dhangu	'this/here' in the *dhangu'mi* 'this-category' or 'dialect-group' of languages
dhapi	'foreskin', circumcision ceremony
dhawu	story, news, information, 'word'
dhin̲'thun	follow tracks (of an animal), follow ancestral precedent
dhiyangu-bala	now
dhud̲akthun	follow, imitate, re-enact
dhukarr	path, track, road, 'way' (of behaving)
dhukuray	shellfish, *Nerita lineata*
dhumun.gur	kin category—father's sister's daughter's daughter's child (reciprocals, *mumalkur, ngathiwalkur*)

dhuwal — 'this/here' in *dhuwal'mirr* tongues
dhuwala — 'this/here' in *dhuwala'mirri* tongues
dhuwala'mirri — '*dhuwala*-having', a this-category or 'dialect group'
dhuwal'mirr — '*dhuwal*-having', a this-category or 'dialect group'
dhuway — kin category—father's sister's child, husband or husband's sibling, person's sister's husband or husband's sibling (reciprocal, *galay*)
dhuyu — 'sacred'
dja:ga, djaka — look after, care for
dja:gamirr/i — custodian, carer
djalathang — south wind, southerly direction
dja:ma — work
dja:mamirr/i — worker
djamarrkuḻi — children
djangu — 'this/here' in *djangu'mi* languages
djapanbani, djapanmani — trestle structure constructed in public domain in Gunapipi ceremony
djarrak — generic category of black-capped tern sp.
djarrka — water monitor, *Varanus mertensi*
djarrpi — bent, crooked, wrong
djinaga — inside, underneath, hidden or secret
djirriḏiḏi — forest or azure kingfisher, *Ceyx azura*
djirrikaymirr — ritual leader who calls *ḻikan* names, Dhuwa moiety—'power-man'
djuḻurrpi — giant threadfin fish, *Eleutheronema tetradactylum*
djunggayarr, djunggayi — position in ceremony of those whose mothers 'own' the sacra through their fathers; the *waku mala* (q.v.)
djutu — paperbark roll used as raft or float
gaḏaku — uncircumcised male, boy
gaḏayka — stringybark tree or its bark, *Eucalyptus tetradonta*
gadjak — subsection or 'skin' name, male
ga:na — separate, distinct
gutjan — subsection or 'skin' name, female
galay — kin category—mother's brother's daughter (and MMBDD etc.), wife, wife's sibling, person's brother's wife or wife's sibling (reciprocal, *dhuway*)
gaḻikaḻi, galiyan — subsection or 'skin' name, female
galka — sorcery, sorcerer
galki — near, nearly

gamanydjan	subsection or 'skin' name, female
gamarrang	subsection or 'skin' name, male
gaminyarr	kin category—woman's own son's child, person's sister's son's child, man's own daughter's child, or person's brother's daughter's child (reciprocals, *mumu*, *ngathi*)
ga:murung	larvae of longicorn and other beetles, 'witchetty grub', symbol of new birth
ganydjarr	power, energy
gapu	water
gapu gulun	water-hole, 'stomach water'
gapu monuk	salt water, sea
garma	public genre of ceremonies
ga:thu	kin category—man's own child, person's brother's child, father's father's father (reciprocals, *ba:pa*, *mukul ba:pa*)
gawal	kin category—mother's brother, syn. *ngapipi* (reciprocal, *waku*)
gaypal	wattle tree, *Acacia auriculiformis*
gudjika	category of songs accompanied by boomerangs
gulu'kulungu	a person's relation to their own patrifilial group, as opposed to mother's, mother's mother's group etc. (see *-pulu*)
gulun	belly, stomach, womb, palm of hand, sole of foot, waterhole
gomili	herb, *Clitoria australis*
gumurr	sternum, chest
gumurr da:l	'strong chested', assertive
gunhu	kin category—father, father's brother, syn. *ba:pa* (reciprocal, *ga:thu*)
gungman	mature woman who has borne children, pregnant or lactating woman
gong-ranggamirr/i	'*rangga*-hand', one whose task it is to refurbish the sacred object in a ceremony
gupa-djuy'yun	send someone on their way, 'say goodbye', funeral ceremony or wake
gupa-ngamathama	'make good', prepare, paint with *likanbuy* (q.v.) designs in a ceremony
gupa-ya:n.guma	send someone on their way, 'say goodbye', funeral ceremony or wake
Gupapuyngu	name of an extended group and its tongue of the *dhuwala'mirri* this-category, 'up-river people', Yirritja moiety

gurrmul young man after the circumcision wound has healed, bachelor (pl. *yawirriny'*)

gurrung kin category—father's sister's daughter's daughter, woman's actual or potential daughter's husband (reciprocals, *mukul rumaru*, *maralkur*)

gurrutu relationship, kinship

gurrutumirr/i mala one's group of close relatives

gutharra kin category—woman's own daughter's children or person's sister's daughter's children (reciprocal, *ma:ri*)

gutharra mala people with the same mother's mother's group

guwak koël cuckoo

guwatjman visit

guya generic term for fish

guyanga remember, think, come to mind

la:rr type of 'coarse' painted body design

lunggurrma north/north-east wind, October–December

likan elbow

likanbuy category of painted designs, 'pertaining to the elbow'

likan ya:ku 'elbow name', see *bundurr*

liya head (first element in many *ba:purru* names)

Liyagawumirr *ba:purru* name, Dhuwa moiety

luku foot, feet, footprints, root, traces (of *wangarr* ancestors), 'foundation' of the religious law

lorrpu sulphur-crested cockatoo, see *dan.gi*

madakarritj fierce, dangerous

madayin sacra, category of things and events associated with *wangarr* ancestors, sacred objects

madayin-watangu holder of the sacra through patrifilial group-identity

makarrata ceremony to settle a dispute

mala group, mob, aggregate

mala wanggany 'one group', united

malgarri style of public ceremony associated with people west of the Blyth River

mali' image, reflection, shadow, spirit

mali'-buthuru-bitjun 'listen-to-the-spirit' ceremony (see *ringgitjthirri*)

ma:lk subsection names, 'skins'

malagatj ancestral being, *wangarr*

man'tjarr leaf, purification ceremony incorporating leaves

Manggalili *ba:purru* name, Yirritja moiety

manikay	song, especially those accompanied by clapsticks and digeridoo
ma<u>n</u>awiny	kin term, used by female *gurrung* to address her *maralkur* or by female *dhumun.gur* to address her *ngathiwalkur*; these are 'avoidance' relations
maralkur	kin category—mother's mother's brother's son, wife's father, brother's wife's father (reciprocal, *gurrung*)
mari	trouble, fighting
ma:ri	kin category—mother's mother, mother's mother's sister, mother's mother's brother (reciprocal, *gutharra*); father's father, father's father's brother, father's father's sister in some areas (see *ma:ri'mu*)
ma:ri'manydji	in a relation of mother's mother or mother's mother's brother to woman's (or sister's) daughter's child (*gutharra*)
ma:ri'mu	kin category—father's father, father's father's brother, father's father's sister (reciprocal, *marratja*)
ma:ri-pulu	a person's mother's mother's group, country, and sacra
ma:ri-wa<u>t</u>angu	'mother's mother holder', person in relation to their mother's mother group, land, and sacra, see *gutharra mala*
marnggi	to be knowledgeable, to know
ma:rr	power, especially of and deriving from *wangarr* ancestors
ma:rr-gangga	somewhat, a little, rather
ma:rr-yuwalkthirr/i	believe, follow, be loyal to
marradjirri	a hair cord or pole presented in ceremony, ceremony incorporating this
marratja	kin category—man's own son's child or person's brother's son's child (reciprocal, *mari'mu* or *ma:ri*)
marrawinydjun	style of 'chorus' performed at climax of ceremonies, usually to accompany calling of *<u>l</u>ikan* names (q.v.)
marrngitj	healer, 'doctor'
marrtji	go, walk
matha	tongue, way of speaking, language
Ma:tjarra	*ba:purru* name, extended group or string of

groups created by Djang'kawu sisters, Dhuwa moiety

mayali' — meaning, significance
mayali'mirr/i — meaningful
maypal — invertebrates, including shellfish, crabs, and larvae
miku — red ochre, paint made from this pigment
milminydjarrk — freshwater springs
miny'tji — painted design, pattern
minggathun — style of 'chorus' performed at climax of ceremonies, usually to accompany calling of *likan* names (q.v.)
mirithirr/i — very
-mirringu — suffix applied to kin categories denoting a close relation
miyalk — female, woman, wife
moda' — circumcised boy, before the wound has healed (see *gurrmul*)
mukul ba:pa — kin category—father's sister, man's own (or person's brother's) son's son's daughter (reciprocal, *ga:thu*)
mukul rumaru — kin category—mother's mother's brother's daughter, wife's mother, brother's wife's mother (reciprocal, *gurrung*)
mokuy — corpse, dangerous spirit of the dead, category of malign and libidinous spirits of the forest celebrated in songs
mokuy-birrimbirr — spirit of the dead
mulkuru — stranger, foreigner
mumalkur — kin category—mother's mother's mother's brother's daughter, wife's mother's mother (reciprocal, *dhumun.gur*)
mum'mumdhun — style of 'chorus' performed at climax of ceremonies, usually to accompany calling of *likan* names (q.v.)
mumu — kin category—father's mother, father's mother's sister, mother's father's sister (reciprocal, *gaminyarr*)
morri'murri — very old person (see *norrutj*)
murrmurryun — style of 'chorus' performed at climax of ceremonies, usually to accompany calling of *likan* names (q.v.)

motj — category of *wangarr* ancestor in the form of snakes, associated with lightning and thunderstorms

natcha — kin category used by hinterland people, father's mother, mother's father and mother's father's sister, mother's mother's mother's brother's son (see *ngathi*, *ngathiwalkur*)

norrutj — very old person, old man

n̲iwud̲a — honeybee, see *birrkud̲a*

n̲ond̲a — shellfish, *Telescopium* sp.

nha:ma — see, look

nhanngu — possessive pronoun, his/her, his/hers

nhe — you (singular)

nyira — unaccompanied style of song used in Gunapipi ceremony, song used in sorcery

nga:ma — hear, listen

ngamin̲i-wat̲angu — 'holder of the breast/milk' (see *ngand̲i-wat̲angu*)

nganakal — indirect kin-term, 'your mother's brother'

ngand̲i — kin category—mother, mother's sister, woman's or sister's daughter's daughter's daughter (reciprocal, *waku*)

ngand̲i-pulu — person's mother's group, country and sacra

ngand̲i-wat̲angu — 'mother-holder', person in relation to mother's group, country and sacra (see *waku mala*)

ngan̲marra — conical mat woven from pandanus formerly used by women as a pubic covering and to protect a baby

ngapawanditj — a person's own patrifilial group (*ngapa*, spine) (see *-pulu* relations)

ngapipi — kin category—mother's brother (and MMBDS etc.), woman's own (or person's sister's) daughter's daughter's son (reciprocal *waku*)

ngaraka — bone

ngaraka wa:nga — 'bone country', country associated with *rangga* sacred object and major ancestor

ngarra — first person singular pronoun, I

nga:rra — category of major revelatory ceremony, the shelter in which *rangga* sacred objects are placed, Mad̲ayin ceremony

ngarraku — pronoun—my, mine

ngarritj — subsection or 'skin' name, male

ngarritjan — subsection or 'skin' name, female

ngathi	kin category—mother's father, mother's father's brother, father's mother's brother, (reciprocal, *gaminyarr*)
nga:thi	weep, cry, mourn, sing in weeping style
ngathiwalkur	kin category—mother's mother's mother's brother's son (reciprocal, *dhumun.gur*)
ngathu	bread made with the nuts of cycad palm (*Cycas media*) (see *warraga*)
ngayathama	hold, look after
ngayi	personal pronoun, he/she
-ngura	suffix denoting 'place'
ngurrud̲awalangu	first-born, leader
ngurrunanggal	ancestor, forerunner
ra:kay	edible rush-corms, *Eleocharis dulcis*
rangga	secret sacred objects normally seen only by men
rapam	generic term for 'big' name of country, incorporating many small named places
ratjpa	haematite, a crimson ochre with a sheen
rebal'yun	clean up, purify, clear country by burning off
renggitj, ringitj	place at confluence of several groups' countries, associated with ceremonies
ringgitjthirr/i	announce a death with a special song (see *mali'-buthuru-bitjun*)
rom	'the proper way', religious 'law', custom
rrambangi	together, level
wakinngu	illegitimate, wild
waku	kin category—woman's own child, person's sister's child, mother's mother's mother, mother's mother's mother's brother (reciprocals, *nga:nd̲i*, *ngapipi*)
waku mala	people with the same mother's group
waku-pulu	a person's mother's mother's mother's group, country and sacra
waku-wat̲angu	'mother's mother's mother-holder', person in relation to mother's mother's mother's group, country and sacra
wa:l̲awun	shellfish, *Telescopium telescopium*
wal̲i	salmon catfish (*wed̲u*) (song-name)
wan'tjirr	'leaf' purification ceremony, see *bukul̲up*
wan̲a	arm, wing, front leg of animal
wa:n̲d̲uwa	bloodwood, *Eucalyptus polycarpa*
wa:nga	place, country

wa:nga-waṯangu	country-holder, owner of country through the father
wangarr	totemic ancestor
wangarryu	in ancestral times
wanggany	one, united
warraga	cycad palm, *Cycas media* (see *ngathu*)
warrakan	animal, game, meat
warrangul	outside, in the open, public
-waṯangu	suffix denoting 'holder' or 'owner'
wa:wa	kin category—older brother of male or female (reciprocals, *yukuyuku, yapa*)
wayirri-waṯangu	see *ma:ri-waṯangu*
weḏu	salmon catfish (Dhuwa moiety)
winikamu	generic term for several species of fern
wiripu	other, another, different
wirrkuḻ	woman between puberty and first pregnancy
wititj	olive python (*Liacis olivaceus*), identity of major *wangarr* ancestor of Dhuwa-moiety groups
wukuṉḏi	subject to restrictions due to pollution or avoidance relation, 'taboo'
wuḻma	thunder, thundercloud
wuḻman	old person, from English 'old man'
-wurru	through, along (suffix)
worrungu	old person
wurrwiliny	middle-aged man, 'mature man'
-wuy	pertaining to (suffix)
yaka	no
ya:ku	name
ya:kumirr/i	name-having, important (of person's, places)
ya:n	tongue, language (see *matha*)
yapa	kin category—older sister of female, sister of male (reciprocals, *yukuyuku*, *wa:wa*)
yapa-pulu	a person's sister's group, country and sacra
yapa-waṯangu	'sister-holder', person in relation to sister's group, country and sacra
yarangu	open, not *dhuyu* (q.v.)
ya:tj, ya:tjkurru	bad, evil, ill, drunk
yawirriny'	bachelors, unmarried but circumcised males (sing. *gurrmul*)
yiḏaki	drone-pipe, didgeridoo
yindi	big, important
yindi-pulu	see *nga:ṉḏi-pulu*
yindi yolngu	'big man', man with grey hair

-yu/thu	suffix denoting agent or instrument of an action, or a temporal location
yukuyuku	kin category—younger brother of male, younger sibling of female (reciprocals, *wa:wa*, *yapa*)
yol	who, someone, whoever, 'person' in *djinang* dialect
yolngu	person, person with black skin
yolngu rom	'people's law'—sorcery
yuṯa	new
yothu	child
yothumirr/i	'child-having', woman who has borne a child
yuw	yes
yuwalk	true

BIBLIOGRAPHY

BOOKS, ARTICLES, AND UNPUBLISHED PAPERS

Abercrombie, N., Hill, S., and Turner, B. S. (1980), *The Dominant Ideology Thesis* (London, George Allen & Unwin).

Alland, A. (1988), 'Phallic Symbolism and Reproductive Expropriation: Sexual Politics in Cross Cultural Perspectives', in R. R. Randolph, D. M. Schneider, and M. N. Diaz (eds.) *Dialectics and Gender: Anthropological Approaches* (Boulder, Col., Westview Press), 20–37.

Allen, M. (1967), *Male Cults and Secret Initiations in Melanesia* (Melbourne, Melbourne University Press).

Althusser, L., and Balibar, E. (1970), *Reading Capital* (London, New Left Books).

Altman, J. C. (1987), *Hunter-gatherers Today: An Aboriginal Economy in North Australia* (Canberra, Australian Institute of Aboriginal Studies).

Barth, F. (1975), *Ritual and Knowledge Among the Baktaman of New Guinea* (New Haven, Conn., Yale University Press).

—— (1987), *Cosmologies in the Making: A Generative Approach to Cultural Variation in Inner New Guinea* (Cambridge, Cambridge University Press).

Bauer, F. H. (1964), *Historical Geography of White Settlement in Part of Northern Australia: Part 2—Katherine Darwin Region* (Canberra, CSIRO Division of Land Research and Regional Survey Report, no. 64/1).

Beckett, J. (1987), *Torres Strait Islanders: Custom and Colonialism* (Cambridge, Cambridge University Press).

Bern, J. (1979), 'Ideology and Domination: Towards a Reconstruction of the Australian Social Formation', *Oceania*, 50, 118–32.

Berndt, C. H. (1950), *Women's Changing Ceremonies in Northern Australia* (*L'Homme*; Paris, Hermann et Cie).

—— (1970*a*), 'Digging Sticks and Spears, or, the Two-sex Model', in F. Gale (ed.), *Women's Role in Aboriginal Society* (Canberra, Australian Institute of Aboriginal Studies), 64–80.

—— (1970*b*), 'Monsoon and Honey Wind', in J. Pouillon and P. Maranda (eds.), *Échanges et Communications: Mélanges offerts à Claude Lévi-Strauss* (Paris, Mouton).

Berndt, R. M., (1948*a*), 'A Wonguri-Mandzikai Song-cycle of the Moon-bone', *Oceania*, 19 (1), 16–50.

—— (1948*b*), 'Badu, Islands of the Spirits', *Oceania*, 19 (2), 93–103.

—— (1951), *Gunapipi* (Melbourne, Cheshire).

Berndt, R. M. (1952), *Djanggawul* (London, Routledge & Kegan Paul).
—— (1955), '"Murngin" (Wulamba) Social Organization', *American Anthropologist*, 57 (1), 84–106.
—— (1962), *An Adjustment Movement in Arnhem Land* (Cahiers de L'Homme; Paris and The Hague, Mouton).
—— (1965), 'Marriage and the Family in North-eastern Arnhem Land', in M. Nimkoff (ed.), *Comparative Family Systems* (Boston, Houghton Mifflin), 77–104.
—— (1971), 'Social Relationships Among Two Australian Aboriginal Societies of Arnhem Land: Gunwinggu and "Murngin"', in F. L. K. Hsu (ed.), *Kinship and Culture* (Chicago, Aldine).
—— (1976), 'Territoriality and the Problem of Defining Sociocultural Space', in N. Peterson (ed.), *Tribes and Boundaries in Australia* (Canberra, Australian Institute of Aboriginal Studies), 50–71.
Berndt, R. M. and Berndt, C. H. (1954), *Arnhem Land: Its History and its People* (Melbourne, Cheshire).
—— and Berndt, C. H. (1970), *Man, Land and Myth in North Australia: The Gunwinggu People* (East Lansing, Mich., Michigan State University Press).
—— (1980), 'Concerning a Document Distributed by the A.I.A.S. and Entitled Social Anthropology Research Priority: Aboriginal "Revitalization" Movements', typescript.
Bettelheim, B. (1954), *Symbolic Wounds: Puberty Rites and the Envious Male* (Glencoe, Ill., Free Press).
Bloch, M. (1985), 'From Cognition to Ideology', in R. Fardon (ed.), *Knowledge and Power: Anthropological and Sociological Approaches* (Edinburgh, Scottish Academic Press), 21–48.
Borsboom, A. (1978*a*), *Marradjirri: A Modern Ritual Complex in Arnhem Land, North Australia* (Nijmegen, Katholieke Universiteit).
—— (1978*b*), 'Dreaming Clusters Among Marangu Clans', in L. R. Hiatt (ed.), *Australian Aboriginal Concepts* (Canberra, Australian Institute of Aboriginal Studies), 106–20.
Bos, Revd. R. (1979*a*), 'Report on Field Work, July 1979' (Department of Sociology and Anthropology, University of Queensland).
—— (1979*b*), 'Report on Field Work, August 1979' (Department of Sociology and Anthropology, University of Queensland).
—— (1988*a*), 'Jesus and the Dreaming: Religion and Social Change in Arnhem Land' Ph.D. thesis (University of Queensland).
—— (1988*b*), 'The Dreaming and Social Change in Arnhem Land', in T. Swain and D. B. Rose (eds.), *Aboriginal Australians and Christian Missions* (Bedford Park, South Australia, The Australian Association for the Study of Religions), 422–37.
Boudon, R. (1989), *The Analysis of Ideology* (Cambridge, Polity Press).

Bourdieu. P. (1972), *Esquisse d'une Théorie de la Pratique* (Geneva, Librairie Droz).

—— (1977), *Outline of a Theory of Practice* (Cambridge, Cambridge University Press).

Boyd, J. (1986), 'The Arnhem Land Revival of 1979: An Australian Religious Movement' (Fuller School of World Mission).

Burton, R. V., and Whiting, J. W. M. (1961), 'The Absent Father and Cross-sex Identity', *Merrill-Palmer Quarterly of Behaviour and Development*, 7 (2), 85–95.

Chapple, E. D., and Coon, C. (1947), *Principles of Anthropology* (London, Jonathan Cape).

Christie, M. (1984), 'The Classroom World of the Aboriginal Child', Ph.D thesis (University of Queensland).

Clunies Ross, M., and Hiatt, L. R. (1977), 'Sand Sculptures at a Gidjingali Burial Site', in P. J. Ucko (ed.), *Form in Indigenous Art: Schematisation in the Art of Aboriginal Australia and Prehistoric Europe* (Canberra, Australian Institute of Aboriginal Studies), 131–46.

Cohen, I. J. (1989), *Structuration Theory: Anthony Giddens and the Constitution of Social Life* (London, Macmillan).

Cohen, Y. A. (1964), *The Transition from Childhood to Adolescence* (Chicago, Aldine).

Cole, K. (1979), *The Aborigines of Arnhem Land* (Adelaide, Rigby).

Comaroff, J., and Comaroff, J. (1991), *Of Revelation and Revolution: Christianity, Colonialism and Consciousness in South Africa* (Chicago, University of Chicago Press).

Cowlishaw, G. (1978), 'Infanticide in Aboriginal Australia', *Oceania*, 48, 262–83.

—— (1987), 'Colour, Culture and Aboriginalists', *Man*, 2 (2), 221–37.

Donovan, P. F. (1984), *At the Other End of Australia: The Commonwealth and the Northern Territory 1911–1978* (St Lucia, University of Queensland Press).

Drummond, L. (1980), 'The Cultural Continuum: A Theory of Intersystems', *Man* (NS), 15, 352–74.

Duffy, E. A. J. (1960), *A Monograph on the Immature Stages of Neotropical Timber Beetles (Cerambycidae)* (London, Trustees of the British Museum).

Eades, D. (1982), 'You Gotta Know How to Talk . . . Information Seeking in South-east Queensland Aboriginal Society', *Australian Journal of Linguistics*, 2 (1), 61–82.

Eco, U. (1983), *The Name of the Rose* (San Diego, Harcourt Brace Jovanovich.)

—— (1986), *Semiotics and the Philosophy of Language* (Bloomington, Ind., University Press).

Elkin, A. P. (1972), *Two Rituals in South and Central Arnhem Land* (The Oceania Monographs no. 19; Sydney, University of Sydney).

——, Berndt, R. M., and Berndt, C. H. (1950), Art in Arnhem Land (Melbourne, Cheshire).

Elliott, C. (1991), 'Mewal is Merre's Name: Form and Ambiguity in Marrangu Cosmology', Master's thesis (Australian National University).

Falk Moore, S. (1987), 'Explaining the Present: Theoretical Dilemmas in Processual Ethnography', *American Ethnologist*, 14 (4), 727–36.

Fish, S. (1989), *Doing What Comes Naturally: Change, Rhetoric, and the Practice of Theory in Literary and Legal Studies* (Durham and London, Duke University Press).

Fogelin, R. (1967), *Evidence and Meaning: Studies in Analytic Philosophy* (London, Routledge & Kegan Paul).

Foucault, M. (1980), *Power/knowledge: Selected Interviews and Other Writings, 1972–1977* (Brighton, Harvester Press).

—— (1986), 'The Subject and Power', in *Michel Foucault: Beyond Structuralism and Hermeneutics*, ed. H. L. Dreyfus and P. Rabinow (Brighton, Harvester Press), 208–26.

Freud, S. (1922), *Introductory Lectures on Psychoanalysis* (London, George Allen & Unwin).

Geertz, C. (1966), 'Religion as a Cultural System', in M. Banton (ed.), *Anthropological Approaches to the Study of Religion* (ASA Monographs; London, Tavistock Publications), 1–46.

Gerard, G. (n.d.), 'Rape in Arnhem Land', Anthropology seminar paper (University of Sydney).

Gerholm, T. (1988), 'On Ritual: A Postmodernist View', *Ethnos*, 53 (3–4), 190–203.

Gerritsen, R. (1982), 'Outstations: Differing Interpretations and Policy Implications', in P. Loveday (ed.), *Service Delivery to Outstations* (Darwin, Australian National University North Australia Research Unit), 57–69.

Giddens, A. (1979), *Central Problems in Social Theory: Action, Structure and Contradiction in Social Analysis* (London, Macmillan).

Gondarra, Revd. Dj. (1980), 'The Land is My Mother . . . ', in I. R. Yule (ed.), *My Mother the Land* (Galiwin'ku, Christian Action Group), 8–9.

Grace, G. W. (1987), *The Linguistic Construction of Reality* (London, Croom Helm).

Gramsci, A. (1971), *Selections from the Prison Notebooks of Antonio Gramsci*, ed. and trans. Q. Hoare and G. N. Smith (London, Lawrence and Wishart).

Groger-Wurm, H. (1973), *Australian Aboriginal Bark Paintings and their Mythological Interpretation: Volume 1 Eastern Arnhem Land* (Canberra, Australian Institute of Aboriginal Studies).

Guha, R. (1983), *Elementary Aspects of Peasant Insurgency in Colonial India* (Delhi, Oxford University Press).

Hamilton, A. (1980), 'Dual Social Systems: Technology Labour and Women's Secret Rites in the Eastern Western Desert of Australia', *Oceania*, 51, 4–19.

—— (1981*a*), 'A Complex Strategical Situation: Gender and Power in Aboriginal Australia', in N. Grieve and P. Grimshaw (eds.), *Australian Women: Feminist Perspectives* (Melbourne, Oxford University Press), 69–85.

—— (1981*b*), *Nature and Nurture: Aboriginal Child-rearing in North-Central Arnhem Land* (Canberra, Australian Institute of Aboriginal Studies).

Harrington, C. (1968), 'Sexual Differentiation in Socialization and Some Male Genital Mutilations', *American Anthropologist*, 70, 952–6.

Harris, Stephen (1984), *Culture and Learning: Tradition and Education in North-east Arnhem Land* (Canberra, Australian Institute of Aboriginal Studies).

Harris, Stewart (1972), *This Our Land* (Canberra, Australian National University Press).

Heath, J. (1978), *Linguistic Diffusion in Arnhem Land* (Canberra, Australian Institute of Aboriginal Studies).

—— (1980), *Dhuwal (Arnhem Land) Texts on Kinship and Other Subjects, with Grammatical Sketch and Dictionary* (Oceania Linguistic Monographs no. 23).

—— (1982), 'Where is That (Knee)?: Basic and Supplementary Kin Terms in Dhuwal (Yuulngu/Murngin)', in J. Heath, F. Merlan, and A. Rumsey (eds.), *Languages of Kinship in Aboriginal Australia* (Oceania Linguistic Monographs no. 24), 40–63.

Herdt, G. (1982), 'Fetish and Fantasy in Sambia Initiation', in G. Herdt (ed.), *Rituals of Manhood: Male Initiation in Papua New Guinea* (Berkeley, Calif., University of California Press), 44–98.

—— (1990), 'Secret Societies and Secret Collectives', *Oceania*, 60 (4), 361–81.

Hiatt, L. R. (1965), *Kinship and Conflict: A Study of an Aboriginal Community in Northern Arnhem Land* (Canberra, Australian National University Press).

—— (1971), 'Secret Pseudo-procreation Rites Among the Australian Aborigines', in L. R. Hiatt and C. Jayawardena (eds.), *Anthropology in Oceania* (Sydney, Angus & Robertson), 77–88.

—— (1975), Introduction in L. R. Hiatt (ed.), *Australian Aboriginal Mythology* (Canberra, Australian Institute of Aboriginal Studies), 1–23.

—— (1986), 'Maidens, Males and Marx: Some Contrasts in the Work of Frederick Rose and Claude Meillasoux', *Oceania*, 56 (1), 34–46.

Hockett, C. F. (1958), *A Course in Modern Linguistics* (New York,

Macmillan).

Johnson-Laird, P. N. (1983), *Mental Models: Towards a Cognitive Science of Language, Inference, and Consciousness* (Cambridge, Cambridge University Press).

Jorgenson, D. (1990), 'Secrecy's Turns', *Canberra Anthropology*, 13 (1), 40–7.

Keen, I. (forthcoming) 'Metaphor and the Meta-language: The Constitution of Yolngu "Groups"', *American Ethnologist.*

—— (1977), 'Ambiguity in Yolngu Religious Language', *Canberra Anthropology*, 1 (1), 33–50.

—— (1978), 'One Ceremony, One Song: An Economy of Religious Knowledge Among the Yolngu of Northeast Arnhem Land', Ph.D. thesis (Australian National University).

—— (1982), 'How Some Murngin Men Marry Ten Wives', *Man* (NS), 17 (4), 620–42.

—— (1985), 'Definitions of Kin', *Journal of Anthropological Research*, 41 (1), 62–90.

—— (1986*a*), 'New Perspectives on Yolngu Affinity: A Review Article', *Oceania*, 56 (3), 219–30.

—— (1986*b*), Stanner on Aboriginal Religion', *Canberra Anthropology*, 9 (2), 26–50.

—— (1989*a*), 'Aboriginal Governance', in J. C. Altman (ed.), *Emergent Inequalities in Aboriginal Australia* (Oceania Monographs no. 38; Sydney, University of Sydney), 17–42.

—— (1989*b*), 'Ecology and Species Attributes in Yolngu Religious Symbolism', in R. Willis (ed.), *Signifying Animals: Human Meaning in the Natural World* (London, Unwin Hyman), 85–102.

—— (1991), 'Images of Reproduction in the Yolngu Maḏayin Ceremony', in W. Shapiro (ed.), *On the Generation and Maintenance of Person: Essays in Honour of John Barnes* (*Australian Journal of Anthropology*, Special Issue 1; Sydney, Anthropological Society of New South Wales), 192–207.

Keen, I., and Merlan, F. (1990), *The Significance of the Conservation Zone to Aboriginal People*, Report to the Kakadu Conservation Zone Inquiry, Resource Assessment Commission (Canberra, Australian Government Publishing Service).

Keesing, R. (1987), 'Anthropology as Interpretive Quest', *Current Anthropology*, 28 (2), 161–76.

Kendon, A. (1990), *Sign Languages of Aboriginal Australia: Cultural Semiotic and Communicative Perspectives* (Cambridge, Cambridge University Press).

Kiefer, C. W., and Cowan, J. (1979), 'State-context Dependence and Theories of Ritual', *Ethos*, 2 (10), 53–84.

Kolig, E. (1981), *The Silent Revolution: The Effects of Modernization on Australian Aboriginal Religion* (Philadelphia, Institute for the Study of Human Issues).

La Fontaine, J. S. (1985), *Initiation: Ritual Drama and Secret Knowledge Across the World* (Harmondsworth, Penguin).

Lakoff, G., and Johnson, M. (1980), *Metaphors We Live By* (Chicago, The University of Chicago Press).

Lalley, E. (1988), 'Yolngu Marriage: An Empirical Analysis', Master's thesis (Australian National University).

Lattas, A. (1989), 'Trickery and Sacrifice: Tambarans and the Appropriation of Female Reproductive Powers in Male Initiation Ceremonies in West New Britain', *Man* (NS), 24 (3), 451–69.

Leach, E. R. (1954), *Political Systems of Highland Burma* (London, G. Bell & Sons).

Lévi-Strauss, C. (1964), *Totemism*, trans. Rodney Needham (London, Merlin Press).

Lewis, G. (1980), *Day of Shining Red: An Essay on the Understanding of Ritual* (Cambridge, Cambridge University Press).

Lindstrom, L. (1990), *Knowledge and Power in a South Pacific Society* (Washington, DC, Smithsonian Institution Press).

Lowe, B. (1976), *Gupapuyngu Dictionary* (Milingimbi, Milingimbi Literacy Production Centre).

Lurhmann T. M. (1989), 'The Magic of Secrecy', *Ethos*, 17, 131–65.

Lyons, J. (1977), *Semantics*, 1 (Cambridge, Cambridge University Press).

McConvell, P. (1985), 'The Origin of Subsections in Northern Australia', *Oceania*, 56 (1), 1–33.

McGowan, F. (n.d.), 'A Musical Education for Life', unpublished paper (University of Oxford).

McKenzie, M. (1976), *Mission to Arnhem Land* (Adelaide, Rigby).

MacKnight, C. C. (1976), *The Voyage to Marege': Macassan Trepangers in Northern Australia* (Melbourne, Melbourne University Press).

McLuhan, M. (1964), *Understanding Media* (London, Routledge & Kegan Paul).

Maddock, K. (1972) *The Australian Aborigines: A Portrait of Their Society* (Ringwood, Victoria, Penguin Books).

—— (1976), 'Communication and Change in Mythology', in N. Peterson (ed.), *Tribes and Boundaries in Australia* (Canberra, Australian Institute of Aboriginal Studies), 162–79.

—— (1982), *The Australian Aborigines: A Portrait of Their Society*, rev. edn. (Ringwood, Victoria, Penguin Books).

Marika, R., and Christie, M. (forthcoming), 'Yolngu Metaphors for Learning', *International Journal of the Sociology of Language*.

Meehan, B. (1982), *Shell Bed to Shell Midden* (Canberra, Australian

Institute of Aboriginal Studies).

Meggitt, M. (1966), *Gadjari Among the Walbiri Aboriginal of Central Australia* (Oceania Monographs no. 14; Sydney, The University of Sydney).

Merlan, F. (1981), 'Land, Language and Social Identity in Aboriginal Australia', *Mankind*, 13, 133–48.

—— (1988), 'Gender', in R. M. Berndt and R. Tonkinson (eds.), *Social Anthropology and Australian Aboriginal Studies: A Contemporary Overview* (Canberra, Aboriginal Studies Press), 15–76.

Michaels, E. (1988), 'Para-ethnography', *Art and Text*, 30, 42–51.

Mirritji, J. (1976), *My People's Life: An Aboriginal's Own Story* (Milingimbi, Milingimbi Literacy Centre).

Morinis, E. A. (1985), 'The Ritual Experience: Pain and the Transformation of Consciousness in Ordeals of Initiation', *Ethos*, 13 (2), 150–74.

Morphy, F. (1977), 'Language and Moiety: Sociolectal Variation in a Yu:lngu Language of North-east Arnhem Land', *Canberra Anthropology*, 1 (1), 51–60.

—— (1983), 'Djapu: A Yolngu dialect', in R. Dixon (ed.), *Handbook of Australian Languages*, iii (Canberra, The Australian National University Press), 1–190.

Morphy, H. (1977), 'Too Many Meanings: An Analysis of the Artistic System of the Yolngu of North-east Arnhem Land', unpublished Ph.D. thesis (Australian National University, Canberra).

—— (1978), 'Rights in Paintings and Rights in Women: A Consideration of Some of the Basic Problems Posed by the Asymmetry of the "Murngin" System', in J. Specht and P. White (eds.), *Trade and Exchange in Oceania and Australia* (*Mankind*, 11, Special Issue), 208–11.

—— (1983), 'Aboriginal Fine Art, the Creation of Audiences and the Marketing of Art', in P. Loveday and P. Cook (eds.), *Aboriginal Arts and Crafts and the Market* (Darwin, Australian National University North Australia Research Unit), 37–43.

—— (1984), *Journey to the Crocodile's Nest: An Accompanying Monograph to the Film Ma̲darrpa Funeral at Gurka'wuy* (Canberra, Australian Institute of Aboriginal Studies).

—— (1988), 'Maintaining Cosmic Unity: Ideology and the Reproduction of Yolngu Clans', in T. Ingold, D. Riches, and J. Woodburn (eds.), *Hunters and Gatherers: Property, Power and Ideology* (Oxford, Berg), 249–71.

—— (1990), 'Myth, Totemism and the Creation of Clans', *Oceania*, 60 (4), 312–29.

—— (1991), *Ancestral Connections: Art and an Aboriginal System of Knowledge* (Chicago, Chicago University Press).

Munn, N. (1969), 'The Effectiveness of Symbols in Murngin Rite and Myth', in R. F. Spencer (ed.), *Forms of Symbolic Action*, Proceedings of the 1969 Annual Spring Meeting of the American Ethnological Society (Seattle, University of Washington Press), 178–207.

—— (1973), *Walbiri Iconography: Graphic Representation and Cultural Symbolism in a Central Australian Society* (Ithaca, NY, Cornell University Press).

Myers, F. R. (1980), 'The Cultural Basis of Politics in Pintupi Life', *Mankind*, 12 (3), 197–214.

—— (1982), 'Always Ask: Resource Use and Land Ownership Among Pintupi Aborigines of the Australian Western Desert', in N. Williams and E. Hunn (eds.), *Resource Managers: North American and Australian Hunter-Gatherers* (Canberra, Australian Institute of Aboriginal Studies), 173–96.

—— (1986), *Pintupi Country, Pintupi Self: Sentiment, Place and Politics Among Western Desert Aborigines* (Washington, DC, Smithsonian Institution).

Nadel, S. F. (1951), *The Foundations of Social Anthropology* (London, Cohen & West).

Norbeck, E., Walker, D. E., and Cohen, M. (1962), 'The Interpretation of Data: Puberty Rites', *American Anthropologist*, 64 (3), 463–85.

Paige, K. E., and Paige, J. M. (1981), *The Politics of Reproductive Ritual* (Berkeley, Calif., University of California Press).

Parkin, D. (1982), Introduction in id. (ed.), *Semantic Anthropology* (London, Academic Press), pp. xi–li.

—— (1985), 'Controlling the U-turn of Knowledge', in R. Fardon (ed.), *Knowledge and Power: Anthropological and Sociological Approaches* (Edinburgh, Scottish Academic Press), 49–60.

Pelto, P. J., and Pelto, G. H. (1975), 'Intra-cultural Diversity: Some Theoretical Issues', *American Ethnologist*, 2 (1), 1–18.

Peterson, N. (n.d.), 'Notes on the Gunapipi Film', (Australian Institute of Aboriginal and Torres Strait Islander Studies).

—— (1971), 'The Structure of Two Aboriginal Ecosystems', Ph.D. thesis (University of Sydney).

—— (1972), 'Totemism Yesterday: Sentiment and Local Organisation Among the Australian Aborigines', *Man* (NS), 7 (1), 12–32.

—— (1973), 'Camp-site Location Amongst Australian Hunter-gatherers: Archaeological and Ethnographic Evidence for a Key Determinant', *Archaeology and Physical Anthropology in Oceania*, 8 (3), 173–93.

—— (1976), 'Mortuary Customs of Northeast Arnhem Land: An Account Compiled from Donald Thomson's Fieldnotes', *Memoirs of the National Museum of Victoria*, 37, 97–108.

—— (1978), 'The Importance of Women in Determining the Composition of Residential Groups in Aboriginal Australia', in F. Gale (ed.), *Women's Role in Aboriginal Society* (Canberra, Australian Institute of Aboriginal Studies), 9–16.

Peterson, N. (1988), 'Is Secular Assimilation Possible? Cash, Commoditisation and the Persistence of Aboriginal Culture'. Paper presented to the Twelfth International Symposium of Ethnology of the Taniguchi Foundation at the National Museum of Ethnology, 20–7 November 1988.

—— Keen, I., and Sansom, B. (1977), 'Succession to Land: Primary and Secondary Rights to Aboriginal Estates', in *Official Hansard Report of the Joint Select Committee on Aboriginal Land Rights in the Northern Territory*, 19 Apr. 1977 (Canberra, Government Printer), 1002–14.

Powell, A. (1982), *A Far Country: A Short History of the Northern Territory* (Melbourne, Melbourne University Press).

Reid, J. (1983), *Sorcerers and Healing Spirits: Continuity and Change in an Aboriginal Medical System* (Canberra, Australian National University).

Roheim, G. (1925), *Australian Totemism: A Psychoanalytic Study in Anthropology* (London, Frank Cass & Co. Ltd.).

—— (1945), *The Eternal Ones of the Dream: A Psychoanalytic Interpretation of Australian Myth and Ritual* (New York, International Universities Press).

Rose, Frederick (1968), 'Australian Marriage, Land-owning Groups, and Initiations', in R. B. Lee and I. Devore (eds.), *Man the Hunter* (Chicago, Aldine Publishing Company), 200–8.

Rudder, J. (1977), *Introduction to Yolngu Science* (Galiwin'ku, Galiwin'ku Adult Education Centre).

—— (1991), 'Yolngu Time', manuscript (Australian National University).

Rumsey, A. (1981), 'Kinship and Context among the Ngarinyin', *Oceania*, 51 (3), 181–92.

Sahlins, M. (1985), *Islands of History* (Chicago, Chicago University Press).

Sansom, B. (1980), *The Camp at Wallaby Cross: Aboriginal Fringe Dwellers in Darwin* (Canberra, Australian Institute of Aboriginal Studies), 159–78.

—— (1982), 'The Aboriginal Commonality', in R. M. Berndt (ed.), *Sites, Rights and Resource Development* (Nedlands, University of Western Australia Press), 117–38.

—— (1988), 'A Grammar of Exchange', in I. Keen (ed.), *Being Black: Aboriginal Cultures in "Settled" Australia* (Canberra, Aboriginal Studies Press), 159–78.

Sapir, E. (1934), 'Symbolism', *Encyclopedia of the Social Sciences* (New York, Macmillan), 492–5.

Sargant, W. W. (1957), *Battle for the Mind: A Physiology of Conversion and Brainwashing* (Melbourne, Heinemann).

Schebeck, B. (1968), 'Dialect and Social Groupings in North-east Arnhem Land', typescript (Australian Institute of Aboriginal and Torres Strait Islander Studies, Canberra).

Scheffler, H. (1973), 'Kinship, Descent and Alliance', in J. J. Honigmann (ed.), *Handbook of Social and Cultural Anthropology* (Chicago, Rand McNally College Publishing Company), 747–94.

—— (1978), *Australian Kin Classification* (Cambridge, Cambridge University Press).

Schlegel, A. and Barry, H. (1980), 'The Evolutionary Significance of Adolescent Initiation Ceremonies', *American Ethnologist*, 7 (4), 696–715.

Searle J. (1969), *Speech Acts: An Essay in the Philosophy of Language* (Cambridge, Cambridge University Press).

Shapiro, W. (1967), 'Relational Affiliation in "unilineal" Descent Systems', *Man*, 12, 161–3.

—— (1968), 'The Exchange of Sisters' Daughters' Daughters in Northeast Arnhem Land', *South Western Journal of Anthropology*, 24, 346–53.

—— (1969), 'Semi-moiety Organization and Mother-in-law Bestowal in Northeast Arnhem Land', *Man*, 4, 629–40.

—— (1973), 'Residential Groupings in Northeast Arnhem Land', *Man* (NS), 8, 365–83.

—— (1979), *Social Organization in Aboriginal Australia* (Canberra, Australian National University Press).

—— (1981), *Miwuyt Marriage: The Cultural Anthropology of Affinity in Northeast Arnhem Land* (Philadelphia, Institute for the Study of Human Issues).

—— (1988), 'Ritual Kinship, Ritual Incorporation and the Denial of Death', *Man* (NS), 23 (2), 275–97.

Sider, G. M. (1986), *Culture and Class in Anthropology and History: A Newfoundland Illustration* (Cambridge, Cambridge University Press).

Simmel, G. (1950), *The Sociology of Georg Simmel*, ed. and trans. K. H. Wolff (Glencoe Ill., The Free Press).

Spencer, B. (1914), *Native Tribes of the Northern Territory of Australia* (London, Macmillan).

Spencer, P. (1965), *The Samburu: A Study of Gerontocracy in a Nomadic Tribe* (London, Routledge & Kegan Paul).

—— (1970), 'The Function of Ritual in the Socialization of the Samburu Moran', in P. Mayer (ed.), *Socialization: the Approach from Social Anthropology* (ASA Monograph 8; London, Tavistock Publications), 127–58.

Sperber, D. (1975), *Rethinking Symbolism* (Cambridge, Cambridge University Press).

Stanner, W. E. H. (1963), *On Aboriginal Religion* (Oceania Monographs no. 11; Sydney, The University of Sydney).

—— (1965), 'Religion, Totemism and Symbolism', in R. M. and C. H. Berndt (eds.), *Aboriginal Man in Australia* (Sydney, Angus and Robertson), 207–37.

—— (1979), *White Man Got No Dreaming: Essays 1938–1973* (Canberra, Australian Institute of Aboriginal Studies).

Steiner (1975), *After Babel: Aspects of Language and Translation* (London, Oxford University Press).

Strathern, M. (1988), *The Gender of the Gift: Problems with Women and Problems with Society in Melanesia* (Berkeley, Calif., University of California Press).

Strecker, I. (1988), *The Social Practice of Symbolization: An Anthropological Analysis* (LSE Monographs on Social Anthropology, 60; London, Athlone Press).

Strehlow, T. G. H. (1970), 'Geography and the Totemic Landscape in Central Australia: A Functional Study', in R. M. Berndt (ed.), *Australian Aboriginal Anthropology* (Nedlands, University of Western Australia Press, for the Australian Institute of Aboriginal Studies), 92–140.

Sutton, P. (1978), 'Wik: Aboriginal Society, Territory and Language at Cape Keerweer, Cape York Peninsula, Australia', Ph.D. thesis (University of Queensland).

—— (1987), 'Mystery and Change', in M. Clunies Ross, T. Donaldson, and S. A. Wild (eds.), *Songs of Aboriginal Australia* (Oceania Monograph No. 32; Sydney, University of Sydney), 77–96.

Taylor, L. (1987), '"The same but different": Social Reproduction and Innovation in the Art of the Kunwinjku of Western Arnhem Land', Ph.D. thesis (Australian National University).

—— (1990), 'The Rainbow Serpent as Visual Metaphor in Western Arnhem Land', *Oceania*, 60, 329–44.

Thomson, D. F. (n.d.), Unpublished field-notes 9.1.1937 (National Museum of Victoria, Melbourne).

—— (1938), 'A New Type of Fish Trap from Arnhem Land, Northern Territory of Australia', *Man*, 39, 216, 193–8.

—— (1939*a*), 'Proof of Indonesian Influence upon the Aborigines of North Australia: The Remarkable Dog Ngarra of the Mildjingi Clan', *Illustrated London News*, 12 Aug. 1939, 277–9.

—— (1939*b*), 'Two Painted Skulls from Arnhem Land, with Notes on the Totemic Significance of the Designs', *Man*, 39, 1–3.

—— (1948), 'An Arnhem Land Adventure', *National Geographic*, 93 (3), 403–30.

—— (1949), *Economic Structure and the Ceremonial Exchange Cycle in Arnhem Land* (Melbourne, Macmillan).

—— (1975), 'The Concept of "marr" in Arnhem Land', *Mankind*, 10 (1), 1–10.

—— (1983), *Donald Thomson in Arnhem Land*, ed. Nicolas Peterson (South Yarra, Victoria, Curry O'Neil).

Toohey, J. (1982), *Kaytej, Warlpiri and Warlmanpa Land Claim* (Canberra, Australian Government Publishing Service).

Trigger, D. (1992), *Whitefella Comin': Aboriginal Responses to Colonialism in Northern Australia* (Sydney, Cambridge University Press).

Turner, D. (1974), *Tradition and Transformation: A study of Aborigines in the Groote Eylandt Area, Northern Australia* (Canberra, Australian Institute of Aboriginal Studies).

Turner, V. (1967), *The Forest of Symbols: Aspects of Ndembu Ritual* (Ithaca, NY, Cornell University Press).

United Church in North Australia (1974), 'Free to Decide' (Darwin).

Urry, J. and Walsh, M. (1982), 'The Lost "Macassar Language" of Northern Australia', *Aboriginal History*, 5 (1), 91–108.

Wagner, R. (1975), *The Invention of Culture* (Englewood Cliffs, NJ, Prentice Hall).

Warner, W. L. (1937), *A Black Civilization: A Social Study of an Australian Tribe* (New York, Harper & Brothers).

Webb, T. T. (1933), 'Tribal Organization in Eastern Arnhem Land', *Oceania*, 3 (4), 406–17.

Wedgewood, C. H. (1930), 'The Nature and Function of Secret Societies', *Oceania*, 1, 130–45.

Weiner, J. F. (1988), *The Heart of the Pearlshell: The Mythological Dimension of Foi Sociality* (Berkeley, Calif., University of California Press).

White, N. (1976), 'A Preliminary Account of the Correspondence Among Genetic, Linguistic, Social and Topographic Divisions in Arnhem Land, Australia', *Mankind*, 10, 240–7.

White, N., Meehan, B., Hiatt, L. R., and Jones, R. (1990), 'Demography of Contemporary Hunter-gatherers: Lessons from Central Arnhem Land', in B. Meehan and N. White (eds.), *Hunter-gatherer Demography: Past and Present* (Oceania Monographs no. 39; Sydney, University of Sydney), 171–85.

Whiting, J. W. M., Kluckhohn, R., and Anthony, J. (1958), 'The Function of Male Initiation Ceremonies at Puberty', in E. E. Maccoby, T. M. Newcombe, and E. L. Hartley (eds.), *Readings in Social Psychology* (New York, Henry Holt & Co.), 359–70.

Williams, N. (1976), 'Australian Aboriginal Art at Yirrkala: The Introduction and Development of Marketing', in N. H. H. Graburn (ed.),

Ethnic and Tourist Arts: Cultural Expression from the Fourth World (Berkeley, Calif., University of California Press), 266–84.

Williams, N. (1986), *The Yolngu and their Land: A System of Land Tenure and the Fight for its Recognition* (Canberra, Australian Institute of Aboriginal Studies).

—— (1987), *Two Laws: Managing Disputes in a Contemporary Aboriginal Community* (Canberra, Australian Institute of Aboriginal Studies).

Yallop, C. (1982), *Australian Aboriginal Languages* (London, André Deutsch).

Yule, I. R. (ed.) (1980), *My Mother the Land* (Galiwin'ku, Christian Action Group).

Yunupingu, G. (1982), 'Putting Things Back How They Belong', *Sunday Territorian*, 7 July 1988.

Zorc, R. D. (1986), *Yolngu-Matha Dictionary* (Batchelor, School of Australian Linguistics).

FILMS

Dunlop, I. (1984), *Memorial to Mawalan* (16mm film) (Sydney, Film Australia).

McKenzie, K. (1980), *Waiting for Harry* (16mm film) (Canberra, Australian Institute of Aboriginal Studies).

NEWSPAPERS

Northern Standard 1938

PARLIAMENTARY AND ARCHIVAL SOURCES

AAD: Australian Archives, Darwin, series F1 and F315.

NTAS: Northern Territory Archive Service, series NTRS 40.

The Northern Territory, Annual Report for 1 July 1957 to 30 June 1958.

The Parliament of the Commonwealth of Australia, The Northern Territory, Report for 1964–5.

INDEX